TELECOMMUNICATION TRANSMISSION HANDBOOK

ROGER L. FREEMAN

A WILEY-INTERSCIENCE PUBLICATION

JOHN WILEY & SONS, New York · London · Sydney · Toronto

Library of Congress Cataloging in Publication Data:

Freeman, Roger L.
  Telecommunication transmission handbook.
  "A Wiley-Interscience publication."
  Includes index.
  1. Telecommunication.   I. Title.

TK5101.F66   621.38   75-1134
ISBN 0-471-27789-4

Printed in the United States of America

10 9 8 7 6 5 4 3 2 1

FOR

CRISTI, BOBBY, and ROSALIND

PREFACE

In writing this book I have sought to reach a varied readership. It is addressed to telecommunication engineering students, and to transmission engineers specializing in one discipline such as data transmission, who wish to have an appreciation of other disciplines such as radio, telephony, or video. The book also should serve as an introduction to the problems of transmission engineering for the nonengineer. This latter group encompasses telecommunication managers, other corporate and military staff responsible for communications, as well as technicians.

To carry out this aim I have made every effort to provide explanation. It has been my intention to express, not to impress. Sufficient material and background are given so that at least a first-cut engineering effort can be made using this text as the only reference; the basic aim is how-to-do-it and why.

It must be kept in mind that a book of this size cannot treat each subdiscipline exhaustively. If more depth is desired in a specific area, a bibliography is provided at the end of each chapter to assist the reader in his search for more information as well as to document the sources used in the preparation of this work.

Engineering today is highly stratified in disciplines; and this is particularly true in the broad field of telecommunications. We have traffic engineers and radiomen, and radio itself is broken down into earth stations, radio links, high frequency, troposcatter, and so forth. We have switching engineers, experts in signaling, and telephone transmission engineers who may or may not know anything about carrier, PCM, or the like. There are experts in outside plants, plant extension, plant operations, and data communications. We lack guidance in how to take a grouping of these fields and work them into an operational system.

Thus it was my first concern to emphasize the systems approach. So that we do not get bogged down in semantics, systems in this text always means the interworking of one discipline with another to reach a definite, practical end product that serves the needs of a specific user. Only point-to-point telecommunication systems are discussed in order to keep the field of discussion within manageable proportions. Television signals are treated only in that area of special system considerations necessary for the transport of a video signal on a point-to-point basis. Yet I have kept the end user always in mind. For example, how will the change of several decibels of signal-to-noise ratio at a radiolink repeater site change the picture reception for a video viewer? It should be noted that I have purposely avoided involvement with broadcast, CATV, air-ground, telemetry, and marine transmission systems. This allows more emphasis to be placed on the basic aim of the work.

To do justice to the book, the typical reader should have a good background in algebra, trigonometry, logarithms, statistics, and electricity. He should also have some knowledge of electrical communications, particularly modulation.

The chapter order of the book is purposeful. I try to avoid jumping into a complex concept without having introduced it previously. The reader will have a grasp of what "noise" is before he has to tangle with radio, telephony before multiplex, and multiplex before broadband systems.

The object of the first chapter is to provide a leveling process. For many, admittedly, it is a review; for me it provides a useful tool to which I can refer further on in the book.

Where desirable I have referred to the appropriate CCITT (International Consultive Committee for Telephone and Telegraph), CCIR (International Consultive Committee for Radio), EIA (Electronics Industry Association), and Defense Communications Agency (U.S.) standards. Appendix A provides a guide to CCITT and CCIR recommendations and, in particular, the G. Recommendations of CCITT. Appendix B contains a short, introductory discussion of static no-break power systems. An overview of telephone signaling systems is given in Appendix C, and Appendix D is a short glossary of acronyms, abbreviations, and some of the more confusing terms used in the text.

ROGER L. FREEMAN

September 1974
Madrid, Spain

ACKNOWLEDGMENTS

Two who have contributed to the final draft of the text were my principal reviewers: John Reynolds, who recently retired from ITT, and Robert J. Ness of the U.S. Army Strategic Communications Command and TriTac Committee. They freely gave much of their time to see this job through. Their suggestions were particularly helpful. Others reviewed parts of the draft which were more in their area of specialty. Among these I would like to mention John Lawlor of ITT's Telecommunication Technology Center in Stamford, Connecticut, Jordan (Jerry) Brilliant, Technical Director of MCI, Washington, D.C., and Dr. Enrique Vilar, on loan to ESRO from ITT Laboratories Spain (ITTLS).

A number of technical staff members of ITTLS were very helpful with suggestions and review of material and, most important, kept me motivated. Among this group were Frank Drake, John Emerson, Luis Alvarez Mazo, Norman Doving, and Stanley Boylls.

R.L.F.

CONTENTS

1. Introductory Concepts   **1**

  1. The Transmission Problem, 1
  2. A Simplified Transmission System, 1
  3. The Decibel, 2
  4. Basic Derived Decibel Units, 5

    4.1 The dBm, 5
    4.2 The dBW, 6
    4.3 The dBmV, 7

  5. The Neper, 8
  6. Addition of Power Levels in dB(dBm/dBW or similar absolute logarithm units), 8
  7. Normal Distribution—Standard Deviation, 10
  8. The Simple Telephone Connection, 11
  9. The Practical Transmission of Speech, 13

    9.1 Bandwidth, 13
    9.2 Speech Transmission—The Human Factor, 13
    9.3 Attenuation Distortion, 14
    9.4 Phase Shift (Envelope Delay Distortion), 17
    9.5 Level, 18
    9.6 Noise, 20

  10. Signal-to-Noise Ratio, 27
  11. Noise Figure, 27
  12. Relating Noise Figure to Noise Temperature, 30
  13. EIRP (Effective Isotropically Radiated Power), 31
  14. Some Common Conversion Factors, 32

2. Telephone Transmission   **34**

   1. General, 34
   2. The Telephone Instrument, 34

      2.1 Transmitters, 34
      2.2 Receivers, 35

   3. The Telephone Loop, 35
   4. Telephone Loop Length Limits, 36
   5. The Reference Equivalent, 38

      5.1 Definition, 38
      5.2 Application, 39

   6. Telephone Networks, 42

      6.1 General, 42
      6.2 Basic Considerations, 42
      6.3 Two-Wire/Four-Wire Transmission, 43
      6.4 Echo, Singing, and Design Loss, 45
      6.5 Via Net Loss (VNL), 53

   7. Notes on Network Hierarchy, 59
   8. Design of Subscriber Loop, 61

      8.1 Introduction, 61
      8.2 Basic Resistance Design, 62
      8.3 Transmission Design, 63
      8.4 Loading, 65
      8.5 Other Approaches to Subscriber Loop Design, 68

   9. Design of Local Area Trunks (Junctions), 72
  10. VF Repeaters, 74
  11. Transmission Considerations of Telephone Switches in the Long-Distance Network (Four-Wire), 75
  12. CCITT Interface, 75

      12.1 Introduction, 75
      12.2 Maximum Number of Circuits in Tandem, 77
      12.3 Noise, 77
      12.4 Variation of Transmission Loss with Time, 77
      12.5 Crosstalk, 79
      12.6 Attenuation Distortion, 79
      12.7 Reference Equivalent, 80
      12.8 Propagation Time, 80
      12.9 Echo Suppressors, 80

3. Frequency Division Mutliplex **82**

1. Introduction, 82
2. Mixing, 82
3. The CCITT Modulation Plan, 86

   3.1 Introduction, 86
   3.2 Formation of the Standard CCITT Group, 86
   3.3 Alternate Method of Formation of the Standard CCITT Group, 88
   3.4 Formation of the Standard CCITT Supergroup, 88
   3.5 Formation of the Standard CCITT Basic Mastergroup and Supermastergroup, 89
   3.6 The "Line" Frequency, 90

4. Loading of Multichannel FDM Systems, 91

   4.1 Introduction, 91
   4.2 Speech Measurement, 92
   4.3 Overload, 98
   4.4 Loading, 98
   4.5 Single Channel Loading, 99
   4.6 Loading with Constant Amplitude Signals, 99

5. Pilot Tones, 101

   5.1 Introduction, 101
   5.2 Level Regulating Pilots, 101
   5.3 Frequency Synchronization Pilots, 102

6. Frequency Generation, 103
7. Noise and Noise Calculations, 105

   7.1 General, 105
   7.2 CCITT Approach, 105
   7.3 U.S. Military Approach, 106

8. Other Characteristics of Carrier Equipment, 108

   8.1 Attenuation Distortion, 108
   8.2 Envelope Delay Distortion (Group Delay), 108

9. Through-Group, Through-Supergroup Techniques, 109
10. Line Frequency Configurations, 110

    10.1 General, 110
    10.2 Twelve-Channel Open-Wire Carrier, 112

10.3 Carrier Transmission over Nonloaded Cable Pairs (K-Carrier), 112

10.4 Type N-Carrier for Transmission over Nonloaded Cable Pairs, 113

10.5 Carrier Transmission on Star and Quad Type Cables, 113

10.6 Coaxial Cable Carrier Transmission Systems, 115

10.7 The L-Carrier Configuration, 121

11. Subscriber Carrier-Station Carrier, 121

12. Economics of Carrier Transmission, 124

13. Compandors, 126

14. Signaling on Carrier Systems, 129

4. High Frequency Radio   131

1. General, 131

2. Basic HF Propagation, 131

2.1 Skywave Transmission, 131

3. Operating Frequency, 135

4. HF Radio Systems, 139

5. Practical HF Point-to-Point Communication, 140

5.1 General, 140

5.2 Emission Types, 140

5.3 Recommended Modulation (Emission Type), 143

5.4 Single Sideband Suppressed Carrier, 143

5.5 SSB Operation and its Comparison with Conventional AM, 147

5.6 Key to SSB Transmission, 149

5.7 Synthesizer Application Block Diagrams, 150

6. Linear Power Amplifiers in Point-to-Point SSB Service, 152

7. Intermodulation Distortion, 153

8. HF Antennas, 154

8.1 General, 154

8.2 Basic Antenna Considerations, 154

8.3 Rhombic Antennas, 154

8.4 Basic Rhombic Design, 156

8.5 Terminating Resistances, 159

8.6 Transmission Lines, 161

8.7 Log Periodic Antennas, 162

8.8 Horizontal and Vertical Polarization, 166

9. HF Facility Layout, 167
10. Great-Circle Bearing and Distance, 167
11. Independent Sideband (ISB) Transmission, 168
12. Transmitter Loading for ISB Operation, 171
13. Diversity Techniques on HF, 171
14. Ionospheric Sounders, 173

5. **Radiolink Systems (Line-of-Sight Microwave)** 177

1. Introduction, 177
2. Link Engineering, 177
3. Propagation, 181

   3.1 Free-Space Loss, 181
   3.2 Bending of Radio Waves above 100 MHz from Straight-Line Propagation, 183
   3.3 Path Profiling—Practical Application, 187
   3.4 Reflection Point, 191

4. Path Calculations, 193

   4.1 General, 193
   4.2 Basic Path Calculations, 193
   4.3 The Mechanism of Fading—An Introductory Discussion, 197

5. FM Radiolink Systems, 199

   5.1 General, 199
   5.2 Preemphasis-Deemphasis, 199
   5.3 The FM Transmitter, 201
   5.4 The Antenna System, 207
   5.5 The FM Receiver, 210
   5.6 Diversity Reception, 210
   5.7 Transmission Lines and Related Devices, 219

6. Loading of a Radio Link System, 224

   6.1 General, 224
   6.2 Noise Power Ratio (NPR), 225
   6.3 Basic NPR Measurement, 227
   6.4 Derived Signal-to-Noise Ratio, 230
   6.5 Conversion of Signal-to-Noise Ratio to Channel Noise, 230

7. Other Testing Techniques, 231

   7.1 Out-of-band Testing, 231

8. Detailed Path Calculations, 232
9. Determination of Fade Margin, 232
10. Path and Link Reliability, 238
11. Rainfall and Other Precipitation Attenuation, 238
12. Hot-Standby Operation, 240
13. Radiolink Repeaters, 241
14. Frequency Planning, 244

14.1 General, 244
14.2 Spurious Emission, 245
14.3 Radio Frequency Interference (RFI), 245
14.4 Overshoot, 246
14.5 Transmit-Receive Separation, 246
14.6 Basis of Frequency Assignment, 246
14.7 IF Interference, 247
14.8 CCIR Recommendations, 247

15. Alarm and Supervisory Systems, 248

15.1 General, 248
15.2 Monitored Functions, 248
15.3 Transmission of Fault Information, 249
15.4 Remote Control, 250

16. Antenna Towers and Masts, 250

16.1 General, 250
16.2 Tower Twist and Sway, 251

17. Plane Reflectors as Passive Repeaters, 253
18. Noise Planning on Radiolinks, 258

6. Tropospheric Scatter   264

1. Introduction, 264
2. The Phenomenon of Tropospheric Scatter, 265
3. Tropospheric Scatter Fading, 266
4. Path Loss Calculations, 267
5. Aperture-to-Medium Coupling Loss, 270
6. Takeoff Angle, 271
7. Other Siting Considerations, 273

7.1 Antenna Height, 273
7.2 Distance to Radio Horizon, 273
7.3 Other Considerations, 274

8. Path Calculations, 274
9. Equipment Configurations, 279

9.1 *General*, 279
9.2 *Tropo Operational Frequency Bands*, 281
9.3 *Antennas, Transmission Lines, Duplexer, and Other Related Transmission Line Devices*, 281
9.4 *Modulator-Exciter and Power Amplifier*, 284
9.5 *The FM Receiver Group*, 285
9.6 *Diversity and Diversity Combiners*, 287

10. Isolation, 289
11. Intermodulation, 290
12. Maximum Feasible Median Path Loss, 290
13. Typical Tropospheric Scatter Parameters, 291
14. Frequency Assignment, 292

7. Earth Station Technology **295**

1. Introduction, 295
2. The Satellite, 295
3. Earth-Space Window, 295
4. Path Loss, 298
5. Satellite-Earth Link, 299

5.1 *Figure of Merit G/T*, 302
5.2 *The Ratio of Carrier-to-Thermal Noise Power (C/T)*, 306
5.3 *Relating C/T to G/T*, 307
5.4 *Deriving Signal Input from Illumination Levels*, 308
5.5 *Station Margin*, 309

6. Up-Link Considerations, 310
7. Multiple Access, 311
8. Functional Operation of a "Standard Earth Station," 316

8.1 *The Communication Subsystem*, 316
8.2 *Antenna Tracking Subsystem*, 319
8.3 *Multiplex, Orderwire, and Terrestrial Link Subsystems*, 321

9. Intelsat IV, 325
10. Intelsat IVA, 330
11. Television Transmission, 330
12. SPADE, 331
13. Regional Satellite Communication Systems, 334
14. The Coordination Contour, 336

8. The Transmission of Digital Data   **339**

   1. Introduction, 339
   2. The Bit and Binary Convention, 340
   3. Coding, 341

     3.1 Introduction to Binary Coding Techniques, 341
     3.2 Some Specific Binary Codes for Information Interchange, 344

   4. Error Detection and Error Correction, 353
   5. The DC Nature of Data Transmission, 357

     5.1 Loops, 357
     5.2 Neutral and Polar Dc Transmission Systems, 358

   6. Binary Transmission and the Concept of Time, 359

     6.1 Introduction, 359
     6.2 Asynchronous and Synchronous Transmission, 360
     6.3 Timing, 362
     6.4 Distortion, 364
     6.5 Bits, Bauds, and Words per Minute, 366

   7. Data Interface, 367
   8. Data Input/Output Devices, 368
   9. Digital Transmission on an Analog Channel, 370

     9.1 Introduction, 370
     9.2 Modulation-Demodulation Schemes, 371
     9.3 Critical Parameters, 373

  10. Channel Capacity, 382
  11. Voice Channel Data Modems Versus Critical Design Parameters, 384
  12. Circuit Conditioning, 387
  13. Practical Modem Applications, 389

     13.1 Voice Frequency Carrier Telegraph (VFCT), 389
     13.2 Medium Data Rate Modems, 391
     13.3 High Data Rate Modems, 392

  14. Serial-to-Parallel Conversion for Transmission on Impaired Media, 393
  15. Parallel-to-Serial Conversion for Improved Economy of Circuit Usage, 394
  16. Modems for Application to Channel Bandwidths in Excess of 4 kHZ, 395
  17. Data Transmission System—Functional Block Diagram, 397

9. Coaxial Cable Systems 399

1. Introduction, 399
2. Basic Construction Design, 400
3. Cable Characteristics, 401
4. System Design, 403
5. Repeater Design—An Economic Trade-off from Optimum, 407

   5.1 General, 407
   5.2 Thermal Noise, 409
   5.3 Overload and Margin, 409
   5.4 Intermodulation Noise, 412
   5.5 Total Noise and its Allocation, 414

6. Equalization, 414

   6.1 Introduction, 414
   6.2 Fixed Equalizers, 415
   6.3 Variable Equalizers, 416

7. Level and Pilot Tones, 418
8. Supervisory, 419
9. Powering the System, 420
10. 60-MHz Coaxial Cable Systems, 420
11. The L5 Coaxial Cable Transmission System, 422
12. Coaxial Cable or Radiolink—The Decision, 424

   12.1 General, 424
   12.2 Land acquisition as a limitation to Coaxial Cable Systems, 425
   12.3 Fading, 426
   12.3 Noise Accumulation, 426
   12.5 Group Delay—Attenuation Distortion, 426
   12.6 Radio Frequency Interference (RFI), 427
   12.7 Maximum VF Channel Capacity, 427
   12.8 Repeater Spacing, 428
   12.9 Power Considerations, 428
   12.10 Engineering and Maintenance, 428
   12.11 Multiplex Modulation Plans, 428

10. Millimeter Wave Transmission 430

1. General, 430
2. Propagation, 431
3. Rainfall Loss, 434

4. Scintillation Fading, 437
5. Practical Millimeter Wave Systems, 438
6. The Short Hop Concept, 443
7. Earth-Space Communication, 445

11. PCM and Its Applications   **448**

1. What is PCM? 448
2. Development of a PCM Signal, 451

   2.1 Sampling, 451
   2.2 The PAM Wave, 451
   2.3 Quantization, 452
   2.4 Coding, 455
   2.5 The Concept of Frame, 461
   2.6 Quantizing Distortion, 463
   2.7 Idle Channel Noise, 465

3. Practical Application, 466

   3.1 General, 466
   3.2 Practical System Block Diagram, 467
   3.3 The Line Code, 471
   3.4 Signal-to-Gaussian-Noise Ratio on PCM Repeated Lines, 471
   3.5 The Eye Pattern, 472

4. Higher Order PCM Multiplex Systems, 473
5. Long-Distance (Toll) Transmission by PCM, 476

   5.1 General, 476
   5.2 Jitter, 476

6. PCM Transmission by Radiolink, 477

   6.1 General, 477
   6.2 Modulation, RF Bandwidth, and Performance, 478

7. PCM Switching, 479

12. Video Transmission   **482**

1. General, 482
2. An Appreciation of Video Transmission, 482
3. The Composite Signal, 485
4. Critical Video Transmission Impairments, 488
5. Critical Video Parameters, 491

   5.1 General, 491

5.2 *Transmission Standard—Level*, 491
5.3 *Other Parameters*, 492

6. Video Transmission Standards, 494

6.1 *Basic Standards*, 494
6.2 *Color Transmission*, 497
6.3 *Standardized Transmission Parameters*, (*Point-to-Point Television*), 499

7. Methods of Program Channel Transmission for Video, 499
8. Video Transmission Over Coaxial Cable, 500

8.1 *Early System*, 500
8.2 *Modern Broadband Coaxial Cable Systems for Video*, 500

9. Transmission of Video over Radiolinks, 503

9.1 *General*, 503
9.2 *Bandwidth of the Baseband and Baseband Response*, 503
9.3 *Preemphasis*, 504
9.4 *Differential Gain*, 504
9.5 *Differential Phase Distortion*, 504
9.6 *Signal-to-Noise Ratio*, 504
9.7 *Square Wave Tilt*, 504
9.8 *Radio Link Continuity Pilot*, 505

10. Transmission of Video Over Conditioned Pairs, 505

10.1 *General*, 505
10.2 *Cable Description*, 505
10.3 *Terminal Equipment and Repeaters*, 506

11. Basic Tests for Video Quality, 507

11.1 *Window Signal*, 507
11.2 *Sine-Squared Test Signal*, 507
11.3 *Multiburst*, 508
11.4 *Stair Steps*, 508
11.5 *Vertical Interval Test Signals* (*VITS*), 509
11.6 *Test Patterns*, 510
11.7 *Color Bars*, 510

APPENDIX A Guide to CCITT and CCIR Recommendations 512
APPENDIX B Static No-Break Power Systems 540
APPENDIX C Signaling on Telephone Circuits 547
APPENDIX D Glossary 559

Index 571

1 INTRODUCTORY CONCEPTS

1.1 THE TRANSMISSION PROBLEM

The word transmission is often misunderstood as we try to compartmentalize telecommunications into neatly separated disciplines. A transmission engineer in telecommunications must develop a signal from a source and deliver it to the sink to the satisfaction of a customer. In a broad sense we may substitute the words transmitter for source and receiver for sink. Of major concern are those phenomena, conditions, and factors that distort or otherwise make the signal at the sink such that the customer is unsatisfied. To understand the problems of transmission, we must do away with some of the compartmentalization. Besides switching and signaling, transmission engineers must have some knowledge of maps, civil engineering, and power. A familiarity with basic traffic engineering concepts, such as busy hour, activity factor, and so forth, is also helpful.

Transmission system engineering deals with the production, transport, and delivery of a quality signal from source to sink. The following chapters describe methods of carrying out this objective.

1.2 A SIMPLIFIED TRANSMISSION SYSTEM

The simple drawing here illustrates a transmission system. The source may be a telephone mouthpiece (transmitter) and the sink may be the telephone earpiece (receiver). The source converts the human intelligence, such as voice, data information, or video, into an electrical equivalent or electrical signal. The sink accepts the electrical signal and reconverts it to an approximation of the original human intelligence. The source and sink are electrical

1

transducers. In the case of printer telegraphy, the source may be a keyboard, where each key, when depressed, transmits to the sink distinct electrical impulses. The sink in this case may be a teleprinter, which converts each impulse grouping back to the intended character keyed or depressed at the source.

The transmission media can be represented as a network:

or as a series of electrical networks:

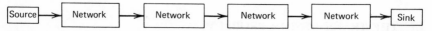

Some networks show a gain in level, others a loss. We must be prepared to discuss these gains and losses as well as electrical signal levels and the level of disturbing effects such as noise, crosstalk, or distortion. To do this we must have a firm and solid knowledge of the decibel and related measurement units.

1.3 THE DECIBEL

The decibel is a unit that describes a ratio. It is a logarithm with a base of 10. Consider first a power ratio. The number of decibels (dB) = $10 \log_{10}$ (the power ratio).

Let us look at the following network:

$$\xrightarrow{\text{1 W}} \boxed{\text{Network}} \xrightarrow{\text{2 W}}$$

The input is 1 W and its output 2 W, in the power domain. Therefore we can say the network has a 3-dB gain approximately. In this case

$$\text{Gain (dB)} = 10 \log \frac{\text{output}}{\text{input}} = 10 \log \frac{2}{1} = 10(0.3013) = 3.0103 \text{ dB}$$

or approximately, a 3-dB gain.

Now let us look at another network:

$$\xrightarrow{\text{1000 W}} \boxed{\text{Network}} \xrightarrow{\text{1 W}}$$

In this case there is a loss of 30 dB.

$$\text{Loss (dB)} = 10 \log \frac{\text{input}}{\text{output}} = 10 \log \frac{1000}{1} = 30 \text{ dB}$$

Or in general we can state

$$\text{Power (dB)} = 10 \log \frac{P_2}{P_1} \qquad (1.1)$$

where P_1 = lower power level, P_2 = higher power level.

A network with an input of 5 W and an output of 10 W is said to have a 3-dB gain.

$$\text{Gain (dB)} = 10 \log \frac{10}{5} = 10 \log \frac{2}{1} = 10 \, (0.30103)$$

$$= 3.0103 \text{ dB or } \sim 3 \text{ dB}$$

That is a good figure to remember. Doubling the power means a 3-dB gain; likewise, halving the power means a 3-dB loss.

Consider another example, a network with a 13-dB gain:

$$\text{Gain (dB)} = 10 \log \frac{P_2}{P_1} = 10 \log \frac{P_2}{0.1} = 13 \text{ dB}$$

Then

$$P_2 = 2 \text{ W}$$

Table 1.1 may be helpful. All values in the power ratio column are $X/1$, or compared to 1.

Table 1.1

Power Ratio		dB	Power Ratio	dB
10^1	(10)	+10	10^{-1} (1/10)	−10
10^2	(100)	+20	10^{-2} (1/100)	−20
10^3	(1,000)	+30	10^{-3} (1/1000)	−30
10^4	(10,000)	+40	10^{-4} (1/10,000)	−40
10^5	(100,000)	+50	10^{-5} (1/100,000)	−50
10^6	(1,000,000)	+60	10^{-6} (1/1,000,000)	−60

To work with decibels without pencil and paper is useful. Relationships of 10 and 3 have been reviewed. Now consider the following:

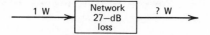

What is the power output of this network? To do this without pencil and paper, we would proceed as follows. Suppose the network attenuated the signal 30 dB. Then the output would be 1/1000 of the input, or 1 mW. 27 dB is 3 dB less than 30 dB. Thus the output would be twice 1 mW, or 2 mW. It really is quite simple. If we have multiples of 10, as in the previous table, or 3 up or 3 down from these multiples, we can work it out in our heads, without pencil and paper.

Look at this next example:

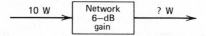

Working it out with pencil and paper, we see that the output is approximately 40 W. Here we have a multiple of 4. A 6-dB gain represents approximately a 4-times power gain. Likewise, a 6-dB loss would represent approximately $\frac{1}{4}$ the power output. Now we should be able to work out many combinations without resorting to pencil and paper.

Consider a network with a 33-dB gain with an input level of 0.15 W. What would be the output? 30 dB represents multiplying the input power by 1000, and 3 additional dB doubles it. In this case the input power is multiplied by 2000. Thus the answer is $0.15 \times 2000 = 300$ W.

The table below may further assist the reader regarding the use of decibels as power ratios.

	Approximate Power Ratio	
Decibels	Losses	Gains
1	0.8	1.25
2	0.63	1.6
3	0.5	2.0
4	0.4	2.5
5	0.32	3.2
6	0.25	4.0
7	0.2	5.0
8	0.16	6.
9	0.125	8.0
10	0.1	10.0

The voltage and current ratios, owing to the squared term, result in twice the decibel level as for power. Remember

$$\text{Power} = I^2 R = \frac{E^2}{R}$$

Thus

$$dB \text{ (voltage)} = 20 \log \frac{E_2}{E_1}$$

or

$$dB \text{ (current)} = 20 \log \frac{I_2}{I_1}$$

when relating current and voltage ratios. E_2 = higher voltage, E_1 = lower voltage; I_2 = higher current, I_1 = lower current.

When using the current and voltage relationships shown above, keep in mind that they must be compared against like impedances. For instance E_2 may not be taken at a point of 600 ohms (Ω) impedance and E_1 at a point of 900 Ω.

Example 1. How many decibels correspond to a voltage ratio of 100?

$$dB = 20 \log \frac{E_2}{E_1}, \text{ where } E_2/E_1 = 100$$

$$= 20 \log 100 = 40 \text{ dB}$$

Example 2. If an amplifier has a 30-dB gain, what voltage ratio does the gain represent? Assume equal impedances at input and output of the amplifier.

$$30 = 20 \log \frac{E_2}{E_1}$$

$$\frac{E_2}{E_1} = 31.6$$

Thus the ratio is 31.6:1.

1.4 BASIC DERIVED DECIBEL UNITS

1.4.1 The dBm

Up to now all reference to decibels has referred to ratios or relative units. We *cannot* say the output of an amplifier is 33 dB. We can say that an amplifier has a gain of 33 dB or that a certain attenuator has a 6-dB loss. These figures or units give no idea whatsoever of absolute level. Several derived decibel units do.

Perhaps the dBm is the most common of these. By definition dBm is a power level related to 1 mW. A most important relationship to remember is : 0 dBm = 1 mW. The dB formula may then be written:

$$\text{Power (dBm)} = 10 \log \frac{\text{power (mW)}}{1 \text{ mW}}$$

Consider the following example: an amplifier has an output of 20 W; what is its output in dBm?

$$\text{Power (dBm)} = 10 \log \frac{20 \text{W}}{1 \text{ mW}}$$

$$= 10 \log \frac{20 \times 10^3 \text{ mW}}{1 \text{ mW}} = +43 \text{ dBm}$$

(The plus sign indicates that the quantity is above the level of reference, 0 dBm.)

Let us try another example: the input to a network is 0.0004 W; what is the input in dBm?

$$\text{Power (dBm)} = 10 \log \frac{0.0004 \text{ W}}{1 \text{ mW}} = 10 \log 4 \times 10^{-1} \text{ mW}$$

$$= -4 \text{ dBm (approximately)}$$

(The minus sign in this case tells us that the level is below reference, 0 dBm or 1 mW.)

1.4.2   The dBW

The dBW is used extensively in microwave applications; it is an absolute decibel unit and may be defined as decibels referred to 1 W.

$$\text{Power level (dBW)} = 10 \log \frac{\text{power (W)}}{1 \text{ W}} \qquad (1.2)$$

Remember the following relationships:

$$+30 \text{ dBm} = 0 \text{ dBW} \qquad (1.3)$$

$$-30 \text{ dBW} = 0 \text{ dBm} \qquad (1.4)$$

Consider this network:

Its output level in dBW is $+20$ dBW. Remember that the gain of the network is 20 dB or 100. The output is 100 W or $+20$ dBW.

Table 1.2, a table of equivalents, may be helpful.

Table 1.2

dBm	dBW	Watts	dBm	dBW	Milliwatts
$+66$	$+36$	4000	$+30$	0	1000
$+63$	$+33$	2000	$+27$	-3	500
$+60$	$+30$	1000	$+23$	-7	200
$+57$	$+27$	500	$+20$	-10	100
$+50$	$+20$	100	$+17$	-13	50
$+47$	$+17$	50	$+13$	-17	20
$+43$	$+13$	20	$+10$	-20	10
$+40$	$+10$	10	$+7$	-23	5
$+37$	$+7$	5	$+6$	-24	4
$+33$	$+3$	2	$+3$	-27	2
$+30$	0	1	0	-30	1
			-3	-33	0.5
			-6	-36	0.25
			-7	-37	0.20
			-10	-40	0.1

1.4.3 The dBmV

The absolute decibel unit dBmV is used widely in video transmission. A voltage level may be expressed in decibels above or below 1 mV across 75 Ω, which is said to be the level in decibel-millivolts or dBmV. In other words,

$$\text{Voltage level (dBmV)} = 20 \log_{10} \frac{\text{voltage (mV)}}{1 \text{ mV}} \tag{1.5}$$

when the voltage is measured at *the 75-Ω impedance level*. Simplified,

$$\text{dBmV} = 20 \log_{10} \text{ (voltage in millivolts at 75-}\Omega \text{ impedance)} \tag{1.6}$$

Table 1.3 may prove helpful.

Table 1.3

Rms voltage across 75 Ω	dBmV
10 V	+80
2 V	+66
1 V	+60
10 mV	+20
2 mV	+ 6
1 mV	0
50 μV	− 6
316 μV	−10
200 μV	−14
100 μV	−20
10 μV	−40
1 μV	−60

1.5 THE NEPER

A transmission unit used in a number of northern European countries as an alternate to the decibel is the neper (N). To convert decibels to nepers, multiply the number of decibels by 0.1151. To convert nepers to decibels, multiply the number of nepers by 8.686. Mathematically,

$$N = \tfrac{1}{2} \log_e \frac{P_2}{P_1} \tag{1.7}$$

where P_2 and P_1 are the higher and lower powers, respectively, and $e = 2.718$, the base of the natural or Naperian logarithm. A common derived unit is the decineper or dN. A decineper is $\frac{1}{10}$ of a neper.

1.6 ADDITION OF POWER LEVELS IN DB
(DBM/DBW OR SIMILAR ABSOLUTE LOGARITHM UNITS)

Adding decibels corresponds to the multiplying of power ratios. Care must be taken when adding or subtracting absolute decibel units such as dBm, dBW, and some noise units. Consider the combining network below, which is theoretically lossless:

What is the resultant output? Answer +33 dBm

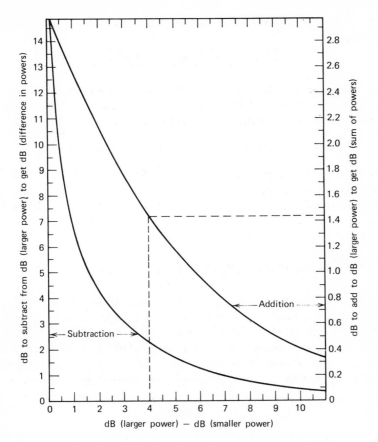

Figure 1.1 Decibels corresponding to the sum or difference of two levels.

Figure 1.1 is a curve for directly determining the level in absolute decibel units corresponding to the sum or difference of two levels, the values of which are known in terms of decibels with respect to some reference.

As an example, let us add two power levels, 10 dBm and 6.0 dBm. Take the difference between them, 4 dB. Spot this value on the horizontal scale (the abscissa) on the curve. Project the point upward to where it intersects the "addition" curve (the upper curve). Take the corresponding number to the right and add it to the larger level. Thus

$$10 \text{ dBm} + 1.45 \text{ dB(m)} = 11.45 \text{ dBm}$$

Suppose we subtract the 6.0 dBm signal from the 10 dBm signal. Again the difference is 4 dB. Spot this value on the horizontal scale as before. Project the point upward to where it meets the "subtraction" curve (the lower curve). Take the corresponding number and subtract it from the larger level. Thus

$$10 \text{ dBm} - 2.3 \text{ dB(m)} = 7.7 \text{ dBm}$$

When it is necessary to add equal absolute levels expressed in decibels, add 10 log (the number of equal powers) to the level value. For example, add four signals of $+10$ dBm each. Thus

$$10 \text{ dBm} + 10 \log 4 = 10 \text{ dB(m)} + 6 \text{ dB} = +16 \text{ dBm}$$

When there are more than two levels to be added and they are not of equal value, proceed as follows. Pair them and sum the pairs, using Figure 1.1. Sum the resultants of the pairs in the same manner until one single resultant is obtained.

1.7 NORMAL DISTRIBUTION—STANDARD DEVIATION

A normal or Gaussian distribution is a binomial distribution where n, the number of points plotted (number of events), approaches infinity (Ref. 11, pp. 121, 122 and 43). A distribution is an arrangement of data. A frequency distribution is an arrangement of numerical data according to size and magnitude. The normal distribution curve (Figure 1.2) is a symmetrical distribution. A nonsymmetrical frequency distribution curve is one in which the distributions extend further in one direction than the other. This

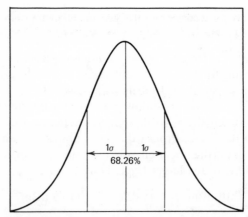

Figure 1.2 A normal distribution curve showing one standard deviation measured off either side of the arithmetic mean.

type of distortion is called skew. The peak of the normal distribution curve is called the point of central tendency, and its measure is its average. This is the point where the group of values tend to cluster.

The dispersion is the variation, scatteration, of data, or the lack of tendency to congregate (Ref. 17). The range is the simplest measure of dispersion and is the difference between maximum and minimum values of a series. The mean deviation is another measure of dispersion. In a frequency distribution, ignoring signs, it is the average distance of items from a measure of the central tendency.

The standard deviation is the root mean square (rms) of the deviations from the arithmetic mean and is expressed by the small Greek letter sigma:

$$\sigma = \sqrt{\frac{\Sigma(X^2)}{N}} \qquad (1.8)$$

where X = deviations from the arithmetic mean $(X - \bar{X})$ and N = total number of items.

The following expressions are useful when working with standard deviations; they refer to a "normal" distribution:

- The mean deviation = 0.7979σ.
- Measure off on both sides of the arithmetic mean one standard deviation; then 68.26% of all samples will be included within the limits.
- For two standard deviations measured off, 95.46% of all values will be included.
- For three standard deviations, 99.73% will be included.

These last three items relate to *exact* normal distributions. In cases where the distribution has moderate skew, approximate values are used, such as 68% for 1σ, 95% for 2σ, etc.

1.8 THE SIMPLE TELEPHONE CONNECTION

Two people may speak to one another over a distance by connecting two telephone subsets together with a pair of wires and a common microphone battery supply. As the wires are extended (i.e., the distance between the talkers is increased), the speech power level decreases until at some point, depending on the distance, the diameter of the wire, and the mutual capacitance between each wire in the pair, communication becomes unacceptable. For example, in the early days of telephony in the United States, it was noted that a telephone connection including as much as 30 mi (48 km) of 19-gauge nonloaded cable was at about the limit of useful transmission.

Suppose several people want to join the network. We could add them in parallel (bridge them together). As each is added, however, the efficiency decreases because we have added subsets in parallel and, as a result, the impedance match between subset and line deteriorates. Besides, each party can overhear what is being said between any two others. Lack of privacy may be a distinct disadvantage at times.

This can be solved by using a switch so that the distant telephone may be selected. Now a signaling system must be developed so that the switch can connect the caller to the distant telephone. A system of monitoring or supervision will also be required so on-hook and off-hook conditions may be known by the switch as well as to permit line seizure by a subscriber. (For a discussion of signaling, see Appendix C.)

Now extend the system again, allowing several switches to be used interconnected by trunks (junctions). Because of extension of the two-wire system without amplifiers, a reduced signal level at many of the subscribers' telephone subsets may be experienced. Now we start to reach into the transmission problem. A satisfactory signal is not being delivered to some subscribers owing to line losses because of excessive wire line lengths. Remember that line loss increases with length.

Before delving into methods of improving subscriber signal level and satisfactory signal-to-noise ratio, we must deal with basic voice channel criteria. In other words, just what are we up against? Consider also that we may want to use these telephone facilities for other types of communication such as telegraph, data, facsimile, and video transmission. The voice channel (telephone channel) criteria covered below are aimed essentially at speech transmission. However, many parameters affecting speech most certainly have bearing on the transmission of other types of signals, and other specialized criteria are peculiar to these other types of transmission. These are treated in depth in later chapters, where they become more meaningful. Where possible, cross reference is made.

Before going on, refer to the simplified sketch (Figure 1.3) of a basic telephone connection. The sketch contains all the basic elements that will deteriorate the signal from source to sink. The medium may be wire or

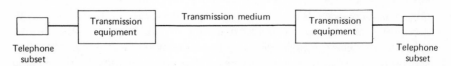

Figure 1.3   A simplified telephone transmission system.

radio or combinations of the two. Other transmission equipment may be used to enhance the medium by extending or expanding it. This equipment might consist of amplifiers, multiplex devices, and other signal processors such as compandors, voice terminals, and so forth.

1.9 THE PRACTICAL TRANSMISSION OF SPEECH

The telephone channel, hereafter called the voice channel, may be described technically using the following parameters:

- Nominal bandwidth.
- Attenuation distortion (frequency response).
- Phase shift.
- Noise and signal-to-noise ratio.
- Level.

Return loss, singing, stability, echo, reference equivalent, and some other parameters deal more with the voice channel in a network and are discussed at length when we look at a transmission network later.

1.9.1 Bandwidth

The range between the lowest and highest frequencies used for a particular purpose may be defined as bandwidth. For our purposes we should consider bandwidth as those frequencies within which a performance characteristic of a device is above certain specified limits. For filters, attenuators, and amplifiers, these limits are generally taken to be where a signal will fall 3 dB below the average level in the passband or below the level at a reference frequency (1000 Hz in the United States and Canada, 800 Hz in Europe). These 3-dB points are by definition half-power points. The nominal bandwidth of a voice channel is often 4 kHz. The actual usable bandwidth is more on the order of 3100 Hz for most modern carrier equipment. The bandwidth of a television transmitter may be defined as 6 MHz and that of a synchronous earth satellite repeater, 500 MHz.

1.9.2 Speech Transmission—The Human Factor

Frequency components of speech may be found between 20 Hz and 20 kHz. The frequency response of the ear (i.e., how it reacts to different frequencies) is a nonlinear function between 30 Hz and 30 kHz, however, the major

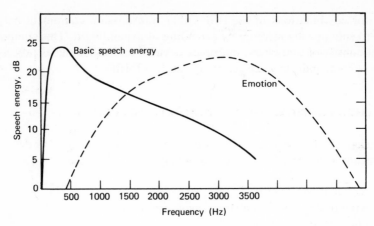

Figure 1.4 Energy and emotion distribution in speech (from *Bell Syst. Tech. J.,* July 1931).

intelligence and energy content exists in a much narrower band. For energy distribution see Figure 1.4. The emotional content, which transfers intelligence, is carried in a band which lies above the main energy portion. Tests have shown that low frequencies up to 600–700 Hz add very little to intelligibility of a signal to the human ear, but in this very band much of the voice energy is transferred (solid line in Figure 1.4). The dotted line in Figure 1.4 shows the portion of the frequency band that carries emotion. From this it can be seen that for economical transfer of speech intelligence, a band much narrower than 20–20,000 Hz is necessary. In fact the standard bandwidth of a voice channel is 300–3400 Hz (CCITT Recs. G.132, G.151A). As is shown later, this bandwidth is a compromise between what a telephone subscriber demands (Figure 1.4) and what can be provided to him economically. However, many telephone subsets have a response range no greater than approximately 500–3000 Hz. This is shown in Figure 1.5, where the response of the more modern Bell System 500 telephone set is compared to the older 302 set.

1.9.3 Attenuation Distortion

A signal transmitted over a voice channel suffers various forms of distortion. That is, the output signal from the channel is distorted in some manner such that it is not an exact replica of the input. One form of distortion is called attenuation distortion and is the result of less than perfect amplitude-frequency response. If attenuation distortion is to be avoided, all frequencies

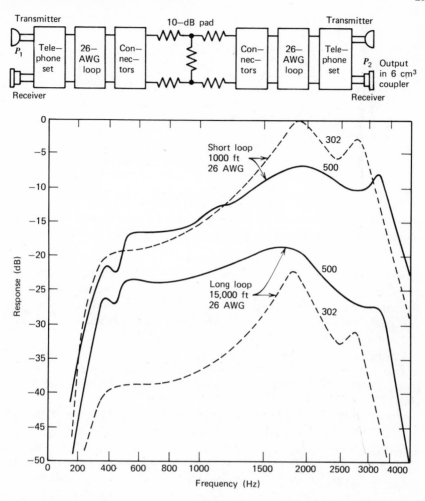

Figure 1.5 Comparison of overall response. W. F. Tuffnell, "500-Type Telephone Set," *Bell Lab. Rec.*, **29**, 414–418 (September 1951). Copyright © 1951, by Bell Telephone Laboratories, Inc.

within the passband should be subjected to the same loss (or gain). On typical wire systems higher frequencies in the passband are attenuated more than lower ones, although inductive loading of voice frequency circuits is used to reduce attenuation distortion. In carrier equipment the filters used tend to attenuate frequencies around band center the least, and attenuation increases as the band edges are approached. Figure 1.6 is a

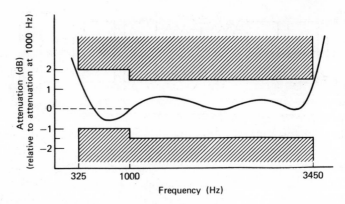

Figure 1.6 A typical attenuation distortion curve for a voice channel. Note that the reference frequency in this case is 1000 Hz.

good example of this. The crosshatched area in the figure expresses the specified limits of attenuation distortion, and the solid line shows measured distortion on typical carrier equipment for the channel band (see Chapter 3). The reader should remember that any practical communication channel will suffer some form of attenuation distortion.

Attenuation distortion across the voice channel is measured compared to a reference frequency. CCITT specifies the reference frequency at 800 Hz. However, 1000 Hz is used more commonly in North America (see Figure 1.6).

For example, one requirement may state that between 600 and 2800 Hz the level will vary by not more than −1, +2 dB, where the plus sign means more loss and the minus sign means less loss. Thus if a signal at −10 dBm is placed at the input of the channel, we would expect −10 dBm at the output at 800 Hz (if there was no overall loss or gain), but at other frequencies we could expect a variation between −1 and +2 dB. For instance, we might measure the level at the output at 2500 Hz at −11.9 dBm and at 1000 Hz at −9 dBm.

CCITT recommendations for attenuation distortion may be found in Volume III, Recs. G.132 and G.151A. Figure 2.15 is taken from Rec. G.132 and shows permissible variation of attenuation between 300 and 3400 Hz. Often the requirement is stated as a slope in decibels. The slope is the maximum excursion that levels may vary in a band of interest about a certain frequency. A slope of 5 dB may be a curve with an excursion from −0.5 to +4.5 dB, 3 to +2 dB, and so forth. As links in a system are added in tandem, to maintain a fixed attenuation distortion across the sytem, the slope requirement for each link becomes more severe.

1.9.4 Phase Shift (Envelope Delay Distortion)

One may look at a voice channel or any bandpass as a bandpass filter. A signal takes a finite time to pass through the filter. This time is a function of the velocity of propagation. The velocity of propagation tends to vary with frequency, increasing toward band center and decreasing toward band edge, usually in the form of a parabola. (See Figure 1.7.)

The finite time it takes a signal to pass through the total extension of a voice channel or any other network is called delay. Absolute delay is the delay a signal experiences passing through the channel at a reference frequency. But we see that the propagation time is different for different frequencies. This is equivalent to phase shift. If the phase shift changes uniformly with frequency, the output signal will be a perfect replica of the input and there will be no distortion, whereas if the phase shift is nonlinear with respect to frequency, the output signal is distorted (i.e., it is not a perfect replica of the input). Delay distortion is usually expressed in milliseconds or microseconds about a reference frequency. Other names for this type of distortion are envelope delay distortion and group delay. In

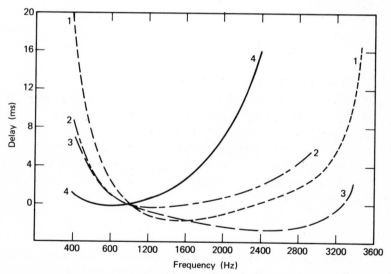

Figure 1.7 Comparison of envelope delay in some typical voice channels. Curves 1 and 3 represent the delay in several thousand miles of a toll-quality carrier system. Curve 2 shows the delay produced by 100 mi of loaded cable. Curve 4 shows the delay in 200 mi of heavily loaded cable. Courtesy *GTE Lenkurt Demodulator,* San Carlos, Calif.

telephone systems the basic contributors to phase delay distortion are filters, particularly those used in carrier equipment.

In commercial telephony the high frequency harmonic components, produced by the discontinuous nature of speech sounds, arrive later than the fundamental components and produce sounds which may be annoying, but do not appreciably reduce intelligibility. With present handset characteristics the evidence is that the human ear is not very sensitive to phase distortions which develop in the circuit. Although a phase delay of 12 ms between the band limits is noticeable, the transmission in commercial telephone systems often contains distortions greatly in excess of this minimum.

Owing to the larger amount of delay distortion in a telephone channel, as measured in its band and relative to a point of minimum delay, the usefulness of the entire telephone channel between its 3-dB cutoff points is severely restricted for the transmission of other-than-voice signals (e.g., data—see Chapter 8).

For the transmission of information which is sensitive to delay distortion, such as medium speed digital signals, it is necessary to restrict occupancy to that part of the telephone channel in which the delay distortion can be tolerated or equalized at reasonable cost.

Applicable CCITT Recommendations are Recs. G.114 and G.133.

1.9.5 Level

GENERAL

In most systems when we refer to level, we refer to a power level which may well be in dBm, dBW, or other power units. One notable exception is video, which uses voltage, usually in dBmV.

Level is an important system parameter. If levels are maintained at too high a point, amplifiers become overloaded, with resulting increases in intermodulation products or crosstalk. If levels are too low, customer satisfaction may suffer.

REFERENCE LEVEL POINTS

System levels usually are taken from a level chart or reference system drawing made by a planning group or as part of an engineered job. On the chart a 0 TLP (zero test level point) is established. A test level point is a location in a circuit or system at which a specified test tone level is expected during alignment. A 0 TLP is a point at which the test tone level should be 0 dBm.

From the 0 TLP other points may be shown using the unit dBr (dB reference). A minus sign shows that the level is so many decibels below reference,
and a positive sign that the level is so many decibels above reference. The
unit dBm0 is an absolute unit of power in dBm referred to the 0 TLP. dBm
can be related to dBr and dBm0 by the following formula:

$$dBm = dBm0 + dBr \qquad (1.9)$$

THE VU (VOLUME UNIT)

One measure of level is the VU, which stands for volume unit. Such a unit
is used to measure the power level (volume) of programs (broadcast) and
certain other types of speech or music. VU meters are usually kept on line
to measure volume levels of program or speech material being transmitted.
If a simple dB meter or voltmeter is bridged across the circuit to monitor
the program volume level, the indicating needle tries to follow every fluctuation of speech or program power and is difficult to read; besides, the
reading will have no real meaning. To further complicate matters, different
meters will probably read differently because of differences in their damping and ballistic characteristics.

The indicating instrument used in VU meters is a dc milliammeter having
a slow response time and damping slightly less than critical. If a steady sine
wave is suddenly impressed on the VU meter, the pointer or needle will
move to within 90% of the steady-state value in 0.3 s and overswing the
steady-state value by no more than 1.5%.

The standard volume indicator (U.S.), which includes the meter and an
associated attenuator, is calibrated to read 0 VU when connected across
a 600-Ω circuit (voice pair) carrying a 1 mW of sine wave power at any
frequency between 35 and 10,000 Hz. For complex waves such as music and
speech, a VU meter will read some value between average and peak of the
complex wave. The reader must remember that there is no simple relationship between the volume measured in volume units and the power of a
complex wave. It can be said, however, that for a continuous sine wave
signal across 600 Ω 0 dBm = 0 VU by definition, or that the reading in
dBm and VU are the same for continuous simple sine waves in the VF
(voice frequency) range. For a complex signal subtract 1.4 from the VU
reading and the result will be approximate talker power in dBm.

Talker volumes, or levels of a talker at the telephone subset, vary over
wide limits, both for long-term average power and peak power. Based on
comprehensive tests by Holbrook and Dixon (Ref. 1) "mean talker" average
power varies between −10 and −15 VU, with a mean of −13 VU.

1.9.6 Noise

GENERAL

"Noise, in its broadest definition, consists of any undesired signal in a communication circuit" (Ref. 2). The subject of noise and its reduction is probably the most important that a transmission engineer must face. It is noise that is the major limiting factor in telecommunication system performance. Noise may be divided into four categories:

1. Thermal noise (Johnson noise).
2. Intermodulation noise.
3. Crosstalk.
4. Impulse noise.

THERMAL NOISE

Thermal noise is that noise occurring in all transmission media and in all communication equipment arising from random electron motion. It is characterized by a uniform distribution of energy over the frequency spectrum and a normal Gaussian distribution of levels.

Every equipment element and the transmission medium contribute thermal noise to a communication system provided the temperature of that element of medium is above absolute zero. Thermal noise is the factor which sets the lower limit for the sensitivity of a receiving system. Often this noise is expressed as a temperature, usually given in degrees referred to absolute zero (degrees Kelvin).

Thermal noise is a general expression referring to noise based on thermal agitations. The term "white noise" refers to the average uniform spectral distribution of energy with respect to frequency. Thermal noise is directly proportional to bandwidth and temperature. The amount of thermal noise to be found in 1 Hz of bandwidth in an actual device is

$$P_n = kT \text{ (W/Hz)} \qquad (1.10)$$

where k = Boltzmann's constant
$\quad$ = $1.3803 \ (10^{-23}) \ \text{J}/°\text{K}$
$\quad T$ = absolute temperature (°K) of thermal noise

At room temperature, $T = 17°\text{C}$ or $290°\text{K}$.

$$P_n = 4.00 \ (10^{-21}) \ \text{W/Hz of bandwidth}$$
$$= -204 \ \text{dBW/Hz of bandwidth}$$
$$= -174 \ \text{dBm/Hz of bandwidth}$$

For a band-limited system (i.e., a system with a specific bandwidth)

$$P_n = kTB \ (\text{W}) \tag{1.11}$$

B refers here to what is called noise bandwidth (Hz). At 0°K

$$P_n = -228.6 \ \text{dBW/Hz of bandwidth} \tag{1.12}$$

For a band-limited system

$$P_n = -228.6 \ \text{dBW} + 10 \log T + 10 \log B \tag{1.13}$$

Example 1. Given a receiver with an effective noise temperature of 100°K and a 10-MHz bandwidth, what thermal noise level may be expect at its output?

$$
\begin{aligned}
P_n &= -228.6 \ \text{dBW} + 10 \log 1 \times 10^2 + 10 \log 1 \times 10^7 \\
&= -228.6 + 20 + 70 \\
&= -138.6 \ \text{dBW}
\end{aligned}
$$

Example 2. Given an amplifier with an effective noise temperature of 10,000°K and a 10-MHz bandwidth, what thermal noise level may we expect at its output?

$$
\begin{aligned}
P_n &= -228.6 \ \text{dBW} + 10 \log 1 \times 10^4 + 10 \log 1 \times 10^7 \\
&= -228.6 + 40 + 70 \\
&= -118.6 \ \text{dBW}
\end{aligned}
$$

From the examples it can be seen that there is little direct relationship between physical temperature and effective noise temperature.

INTERMODULATION NOISE

Intermodulation noise is the result of the presence of intermodulation products. Pass two signals with frequencies F_1 and F_2 through a nonlinear device or medium. Intermodulation products will result which are spurious frequencies. These frequencies may be present either inside or outside the band of interest for the device. Intermodulation products may be produced from harmonics of the signals in question, either as products between harmonics or as one or the other or both signals themselves.

The products result when the two (or more) signals beat together or "mix." Look at the "mixing" possibilities when passing F_1 and F_2 through a nonlinear device. The coefficients indicate first, second, or third harmonics.

- Second order products $F_1 \pm F_2$.
- Third order products $F_1 \pm 2F_2$; $2F_1 \pm F_2$.
- Fourth order products $2F_1 \pm 2F_2$; $3F_1 \pm F_2$.

Devices passing multiple signals, such as multichannel radio equipment, develop intermodulation products that are so varied that they resemble white noise.

Intermodulation noise may result from a number of causes:

- Improper level setting. Too high a level input to a device drives the device into its nonlinear operating region (overdrive).
- Improper alignment causing a device to function nonlinearly.
- Nonlinear envelope delay.

To sum up, intermodulation noise results from either a nonlinearity or a malfunction having the effect of nonlinearity. The cause of intermodulation noise is different from thermal noise; however, its detrimental effects and physical nature are identical to those of thermal noise, particularly in multichannel systems carrying complex signals.

CROSSTALK

Crosstalk refers to unwanted coupling between signal paths. Essentially there are three causes of crosstalk. The first is the electrical coupling between transmission media, for example, between wire pairs on a VF (voice frequency) cable system. The second is poor control of frequency response (i.e., defectivy filters or poor filter design), and the third is the nonlinearity performance in analog (FDM) multiplex systems. Crosstalk has been categorized into two types:

- *Intelligible crosstalk:* at least four words are intelligible to the listener from extraneous conversation(s) in a 7-s period (Ref. 16, 3rd ed., p. 46).
- *Unintelligible crosstalk:* any other form of disturbing effects of one channel upon another. *Babble* is one form of unintelligible crosstalk.

Intelligible crosstalk presents the greatest impairment because of its distraction to the listener. One point of view is that the distraction is caused by fear of loss of privacy. Another is that the annoyance is caused primarily by the user of the primary line consciously or unconsciously trying to understand what is being said on the secondary or interfering circuits; this would be true for any interference that is syllabic in nature.

Received crosstalk varies with the volume of the disturbing talker, the loss from the disturbing talker to the point of crosstalk, the coupling loss between the two circuits under consideration, and the loss from the point of crosstalk to the listener.

As far as this discussion is concerned, the controlling element is the coupling loss between the two circuits under consideration. Talker volume or level is covered in Section 1.9.5. The effects of crosstalk are subjective and other factors also have to be considered when the crosstalk impairment is to be measured. Among these factors are the type of people who use the channel, the acuity of listeners, traffic patterns, and operating practices.

Crosstalk coupling loss can be measured quantitatively with precision between a given sending point on a disturbing circuit and a given receiving point on a disturbed circuit. Essentially, then, it is the simple measurement of transmission loss in decibels between the two points. Between carrier circuits crosstalk coupling in most usually flat. In other words the amount of coupling experienced at one frequency will be nearly the same for every other frequency in the voice channel. For speech pairs the coupling is predominantly capacitive, and coupling loss usually has an average slope of 6 dB/octave. See the annex to CCITT Rec. G.134 which treats crosstalk measurements.

Two other units are also commonly used to measure crosstalk. One expresses the coupling in decibels above "reference coupling," and uses dBx for the unit of measure. The dBx was invented to allow crosstalk coupling to be expressed in positive units. The "reference coupling" is taken as a coupling loss of 90 dB between disturbing and disturbed circuits. Thus crosstalk coupling in dBx is equal to 90 minus the coupling loss in dB. If crosstalk coupling loss is 60 dB, we have 30 dBx crosstalk coupling. Thus, by definition, 0 dBx = −90 dBm at 1000 Hz.

The second unit is the crosstalk unit, abbreviated CU. Where the impedances of the disturbed and disturbing circuits are the same, the number of crosstalk units is one million times the ratio of the induced crosstalk voltage or current to the disturbing voltage or current. When the impedances are not equal,

$$CU = 10^6 \sqrt{\frac{\text{crosstalk signal power}}{\text{disturbing signal power}}} \tag{1.14}$$

Figure 1.8 relates all three units used in measuring crosstalk. CCITT recommends crosstalk criteria in Rec. G.151D (p. 4)

The percentage chance of intelligible crosstalk on a circuit is defined by the crosstalk index. North American practice arbitrarily is to allow that on no more than 1% of calls will a customer hear a foreign conversation, which we have defined as intelligible crosstalk. The design objective is 0.5%. The graph in Figure 1.9 may be used for guidance. It relates customer reaction to crosstalk, and crosstalk index to crosstalk coupling.

All forms of unintelligible crosstalk are covered above; its nature is very similar to intermodulation noise.

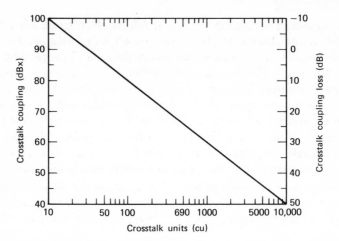

Figure 1.8 Relations between crosstalk measuring units (Ref. 10). Copyright © 1961, by American Telephone and Telegraph Co.

IMPULSE NOISE

This type of noise is noncontinuous, consisting of irregular pulses or noise spikes of short duration and of relatively high amplitude. Often these spikes are called "hits." Impulse noise degrades voice telephony only marginally if at all. However, it may seriously degrade the error rate on a data transmission circuit, and the subject is covered more in depth in Chapter 8.

NOISE MEASUREMENT UNITS

The interfering effect of noise on speech telephony is a function of the response of the human ear to specific frequencies in the voice channel as well as of the type of subset used.

When noise measurement units were first defined, it was decided that it would be convenient to measure the relative interfering effect of noise on the listener as a positive number. The level of a 1000-Hz tone at −90 dBm or 10^{-12} W (1 pW) was chosen by the U.S. Bell System because a tone whose level is less than −90 dBm is not ordinarily audible. Such a negative threshold meant that all noise measurements used in telephony would be greater than this number, or positive. The telephone subset then in early universal use in North America was the Western Electric 144 handset. The noise measurement unit was the dB-rn or dBrn. 0 dBrn = −90 dBm at 1000 Hz, rn standing for reference noise.

With the 144-type handset as a test receiver and with a wide distribution of "average" listeners, it was found that a 500-Hz sinusoidial signal had

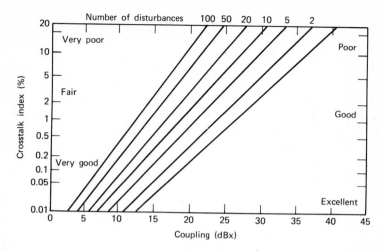

Figure 1.9 Crosstalk judgment curves (Ref. 14). Copyright © 1964, by Bell Telephone Laboratories, Inc.

to have its level increased by 15 dB to have the same interfering effect on the "average" listener over the 1000-Hz reference. A 3000-Hz signal required an 18-dB increase to have the same interfering effect, 6 dB at 800 Hz, etc. A curve showing the relative interfering effects of sinusoidial tones compared to a reference frequency is called a weighting curve. Artificial filters are made with a response resembling the weighting curve. These filters, normally used on noise measurement sets, are called weighting networks.

Subsequent to the 144 handset, Western Electric Company developed the F1A handset, which had a considerably broader response than the older handset but was 5 dB less sensitive at 1000 Hz. The reference level for this type of handset was −85 dBm. The new weighting curve and its noise measurement weighting network were denoted F1A (i.e., an F1A line weighting curve, and F1A weighting network). The noise measurement unit was the dBa, or dB adjusted.

A third, more sensitive handset, is now in use in North America, giving rise to the C-message line weighting curve and its companion noise measurement unit, the dBrnC. It is 3.5 dB more sensitive at the 1000-Hz reference frequency than F1A, and 1.5 dB less sensitive than the 144 type weighting. Rather than a new reference power level (−88.5 dBm), the reference power level of −90 dBm was maintained.

Figure 1.10 compares the various noise weighting curves now in use. Table 1.4 compares weighted noise units.

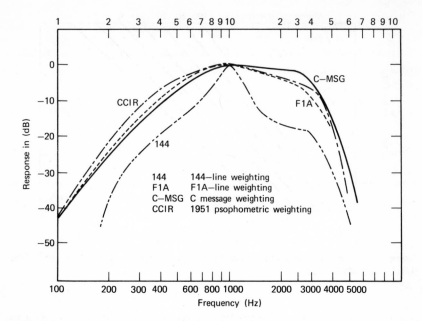

Figure 1.10   Line weightings for telephone (voice) channel noise.

One important weighting curve and noise measurement unit has yet to be mentioned. The curve is the CCIR (CCITT) psophometric weighting curve. The noise measurement units associated with this curve are dBmp and pWp (dBm psophometrically weighted and picowatts psophometrically weighted, respectively). The reference frequency in this case is 800 Hz rather than 1000 Hz.

Consider now a 3-kHz band of white noise (flat, i.e., not weighted). Such a band is attenuated 8 dB when measured by a noise measurement set using a 144 weighting network, 3 dB using F1A weighting, 2.5 dB for CCIR/ CCITT weighting, and 1.5 dB rounded off to 2.0 dB for C-message weighting. Table 1.4 may be used to convert from one noise measurement unit to another.

CCITT states in Rec. G.223 that, "If uniform-spectrum random noise is measured in a 3.1 kHz band with a flat attenuation frequency characteristic, the noise level must be reduced 2.5 dB to obtain the psophometric power level. For another bandwidth, B, the weighting factor will be equal to

$$2.5 + 10 \log B/3.1 \text{ dB} \tag{1.15}$$

When $B = 4$ kHz for example, this formula gives a weighting factor of 3.6 dB."

1.10 SIGNAL-TO-NOISE RATIO

The transmission system engineer deals with signal-to-noise ratio probably more frequently than any other criteria when engineering a telecommunication system.

Signal-to-noise ratio expresses in decibels the amount by which a signal level exceeds its corresponding noise.

As we review the several types of material to be transmitted, each will require a minimum signal-to-noise ratio to satisfy the customer or to make the receiving end instrument function within certain specified criteria. We might require the following signal-to-noise ratios with corresponding end instruments:

$$\left.\begin{matrix} \text{Voice} & \text{30 dB} \\ \text{Video} & \text{45 dB} \end{matrix}\right\} \text{Based on customer satisfaction}$$

$$\text{Data} \quad \text{15 dB} \quad \text{Based on a specified error rate}$$

In Figure 1.11 the 1000-Hz signal has a signal-to-noise ratio (S/N) of 10 dB. The level of the noise is 5 dBm and the signal, 15 dBm. Thus

$$S/N_{dB} = \text{level}_{\text{signal(dBm)}} - \text{level}_{\text{noise(dBm)}} \tag{1.16}$$

1.11 NOISE FIGURE

It has been established that all networks, whether passive or active, and all other forms of transmission media contribute noise to a transmission system. Noise figure is a measure of the noise produced by a practical network

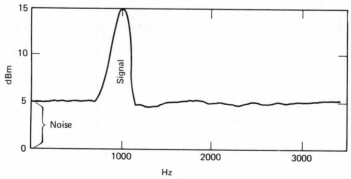

Figure 1.11 Signal-to-noise ratio (S/N).

Table 1.4 Conversion Chart, Psophometric, F1A and C-message Noise Units

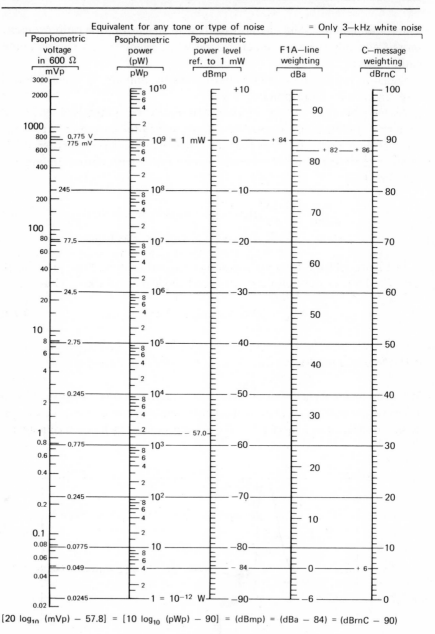

Equivalent for any tone or type of noise = Only 3–kHz white noise

Psophometric voltage in 600 Ω (mVp)	Psophometric power (pWp)	Psophometric power level ref. to 1 mW (dBmp)	F1A–line weighting (dBa)	C–message weighting (dBrnC)

$$[20 \log_{10}(mVp) - 57.8] = [10 \log_{10}(pWp) - 90] = (dBmp) = (dBa - 84) = (dBrnC - 90)$$

compared to an ideal network (i.e., one that is noiseless). For a linear system, the noise figure, *NF*, is expressed by

$$NF = \frac{S/N_{\text{in}}}{S/N_{\text{out}}} \tag{1.17}$$

It simply relates the signal-to-noise ratio of output signal from the network to the signal-to-noise ratio of the input signal. From equation 1.11, thermal noise may be expressed by the basic formula, *kTB*, where $T = 290°\text{K}$ (room temperature). As we can see, the noise figure can be interpreted as the degradation of the signal to-noise ratio (S/N) by the network.

By letting *G*, the gain of the network, $= S_{\text{out}}/S_{\text{in}}$, then

$$F = \frac{N_{\text{out}}}{kTBG} \tag{1.18}$$

It should be noted that we defined the network as fully linear, so *F* has not been degraded by intermodulation noise. *F* more commonly is expressed in decibels where

$$F_{\text{dB}} = 10 \log_{10} F \tag{1.19}$$

Example.   Consider a receiver with a noise figure of 10 dB. Its output signal-to-noise ratio is 50 dB. What is its input equavalent signal-to-noise ratio?

$F_{\text{dB}} = S/N_{\text{dB}}(\text{input}) - S/N_{\text{dB}}(\text{output})$

$10 \text{ db} = S/N_{\text{input}} - 50 \text{ dB}$

$S/N_{\text{input}} = 60 \text{ dB}$

Table 1.4   Conversion Chart, Psophometric, F1A, and C-Message Noise Units

Chart basis:
 dBmp = dBa − 84
 1 mW unweighted 3-kHz white noise reads 82. dBa = 88.5 dBrnC (C-message) rounded off to 88.0 dBrnC. 1 mW into 600 Ω = 775 mV = 0 dBm = 10^9 pW

Readings of noise measuring sets when calibrated on 1 mW of test tone:
 F1A at 1000 Hz reads 85 dBa
 C-message at 1000 Hz reads 90 dBrn
 Psophometer, 800 Hz, reads 0 dBm

1.12 RELATING NOISE FIGURE TO NOISE TEMPERATURE

The noise temperature of a two-port device, a receiver, for instance, is the thermal noise that that device adds to a system. If the device is connected to a noisefree source, its equivalent noise temperature,

$$T_e = \frac{P_{ne}}{Gkdf} \tag{1.20}$$

where G = gain and df = specified small band of frequencies. T_e is referred to as the effective input noise temperature of the network and is a measure of the internal noise sources of the network.

Noise temperature of a device and its noise figure are analytically related. Thus

$$NF = 1 + \frac{T_e}{T_0} \tag{1.21}$$

where T_0 is equivalent room temperature or 290°K.

$$T_e = T_0(F - 1) \tag{1.22}$$

To convert noise figure in decibels to equivalent noise temperature (T_e) in degrees Kelvin, use the following formula:

$$NF_{dB} = 10 \log_{10} \left(1 + \frac{T_e}{290}\right) \tag{1.23}$$

Example 1. Consider a receiver with an equivalent noise temperature of 290°K; what is its noise figure?

$$NF_{dB} = 10 \log \left(1 + \frac{290}{290}\right)$$
$$NF = 10 \log 2$$
$$NF = 3 \text{ dB}$$

Example 2. A receiver has a noise figure of 10 dB; what is its equivalent noise temperature in degrees Kelvin?

$$10 \text{ dB} = 10 \log \left(1 + \frac{T_e}{290}\right)$$
$$10 = 10 \log X \quad \text{where } X = 1 + (T_e/290)$$
$$\text{Log } X = 1, \, X = 10$$
$$T_e = 2900 - 290$$
$$T_e = 2610°K$$

Several noise figures (*NF*) are given with their corresponding equivalent noise temperatures in Table 1.5.

Table 1.5 Noise Figure-Noise Temperature Conversion

NF_{dB}	T (°K) (approx.)	F_{dB}	T (°K) (approx.)
15	8950	6	865
14	7000	5	627
13	5500	4	439
12	4300	3	289
11	3350	2.5	226
10	2600	2.0	170
9	2015	1.5	120
8	1540	1.0	75
7	1165	0.5	35.4

1.13 EIRP (EFFECTIVE ISOTROPICALLY RADIATED POWER)

A useful tool in describing antenna performance is EIRP, or the power in dBm or dBW over an isotropic (antenna) that is radiated. An isotropic antenna does not exist in real life situations. It is an imaginary antenna that is used as a reference. By definition, an isotropic antenna radiates uniformly in *all* directions and therefore has a gain of 1 or 0 dB. The isotropic is a very handy tool to help radio engineers describe how common antennas function with regard to radiated power.

Perhaps an example will best assist in describing how we can use EIRP. Let an antenna have a 10-dB gain in a certain direction and connect a 10-kW transmitter to it via a very efficient transmission line, in effect, lossless (assume a VSWR of 1:1). Now we can describe the power radiated in a desired direction by converting 10 kW to dBW.

$$10 \text{ kW} = 40 \text{ dBW}$$

Add the 10-dB gain of the antenna; the EIRP, the effective radiated power in the desired direction (over an isotropic) is $+50$ dBW. Let us suppose that the first side lobe of the antenna was down 50 dB from the main lobe. What is the EIRP of the side lobe?

$$+50 \text{ dBW} - 50 \text{ dB} = +25 \text{ dBW}$$

Consider the following additional examples.

Example 1. A radiolink transmitter has an output of 1 W; its line losses are negligible, and the antenna gain is 35 dB. What is the EIRP on the main beam?

1 W = 0 dBW (or $+30$ dBm)
0 dBW + 35 dB = $+35$ dBW (or $+65$ dBm)

Example 2. A tropospheric scatter transmitter has a power output of 2 kW, line losses are 2 dB, and its antenna gain is 43 dB. What is the EIRP on the main beam?

$$1 \text{ kW} = +30 \text{ dBW}$$
$$2 \text{ kW} = +33 \text{ dBW}$$
$$+33 \text{ dBW} - 2 \text{ dB} + 43 \text{ dB} = +74 \text{ dBW}$$

Often we hear the term ERP used. ERP, as the reader is probably aware, means effective radiated power. We must beware of the reference used; if it is not an isotropic but a dipole,* then all EIRP values will be somewhat less.

Note that the term dBi is sometimes used when we refer to gain over an isotropic antenna.

1.14 SOME COMMON CONVERSION FACTORS

To convert	into	multiply by	Conversely, multiply by
acres	hectares	0.407	2.471
Btu	kilogram-calories	0.2520	3.969
Centigrade	Fahrenheit	$9C/5 = F - 32$	
		$9(C + 40)/5 = (F + 40)$	
circular mils	square centimeters	5.067×10^{-6}	1.973×10^{5}
circular mils	square mils	0.7854	1.273
degrees (angle)	radians	1.745×10^{-2}	57.30
kilometers	feet	3281	3.048×10^{-4}
kilowatt-hours	Btu	3413	2.930×10^{-4}
liters	gallons (liq. U.S.)	0.2642	3.785
$\log_e$ or ln	$\log_{10}$	0.4343	2.303
meters	feet	3.281	0.3048
miles (nautical)	meters	1852	5.400×10^{-4}
miles (nautical)	miles (statute)	1.1508	0.8690
miles (statute)	feet	5280	1.890×10^{-4}
miles (statute)	kilometers	1.609	0.6214
nepers	decibels	8.686	0.1151
square inches	circular mils	1.273×10^{6}	7.854×10^{-7}
square millimeters	circular mils	1973	5.067×10^{-4}

* A half-wave dipole has a gain of 1.64 or approximately 2.15 dB over an isotropic.

Boltzmann's constant $(1.38044 \pm 0.00007) \times 10^{-16}$ erg/deg
Velocity of light 2.998×10^8 m/s
               186,280 mi/s
               984×10^6 ft/s
degree of longitude
   at the equator   68.703 statute mi or 59.661 nautical mi
1 rad = $180°/\pi$ = 57.2958°
1 m = 39.3701 in. = 3.28084 ft
1° = 17.4533 mrad
e = 2.71828

REFERENCES AND BIBLIOGRAPHY

1. B. D. Holbrook and J. T. Dixon, "Load Rating Theory for Multichannel Amplifiers," *Bell Syst. Tech. J.*, 624–644 (Oct. 1939).
2. *Reference Data for Radio Engineers*, 5th ed., Howard W. Sams & Co., Indianapolis.
3. CCITT, White Books, Mar del Plata, 1968, Vol. III, *G* Recommendations.
4. *DCS Engineering Installation Manual*, DCAC 330-175-1, through Change 9, U.S. Department of Defense, Washington, D.C.
5. Military Standard 188C, U.S. Department of Defense, Washington, D.C.
6. W. Oliver, *White Noise Loading of Multi-channel Communication Systems*, Marconi Instruments.
7. Noah Kramer, "Communication Needs versus Existing Facilities," lecture given at the 1964 Planning Seminar of the North Jersey Section, IEEE.
8. *Lenkurt Demodulator*, Lenkurt Electric Company, San Carlos, Calif.: Dec. 1964, June 1965, Sept. 1965.
9. F. R. Connor, *Introductory Topics in Electronics and Telecommunication—Modulation*, Edward Arnold, London, 1973.
10. *Principles of Electricity Applied to Telephone and Telegraph Work*, American Telephone and Telegraph Company, New York.
11. H. Arkin and R. R. Colton, *Statistical Methods*, 5th ed., Barnes and Noble College Outline Series, New York.
12. Carl E. Smith, *Applied Mathematics for Radio and Communication Engineers*, Dover, New York, 1945.
13. Stanford Goldman, *Information Theory*, Dover, New York, 1953.
14. H. H. Smith, "Noise Transmission Level Terms in American and International Practice," paper, ITT Communication Systems, Paramus, N.J., 1964.
15. M. M. Rosenfeld, "Noise in Aerospace Communication," *Electro-Technology*, May 1965.
16. *Transmission Systems for Communications*, rev. 4th ed., Bell Telephone Laboratories.
17. Robert C. James (James and James), *Mathematics Dictionary*, 3rd ed., D. Van Nostrand, Princeton, N.J.

2 | TELEPHONE TRANSMISSION

2.1 GENERAL

Section 1.8 introduced the reader to the simple telephone connection. This chapter delves into telephony and problems of telephone transmission more deeply; it exclusively treats speech transmission over wire systems. Other transmission media are treated only in the abstract so that we can consider problems in telephone networks. The subscriber loop, an important segment of the telephone network, is also treated.

2.2 THE TELEPHONE INSTRUMENT

The input-output (I/O) device that provides the human interface with the telephone network is the telephone instrument or subset. It converts sound energy into electrical energy, and vice versa. The degree of efficiency and fidelity with which it performs these functions has a vital effect upon the quality of telephone service provided. The modern telephone subset consists of a transmitter (mouthpiece), receiver (earpiece), and electrical network for equalization, sidetone circuitry, and devices for signaling and supervision. All these items are contained in a device that, when mass-produced, sells for about $30.

Let us discuss transmitters and receivers for a moment.

2.2.1 Transmitters

The transmitter converts acoustic energy into electric energy by means of a carbon granule transmitter. The transmitter requires a dc potential, usually on the order of 3–5 V, across its electrodes. We call this the talk battery, and in modern systems it is supplied over the line (central battery) from the switch. (See Section 1.8.) Current from the battery flows through the carbon granules or grains when the telephone is lifted off its cradle (off-hook). When sound impinges on the diaphragm of the transmitter, variations of air pressure are transferred to the carbon and the resistance of the electrical path through the carbon changes in proportion to the pressure. A pulsating direct current results. Frequency response of carbon transmitters peak between 800 and 1000 Hz.

2.2.2   Receivers

A typical receiver consists of a diaphragm of magnetic material, often soft iron alloy, placed in a steady magnetic field supplied by a permanent magnet, and a varying magnetic field, caused by the voice currents flowing through the voice coils. Such voice currents are alternating (ac) in nature and originate at the far-end telephone transmitter. These currents cause the magnetic field of the receiver to alternately increase and decrease, making the diaphragm move and respond to the variations. As a result an acoustic pressure wave is set up, reproducing, more or less exactly, the original sound wave from the distant telephone transmitter. The telephone receiver, as a converter of electrical energy to acoustic energy, has a comparatively low efficiency, on the order of 2–3%.

Sidetone is the sound of the talker's voice heard in his own receiver. Sidetone level must be controlled. When the level is high, the natural human reaction is for the talker to lower his voice. Thus by regulating sidetone, talker levels can be regulated. If too much sidetone is fed back to the receiver, the output level of the transmitter is reduced owing to the talker lowering his voice, thereby reducing the level (voice volume) at the distant receiver, deteriorating performance.

2.3   THE TELEPHONE LOOP

We speak of the telephone subscriber as the user of the subset. As mentioned in Section 1.8, subscribers' telephone sets are interconnected via a switch or network of switches. Present commercial telephone service provides for transmission and reception on the same pair of wires that connect the subscriber to the local switch. Let us now define some terms.

The pair of wires connecting the subscriber to the local switch that serves him is the *subscriber loop*. It is a dc loop in that it is a wire pair typically supplying a metallic path for the following:

1. Talk battery for the telephone transmitter

2. An ac ringing voltage for the bell on the telephone instrument supplied from a special ringing source voltage.

3. Current to flow through the loop when the telephone instrument is taken out of its cradle, telling the switch that it requires "access" and permitting line seizure at the switching center.

4. The telephone dial that, when operated, makes and breaks the dc current on the closed loop, which indicates to the switching equipment the number of the distant telephone with which communication is desired.

The typical subscriber loop is supplied battery by means of a battery feed circuit at the switch. Such a circuit is shown in Figure 2.1; as shown, one important aspect of battery feed is line balance. Telephone battery voltage has been fairly well standardized at −48 V.

2.4 TELEPHONE LOOP LENGTH LIMITS

It is desirable from an economic viewpoint to permit subscriber loop lengths to be as long as possible. Thus the subscriber area served by a single switching center may be much larger. As a consequence, the total number of switches or telephone central offices may be reduced to a minimum. For instance if loops were limited to 4 km in length, a switching center could serve all subscribers within a radius of something less than 4 km. If 10 km were the maximum loop length, the radius of an equivalent area that

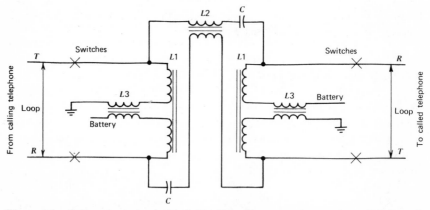

Figure 2.1 Battery feed circuit (Ref. 7). *Note.* Battery and ground are fed through inductors L3 and L1 through switch to loops. Copyright © 1961, by Bell Telephone Laboratories, Inc.

one office could cover would be extended an additional 6 km, out to a total of nearly 10 km. It is evident that to serve a large area, fewer switches (switching centers) are required for the 10-km situation than for the 4-km. The result is fewer buildings, less land to buy, fewer locations where maintenance is required, and all the benefits accruing from greater centralization, which become even more evident as subscriber density decreases, such as in rural areas.

The two basic criteria that must be considered when designing subscriber loops, and which limit their length, are the following:

- Attenuation limits (covered under what we call *transmission design*).
- Signaling limits (covered under what we call *resistance design*).

Attenuation in this case refers to loop loss in decibels (or nepers) at
- 1000 Hz in North America.
- 800 Hz in Europe and many other parts of the world.

As a loop is extended in length, its loss at reference frequency increases. It follows that at some point as the loop is extended, level will be attenuated such that the subscriber cannot hear *sufficiently well*.

Likewise, as a loop is extended in length, some point is reached where signaling and/or supervision (see Appendix C) is no longer effective. This limit is a function of the IR drop of the line. We know that R increases as length increases. With today's modern telephone sets, the first to suffer is usually the "supervision." This is a signal sent to the switching equipment requesting "seizure" of a switch circuit and, at the same time, indicating the line is busy. "Off-hook" is a term more commonly used to describe this signal condition. When a telephone is taken "off-hook" (i.e., out of its cradle), the telephone loop is closed and current flows, closing a relay at the switch. If current flow is insufficient, the relay will not close or it will close and open intermittently (chatter) such that line seizure cannot be effected.

Signaling and supervision limits are a function of the conductivity of the cable conductor and its diameter or gauge. For this introductory discussion, we can consider that the transmission limits are controlled by the same parameters

Consider a copper conductor. The larger the conductor, the higher the conductivity and thus the longer the loop may be for signaling purposes. Copper is expensive so we cannot make the conductor as large as we would wish and extend subscriber loops long distances. These economic limits of loop length are discussed in detail below. First, we must describe what a subscriber considers as hearing sufficiently well, which is embodied in "transmission design" (regarding subscriber loop).

2.5 THE REFERENCE EQUIVALENT

2.5.1 Definition

Hearing "sufficiently well" on a telephone connection is a subjective matter under the blanket heading of "customer satisfaction." Various methods have been devised over the years to rate telephone connections regarding customer (subscriber) satisfaction. Subscriber satisfaction will be affected by the following regarding the received telephone signal:

- Level (see Section 1.9.5).
- Signal-to-noise ratio (see Section 1.10).
- Response or attenuation frequency characteristic (see Section 1.9.3).

A common rating system in use today to grade customer satisfaction is the *"reference equivalent"* system. This system considers only the first criterion mentioned above, namely, level. It must be emphasized that subscriber satisfaction is subjective. To measure satisfaction, the world regulative body for telecommunications, the International Telecommunication Union, devised a system of rating sufficient level to "satisfy," using the familiar decibel as the unit of measurement. It is particularly convenient in that, first disregarding the subscriber telephone subset, essentially we can add losses and gains (measured at 800 Hz) in the intervening network end-to-end, and determine the reference equivalent of a circuit by then adding this sum to a decibel value assigned to the subset, or to a subset plus a fixed subscriber loop length with wire gauge stated.

Let us look at how the reference equivalent system was developed, keeping in mind again that it is a subjective measurement dealing with the likes and dislikes of the "average" human being. Development took place in Europe. A standard for reference equivalent is determined using a team of qualified personnel in a laboratory. A telephone connection was established in the laboratory which was intended to be the most efficient telephone system known. The original reference system or unique master reference consisted of the following:

- A solid-back telephone transmitter.
- Bell telephone receiver.
- Interconnecting these, a "zero decibel loss" subscriber loop.
- Connecting the loop, a manual, central battery, 22-V dc telephone exchange (switch).

The test team, to avoid ambiguity of language, used a test language which consisted of logatoms. A logatom is a one-syllable word consisting of a consonant, a vowel, and another consonant.

At present more accurate measurement methods have evolved. A more modern reference system is now available in the ITU laboratory in Geneva, Switzerland, called the NOSFER. From this master reference, field test standards are available to telephone companies, administrations, and industry to establish the reference equivalent of telephone subsets in use. These field test standards are equivalent to the NOSFER.

The NOSFER is made up of a standard telephone transmitter, receiver, and network. The reference equivalent of a subscriber's subset, together with the associated subscriber line and feeding bridge, is a quantity obtained by balancing the loudness of received speech signals and is expressed relative to the whole or a corresponding part of the NOSFER (or field) reference system.

2.5.2 Application

Essentially, as mentioned earlier, type tests are run on subscriber subsets or on the subsets plus a fixed length of subscriber loop of known characteristics. These are subjective tests carried out in a laboratory to establish the reference equivalent of a specific subset as compared to a reference standard. The microphone or transmitter and the earpiece or receiver are each rated separately and are called

TRE, Transmit reference equivalent
RRE, Receive reference equivalent

respectively. *Note.* Negative values indicate that the reference equivalent is better than the laboratory standard.

Consider the following simplified telephone network:

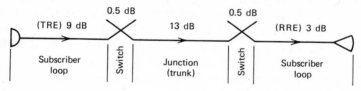

(Simplified telephone network)

The reference equivalent for this circuit is 26 dB including a 0.5-dB loss for each switch. Junction here takes on the meaning of a circuit connecting two adjacent local or metropolitan switches.

The above circuit may be called a small transmission plan. For this discussion we can define a transmission plan as a method of assigning losses end-to-end on a telephone circuit. Later in this chapter we discuss why all telephone circuits that have two-wire telephone subscribers must be lossy. The reference equivalent is a handy device to rate such a plan regarding subscriber satisfaction.

When studying transmission plans or developing them, we usually consider that all sections of a circuit in a plan are symmetrical. Let us examine this. On each end of a circuit we have a subscriber loop. Thus in the plan the same loss is assigned to each loop, which may not be the case at all in real life. From the local exchange to the first long-distance exchange, called variously junctions or toll connecting trunks, a loss is assigned which is identical at each end, and so forth.

To maintain this symmetry regarding reference equivalent of telephone subsets, we use the term $(T + R)/2$. As we see from the above drawing the TRE and the RRE of the subset have different values. We get the $(T + R)/2$ by summing the TRE and the RRE and dividing by 2. This is done to arrive at the desired symmetry. Table 2.1 gives reference equivalent data on a number of standard subscriber sets used in various parts of the world.

The ORE is the overall reference equivalent and equals the sum of the TRE, RRE, and all the intervening losses of a telephone connection end-to-end with reference to 800 Hz. We would arrive at the same figure if we added twice the $(T + R)/2$ and the intervening losses at 800 Hz. On all these calculations we assume that the same telephone set is used on either end.

CCITT Rec. G.121* states that the reference equivalent from the subscriber set to an international connection should not exceed 20.8 dB (TRE) and to the subscriber set at the other end from the same point of reference (RRE) should not exceed 12.2 dB. (*Note.* The intervening losses already are included in these figures.) By adding 12.2 and 20.8 dB, we find 33 dB to be the ORE recommended as a maximum* for an international connection. In this regard Table 2.2 should be of interest. It should also be noted that as the reference equivalent (overall, end-to-end) drops to about 6 dB, the subscriber begins to complain that the call is too loud.

It is noted in Table 2.2 that the 33-dB ORE discussed above is unsatisfactory for more than 10% of calls. Therefore the tendency in many telephone administrations is to reduce this figure as much as possible. As we shall see later, this process is difficult and can prove costly.

---

* For 97% of the connections made in a country of average size.

Table 2.1   Reference Equivalents for Subscriber Sets in Various Countries

Country	Sending (dB)	Receiving (dB)
With limiting subscriber lines and exchange feeding bridges		
Australia	14[a]	6[a]
Austria	11	2.6
France	11	7
Norway	12	7
Germany	11	2
Hungary	12	3
Netherlands	17	4
United Kingdom	12	1
South Africa	9	1
Sweden	13	5
Japan	7	1
New Zealand	11	0
Spain	12	2
Finland	9.5	0.9
With no subscriber lines		
Italy	2	−5
Norway	3	−3
Sweden	3	−3
Japan	2	−1
United States (loop length 1000 ft, 83 Ω)	5	−1[b]

Source: CCITT, *Local Telephone Networks*, ITU, Geneva, July 1968, and *National Telephone Networks for the Automatic Service.*
[a] Minimum acceptable performance.
[b] Ref. 8.

Table 2.2   British Post Office Survey of Subscribers for Percentage of Unsatisfactory Calls

Overall Reference Equivalent (dB)	% of Unsatisfactory Calls
40	33.6
36	18.9
32	9.7
28	4.2
24	1.7
20	0.67
16	0.228

2.6 TELEPHONE NETWORKS

2.6.1 General

The next logical step in our discussion of telephone transmission is to consider the large-scale interconnection of telephones. As we have seen, subscribers within a reasonable distance of one to another can be interconnected by wire lines and we can still expect satisfactory communication. A switch is used so a subsscriber can speak with some other discrete subscriber as he chooses. As we extend the network to include more subscribers and circumscribe a wider area, two technical/economic factors must be taken into account:

1. More than one switch must be used.
2. Wire pair transmission losses on longer circuits must be offset by amplifiers or the pairs must be replaced by other, more efficient means.

Let us accept item 1. The remainder of this section concentrates on item 2. The reason for the second statement becomes obvious to the reader if he reviews the salient point of the previous subsection regarding reference equivalent.

Example.   How far can a two-wire line be run without amplifiers following the rules of reference equivalent? Allow, in this case, no more than an ORE of 33 dB, a high value for a design goal. Referring to Table 2.1 and using Spain as an example the TRE + RRE for telephone subsets sums to 14 dB. This leaves us with a limiting loss of 19 dB (33 − 14) for the remaining network. Use 19-gauge (0.91-mm) telephone cable, typical for "long-distance" communication (Table 2.8). This cable has a loss of 0.71 dB/km. Thus the total extension of the network will be about 26 km allowing no loss for a switch

2.6.2 Basic Considerations

What are some of the more common appraches that may be used to extend the network? We may use:

1. Coarser gauge cable (larger diameter conductors)
2. Amplifiers in the present wire pair system.
3. Carrier transmission techniques (Chapters 3 and also 11).
4. Radio transmission techniques (Chapters 4–7 and Chapter 10).

Items 1 and 2 are used quite widely but often become unattractive from an economic point of view as the length increases. Items 3 and 4 become attractive for multichannel transmission over longer distances. Discussion of the formation of a multichannel signal is left for Chapter 3. Such a multichannel transmission technique is referred to as *carrier* transmission.

For our purposes we can consider carrier as a method of high velocity transmission of bands of frequency above the voice frequency region (i.e., above 4000 Hz) over wire or radio.

2.6.3 Two-Wire/Four-Wire Transmission

TWO-WIRE TRANSMISSION

By its basic nature a telephone conversation requires transmission in both directions. When both directions are carried on the same wire pair, we call it two-wire transmission. The telephones in our home and office are connected to a local switching center by means of two-wire circuits. A more proper definition for transmitting and switching purposes is that when oppositely directed portions of a single telephone conversation occur over the same electrical transmission channel or path, we call this two-wire operation.

FOUR-WIRE TRANSMISSION

Carrier and radio systems require that oppositely directed portions of a single conversation occur over separate transmission channels or paths (or using mutually exclusive time periods). Thus we have two wires for the transmit path and two wires for the receive path, or a total of four wires for a full-duplex (two-way) telephone conversation. For almost all operational telephone systems, the end instrument (i.e., the telephone subset) is connected to its intervening network on a two-wire basis.*

Nearly all long-distance telephone connections traverse four-wire links. From the near-end user the connection to the long-distance network is two wire. Likewise, the far-end user is also connected to the long-distance network via a two-wire link. Such a long-distance connection is shown in Figure 2.2. Schematically, the four-wire interconnection is shown as if it were wire line, single channel with amplifiers. More likely it would be multichannel carrier on cable and/or multiplex on radio. However, the amplifiers in the figure serve to convey the ideas that this chapter considers.

* A notable exception is the U.S. military telephone network, Autovon, where end users are connected on a four-wire basis.

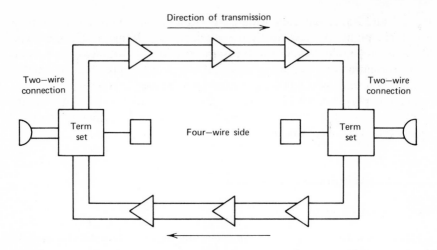

Figure 2.2 A typical long-distance telephone connection.

As shown in Figure 2.2, conversion from two-wire to four-wire operation is carried out by a terminating set, more commonly referred to in the industry as a "term set." This set contains a three-winding balanced transformer (a hybrid) or a resistive network, the latter being less common.

THE OPERATION OF A HYBRID

A hybrid, for telephone work (at voice frequency), is a transformer For a simplified description, a hybrid may be viewed as a power splitter with four sets of wire pair connections. A functional block diagram of a hybrid device is shown in Figure 2.3. Two of these wire pair connections belong to the four-wire path, which consists of a transmit pair and a receive pair. The third pair is the connection to the two-wire link to the subscriber subset. The last wire pair connects the hybrid to a resistance-capacitance balancing network, which electrically balances the hybrid with the two-wire connection to the subscriber's subset over the frequency range of the balancing network. An artificial line may also be used for this purpose.

The hybrid function permits signals to pass from any pair through the transformer to both adjacent pairs but blocks signals to the opposite pair (as shown in Figure 2.3). Signal energy entering from the four-wire side divides equally, half dissipating into the balancing network and half going to the desired two-wire connection. Ideally no signal energy in this path crosses over the four-wire transmit side. This is an important point, which we take up later.

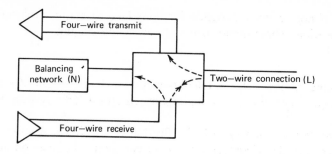

Figure 2.3 Operation of a hybrid transformer.

Signal energy entering from the two-wire subset connection divides equally, half of it dissipating in the impedance of the four-wire side receive path, and half going to the four-wire side transmit path. Here the *ideal* situation is that no energy is to be dissipated by the balancing network (i.e., there is a perfect balance). The balancing network is supposed to display the characteristic impedance of the two-wire line (subscriber connection) to the hybrid.

The reader notes that in the description of the hybrid, in every case, ideally half of the signal energy entering the hybrid is used to advantage and only half is dissipated, wasted. Also keep in mind that any passive device inserted in a circuit such as a hybrid has an insertion loss. As a rule of thumb, we say that the insertion loss of a hybrid is 0.5 dB. Thus there are two losses here than the reader must not lose sight of:

Hybrid insertion loss	0.5 dB	
Hybrid dissipation loss	3.0 dB	(half power)
	3.5 dB	total

As far as this chapter is concerned, any signal passing through a hybrid suffers a 3.5-dB loss. Some hybrids used on short subscriber loops purposely have higher losses, as do special resistance type hybrids.

2.6.4   Echo, Singing, and Design Loss

GENERAL

The operation of the hybrid with its two-wire connection on one end and four-wire connection on the other leads us to the discussion of two phenomena that, if not properly designed for, may lead to major impairments in communication. These impairments are echo and singing.

Echo

As the name implies, echo in telephone systems is the return of a talker's voice. The returned voice, to be an impairment, must suffer some noticeable delay.

Thus we can say that echo is a reflection of the voice. Analogously, it may be considered as that part of the voice energy that bounces off obstacles in a telephone connection. These obstacles are impedance irregularities, more properly called impedance mismatches.

Echo is a major annoyance to the telephone user. It affects the talker more than the listener. Two factors determine the degree of annoyance of echo: its loudness, and how long it is delayed.

Singing

Singing is the result of sustained oscillations due to positive feedback in telephone amplifiers or amplifying circuits. Circuits that sing are unusable and promptly overload multichannel carrier equipment (FDM—see Chapter 3).

Singing may be thought of as echo that is completely out of control. This can occur at the frequency at which the circuit is resonant. Under such conditions the circuit losses at the singing frequency are so low that oscillation will continue even after the impulse that started it ceases to exist.

The primary cause of echo and singing generally can be attributed to the mismatch between the balancing network and its two-wire connection associated with the subscriber loop. It is at this point that the major impedance mismatch usually occurs and an echo path exists. To understand the cause of the mismatch, remember that we always have at least one two-wire switch between the hybrid and the subscriber. Ideally the hybrid balancing network must match each and every subscriber line to which it may be switched. Obviously the impedances of the four-wire trunks (lines) may be kept fairly uniform. However, the two-wire subscriber lines may vary over a wide range. The subscriber loop may be long or short, may or may not have inductive loading (see Section 2.8.4), and may or may not be carrier derived (see Chapter 3). The hybrid imbalance causes signal reflection or signal "return." The better the match, the more the return signal is attenuated. The amount that the return signal (or reflected signal) is attenuated is called the *return loss* and is expressed in decibels. The reader should remember that any four-wire circuit may be switched to hundreds or even thousands of different subscribers. If not, it would be a simple matter to match the four-wire circuit to its single subscriber through the hybrid. This is why the hybrid to which we refer has a compromise balancing network rather than a precision network. A compromise network is usually adjusted for a compromise in the range of impedance that is expected to be encountered on the two-wire side.

Let us consider now the problem of match. For our case the impedance match is between the balancing network (N) and the two-wire line (L) (see Figure 2.3). With this in mind,

$$\text{Return loss}_{dB} = 20 \log_{10} \frac{Z_N + Z_L}{Z_N - Z_L}$$

If the network perfectly balances the line, $Z_N = Z_L$, and the return loss would be infinite.

Return loss may also be expressed in terms of reflection coefficient, or

$$\text{Return loss}_{dB} = 20 \log_{10} \frac{1}{\text{reflection coefficient}}$$

where the reflection coefficient = reflected signal/incident signal.

The CCITT uses the term "balance return loss" (see CCITT Rec. G.122) and classifies it as two types:

1. Balance return loss from the point of view of echo.[*] This is the return loss across the band of frequencies from 500 to 2500 Hz.

2. Balance return loss from the point of view of stability. This is the return loss between 0 and 4000 Hz.

The band of frequencies most important from the standpoint of echo for the voice channel is that between 500 and 2500 Hz. A good value for echo return loss for toll telephone plant is 11 dB, with values on some connections dropping to as low as 6 dB. For the local telephone network, CCITT recommends better than 6 dB with a standard deviation of 2.5 dB (Ref. 18).

For frequencies outside the 500–2500-Hz band, return loss values often are below the desired 11 dB. For these frequencies we are dealing with return loss from the point of view of stability. CCITT recommends that balance return loss from the point of view of stability (singing) should have a value of not less than 2 dB for all terminal conditions encountered during normal operation (CCITT Rec. G.122, p. 3). For further information the reader should consult Appendix A and CCITT Recs. G.122 and G.131.

---

[*] Called echo return loss (ERL) in VNL (north American practice, Section 2.6.5), but uses a weighted distribution of level.

Echo and singing may be controlled by

- Improved return loss at the term set (hybrid).
- Adding loss on the four-wire side (or on the two-wire side).
- Reducing the gain of the individual four-wire amplifiers.

The annoyance of echo to a subscriber is also a function of its delay. Delay is a function of the velocity of propagation of the intervening transmission facility. A telephone signal requires considerably more time to traverse 100 km of a voice pair cable facility, particularly if it has inductive loading, than 100 km of radio facility.

Delay is measured in one-way or round-trip propagation time measured in milliseconds. CCITT recommends that if the mean round-trip propagation time exceeds 50 ms for a particular circuit, an echo suppressor should be used. Bell System practices in North America use 45 ms as a dividing line. In other words, where echo delay is less than that stated above, echo will be controlled by adding loss.

An echo suppressor is an electronic device inserted in a four-wire circuit which effectively blocks passage of reflected signal energy. The device is voice operated with a sufficiently fast reaction time to "reverse" the direction of transmission, depending on which subscriber is talking at the moment. The blocking of reflected energy is carried out by simply inserting a high loss in the return four-wire path.

Figure 2.4 shows the echo path on a four-wire circuit.

TRANSMISSION DESIGN TO CONTROL ECHO AND SINGING

As stated previously, echo is an annoyance to the subscriber. Figure 2.5 relates echo path delay to echo path loss. The curve in Figure 2.5 is a group of points at which the average subscriber will tolerate echo as a function of its delay. Remember that the greater the return signal is delayed, the more annoying it is to the telephone talker (i.e., the more the echo signal must be attenuated). For instance, if the echo path delay on a particular circuit is 20 ms, an 11-dB loss must be inserted to make echo tolerable to the talker. The careful reader will note that the 11 dB designed into the circuit will increase the end-to-end reference equivalent by that amount, something quite undesirable. The effect of loss design on reference equivalents and the trade-offs available are discussed below.

To control singing all four-wire paths must have some loss. Once they go into a gain condition, and we refer here to overall circuit gain, we will have positive feedback and the amplifiers will begin to oscillate or "sing." North American practice calls for a 4-dB loss on all four-wire circuits to

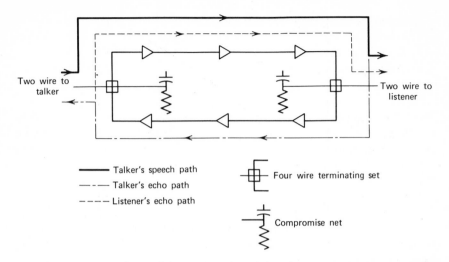

Figure 2.4 Echo paths in a four-wire circuit.

ensure against singing. CCITT recommends a minimum loss for a national network of 7 dB (CCITT Rec. G.122, p. 2).

Almost all four-wire circuits have some form of amplifier and level control. Often such amplifiers are embodied in the channel banks of the carrier equipment. For a discussion of carrier equipment, see Chapter 3.

AN INTRODUCTION TO TRANSMISSION LOSS PLANNING

One major aspect of transmission system design for a telephone network is to establish a transmission loss plan. Such a plan, when implemented, is formulated to accomplish three goals:

- Control singing.
- Keep echo levels within limits tolerable to the subscriber.
- Provide an acceptable overall reference equivalent to the subscriber.

For North America the VNL (via net loss) concept embodies the transmission plan idea. VNL is covered in Section 2.6.5.

From our discussions above we have much of the basic background necessary to develop a transmission loss plan. We know the following:

1. A certain minimum loss must be maintained in four-wire circuits to ensure against singing.

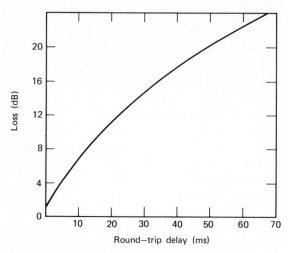

Figure 2.5 Talker echo tolerance for average telephone users.

2. Up to a certain limit of round-trip delay, echo is controlled by loss.

3. It is desirable to limit these losses as much as possible to improve reference equivalent.

National transmission plans vary considerably. Obviously length of circuit is important as well as the velocity of propagation of the transmission media. Two approaches are available in the preparation of a loss plan. These are

> Variable loss plan (i.e., VNL)
> Fixed loss plan (i.e., as used in Europe)

A national transmission loss plan for a small country (i.e., small in extension) such as Belgium could be quite simple. Assume that a 4-dB loss is inserted in all four-wire circuits to prevent singing. Consult Figure 2.5. Here 4 dB allows for 5 ms of round-trip delay. If we assume carrier transmission for the entire length of the connection and use 105,000 mi/s for the velocity of propagation, we can then satisfy Belgium's echo problem. The velocity of propagation used comes out to 105 mi (168 km)/ms. By simple arithmetic we see that a 4-dB loss on all four-wire circuits will make echo tolerable for all circuits extending 210 mi (336 km). This is an application of the fixed loss type of transmission plan. In the case of small countries or telephone companies operating over a small geographical extension, the minimum loss inserted to control singing controls echo as well for the entire country.

Let us try another example. Assume that all four-wire connections have a 7-dB loss. Figure 2.5 indicates that 7 dB permits an 11-ms round-trip delay. Assume that the velocity of propagation is 105,000 mi/s. Remember that we deal with round-trip delay. The talker's voice goes out to the far-end hybrid and is then reflected back. This means that the signal traverses the system twice, as shown.

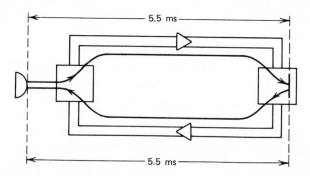

Example of echo–round-trip delay. $5.5 + 5.5 = 11$ ms round-trip delay.

Thus 7 dB of loss for the velocity of propagation given allows about 578 mi of extension or, for all intents and purposes, the distance between subscriber of extension or, for all intents and purposes, the distance between subscribers.

It has become evident by now that we cannot continue increasing losses indefinitely to compensate for echo on longer circuits. Most telephone companies and administrations have set the 45- or 50-ms round-trip delay criterion. This sets a top figure, above which echo suppressors shall be used.

One major goal of the transmission loss plan is to improve overall reference equivalent or apportion more loss to the subscriber plant so that subscriber loops can be longer or allow the use of less copper (i.e., smaller diameter conductors). The question is, what measures can be taken to reduce losses and still keep echo within tolerable limits? One obvious target is to improve return losses at the hydrids. If all hybrid return losses are improved, then the echo tolerance curve gets shifted. This is so because improved return losses reduce the intensity of the echo returned to the talker. Thus he is less annoyed by the echo effect.

One way of improving return loss is to make all two-wire lines out of the hybrid look alike, that is, have the same impedance. The switch at the other end of the hybrid (i.e., on the two-wire side) connects two-wire loops of varying length causing the resulting impedances to vary greatly. One approach is to extend four-wire transmission to the local office such that each hybrid can be better balanced. This is being carried out with success in Japan. The U.S. Department of Defense has its Autovon (automatic voice network) in which every subscriber line is operated on a four-wire basis. Two-wire subscribers connect through the system on a PABX (private automatic branch exchange).

Let us return to standard telephone networks using two-wire switches in the subscriber area; suppose balance return loss could be improved to 27 dB. Thus minimum loss to assure against singing could be reduced to 0.4 dB. Suppose we distributed this loss across 4 four-wire circuits in tandem. Thus each four-wire circuit would be assigned a 0.1-dB loss. If we have gain in the network, singing will result. The safety factor between loss and gain is 0.4 dB. The loss in each circuit or link is maintained by amplifiers. It is difficult to adjust the gain of an amplifier to 0.1 dB, much less keep it there over long periods even with good automatic regulation. *Stability* or gain stability is the term used to describe how well a circuit can maintain a desired level. Of course in this case we refer to a test-tone level. In the example above it would take only one amplifier to shift 0.4 dB, two to shift in the positive direction 0.2 dB and so forth. The importance of stability, then, becomes evident.

The stability of a telephone connection depends on

- The variation of transmission level with time.
- The attenuation-frequency characteristics of the links in tandem.
- The distribution of balance return loss.

Each of these criteria becomes magnified when circuits are switched in tandem. To handle the problem properly we must talk about statistical methods and standard distributions.

Returning to the criteria above, in the case of the first two items, we refer to the tandeming of four-wire circuits. The last item refers to switching subscriber loops/hybrid combinations that will give a poorer return loss than the 11 dB stated above. Return losses on some connections can drop to 3 dB or less.

CCITT Recs. G.122, G.131, and G.151C treat stability. In essence the loss through points *a-t-b* in Figure 2.6 shall have a value not less than $(6 + N)$ dB, where N is the number of four-wire circuits in the national chain. Thus the minimum loss is stated (CCITT Rec. G.122). Rec. G.131 is quoted in part below:

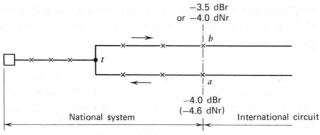

Figure 2.6 Definition of points *a-t-b* (CCITT Rec. G.122); × indicates a switch. Courtesy International Telecommunications Union.

The standard deviation of transmission loss among international circuits routed in groups equipped with automatic regulation is 1 dB . . . . This accords with . . . that the tests . . . indicate that this target is being approached in that 1.1 dB was the standard deviation of the recorded data . . . .

CCITT Rec. G.131 continues

It is also evident that those national networks which can exhibit no better stability balance return loss than 3 dB, 1.5 dB standard deviation, are unlikely to seriously jeopardize the stability of international connections as far as oscillation is concerned. However, the near-singing (rain barrel effect) distortion and echo effects that may result give no grounds for complacency in this matter.

Stability requirements in regard to North American practice are embodied in the VNL concept discussed in the next section.

2.6.5 Via Net Loss (VNL)

Via Net Loss is a concept or method of transmission planning which permits a relatively close approach to an overall zero transmission loss in the telephone network (lowest practicably attainable) and maintains singing and echo within specified limits. The two criteria that follow are basic to VNL design:

1 Customer-to-customer talker echo shall be satisfactorily low on more than 99% of all telephone connections which encounter the maximum delay likely to be experienced.

2. The total amount of overall loss is distributed throughout the trunk segments of the connection by allocation of loss to the echo characteristics of each segment.

One important concept in the development of the discussion on VNL is that of echo return loss (ERL) (See Section 2.6.4). For this discussion we consider ERL as a single valued, weighted figure of return losses in the frequency band 500–2500 Hz. ERL differs from return loss in that it takes into account a weighted distribution of level versus frequency in order to simulate the nonlinear characteristics of the transmitter and receiver of the telephone instrument. By using ERL measurements it is possible to arrive at a basic design factor for the development of the VNL formula. This design factor states that the average return loss at Class 5 offices (local offices) is 11 dB with a standard deviation of 3 dB. Considering then a standard distribution curve and the one, two, or three sigma points on the curve, we could therefore expect practically all measurements of ERL to fall between 2 and 20 dB at Class 5 offices (local offices). VNL also considers that reflection occurs at the far end in relation to the talker where the toll connecting trunks are switched to the intertoll trunks (see Section 2.7).

The next concept in the development of the VNL discussion is that of overall connection loss (OCL), which is the value of one-way trunk loss between two end (local) offices (not subscribers).

Consider that:

$$\text{Echo path loss} = 2 \times \text{trunk loss (one-way)} + \text{return loss (hybrid)}$$

Now let us consider the average tolerance for a particular echo path loss. Average echo tolerance is taken from the curve in Figure 2.5. Therefore,

$$\text{OCL} = \frac{\text{average echo tolerance (loss)} - \text{return loss}}{2}$$

Return loss in this case is the average echo return loss which must be maintained at the distant Class 5 office—the 11 dB given above.

An important variability factor has not been considered in the formula. This is trunk stability. This factor defines how close assigned levels are maintained on a trunk. VNL practice dictates trunk stability to be maintained with a normal distribution of levels and a standard deviation of 1 dB in each direction. For a round-trip echo path the deviation is taken as 2 dB. This variability applies to each trunk in a tandem connection. If there are three trunks in tandem, this deviation must be applied to each of them.

The reader will recall that the service requirement in VNL practice is satisfactory echo performance for 99% of all connections. This may be considered as a cumulative distribution or $+2.33$ standard deviations summing from negative infinity toward the positive direction.

The OCL formula may now be rewritten:

$$\text{OCL} = \frac{\text{avg. echo tolerance} - \text{avg. return loss} + 2.33D}{2}$$

where D = composite standard deviation of all functions.

Notes on the derivation of D, the composite standard deviation of all functions:

$$D = \sqrt{D_t^2 + D_{rl}^2 + ND_1^2}$$

where D_t = standard deviation of the distribution of echo tolerance among a large group of observers, 2.5 dB

D_{rl} = standard deviation of the distribution of return loss, 3 dB

D_1 = Standard deviation of the distribution of the variability of trunk loss for a round trip echo path. Given as 2 dB

N = number of trunks switched in tandem to form a connection Class 5 office to Class 5 office

Consider now several trunks in tandem, and it then can be calculated that at just about any given echo path delay, the OCL increases approximately 0.4 dB for each trunk added. With this simplification, once we have the OCL for one trunk, all that is needed to compute the OCL for additional trunks is to add 0.4 dB times the number of trunks added in tandem. This loss may be regarded as an additional constant needed to compensate for variations in trunk loss in the VNL formula.

Figure 2.7 relates echo path delay (round trip) to overall connection loss (OCL for one trunk, then for a second trunk in tandem and for four and six trunks in tandem). The straight-line curve has been simplified, yet the approximation is sufficient for engineering VNL circuits. On examining the straight-line curve in Figure 2.7 it will be noted that that curve cuts the Y axis at 4.4 dB where round-trip delay is 0. This 4.4 dB is made up of two elements, namely, that all trunks have a minimum of 4 dB to control singing and 0.4 dB protection against negative variation of trunk loss.

Another important point to be defined on the linear curve in Figure 2.7 is that of a round-trip delay of 45 ms, which corresponds to an OCL of 9.3 dB. Empirically it has been determined that for delays greater than 45 ms, echo suppressors must be used.

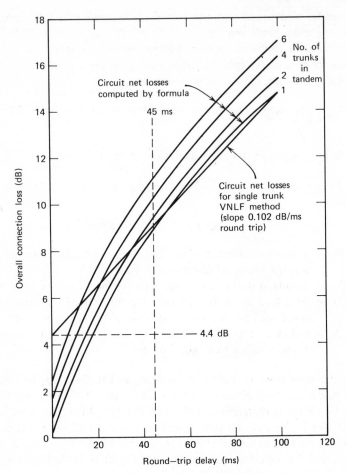

Figure 2.7 Approximate relationship between round-trip echo delay and overall connection loss (OCL).

From the linear curve in the figure the following formula for OCL may be derived:

$$\text{OCL} = (0.102)\,(\text{path delay in ms})$$
$$+ (0.4\ \text{dB})\,(\text{number of trunks in tandem}) + 4\ \text{dB}$$

A word now about the last term in the OCL equation, a constant (4 dB): usually the 4 dB is applied to the extremity of each trunk network, namely, to the toll connecting trunks, 2 dB to each.

OCL deals with the losses of an entire network consisting of trunks in tandem. VNL deals with the losses assigned to one trunk. The VNL formula follows from the OCL formula. The key here is the round-trip delay on the trunk in question. The delay time for a transmission facility employing only one particular medium is equal to the reciprocal of the velocity of propagation of the medium multiplied by the length of the trunk. To obtain round-trip time this figure must be multiplied by 2 Thus

$$VNL = 0.102 \times 2 \times \left(\frac{1}{\text{velocity of propagation}} \right)$$
$$\times \text{ (one-way length of the trunk)} + 0.4 \text{ dB}$$

Often another term is introduced to simplify the equation. That is VNLF (via net loss factor):

$$VNL = VNLF \times \text{(one-way length of trunk in miles)} + 0.4 \text{ dB}$$

$$VNLF = \frac{2 \times 0.102}{\text{velocity of propagation of the medium}} \quad \text{(dB/mi)}$$

It should be noted that the velocity of propagation of the medium used here must be modified by such things as delays caused by repeaters, intermediate modulation points, and facility terminals

VNL factors (VNLF) for loaded two-wire facilities are 0.03 dB/mi with H-88 loading on 19-gauge wire and increased to 0.04 dB/mi on B-88 and H-44 facilities. On four-wire carrier and radio facilities the factor improves to 0.0015 dB/mi.

For connections with round-trip delay times in excess of 45 ms the standard VNL approach must be modified. As mentioned previously, these circuits use echo suppressors that automatically switch about 50 dB into the echo return path, and the switch actuates when speech is received in the "return" path switching the pad into the "go" path.

VNL practice in North America treats long delay circuits with up to a maximum of 45 ms delay in the following manner. Refer to Figure 2.8. Here the total round-trip delay is arbitrarily split in two parts for connections involving regional intertoll trunks. If the regional intertoll delay exceeds 22 ms, echo suppressors are used. If the figure is 22 ms or less echo is controlled by VNL design. Thus, allow 22 ms for the maximum delay for the regional intertoll segment to the connection. This leaves 23-ms maximum delay for the other segments (45 − 22). Now apply the VNL formula for a delay of 22 ms.

Thus VNL $= 0.102 \times 22 + 0.4 = 2.6$ dB. This loss is equivalent to the maximum length of an intertoll trunk without an echo suppressor. What is that length?

$$\text{Length (one-way)} = \frac{\text{VNL} - 0.4}{\text{VNLF}} \text{ mi}$$

$$= \frac{0.102 \times 22 + 0.4 - 0.4}{0.0015} = 1498 \text{ mi}$$

Note. The VNLF indicates carrier and/or radio for the whole trunk.

In summary, in VNL design we have three types of losses that may be assigned to a trunk:

Type	Loss
Toll connecting trunk	VNL + 2 dB
Intertoll trunk (no echo suppressor)	VNL
Intertoll trunk (with echo suppressor)	0 dB

Note. See discussion on echo suppressors for explanation of the 0 dB figure.

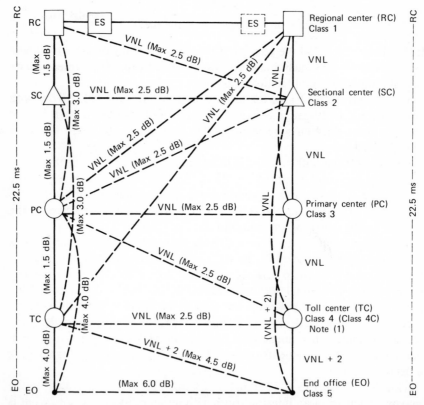

Figure 2.8 Trunk losses with VNL design (from Bell System, *Notes on Distance Dialing*). ES = echo suppressor; Losses include 0.4-dB design loss allowed for maintenance.

VNL PENALTY FACTORS

Much of the toll network using the VNL concept utilizes two-wire switches even though the network is considered four-wire. At each point where the network has a two-wire to four-wire transformation, another source of echo occurs. Again the amount of echo is a function of the echo return loss at each point of transformation. If the return loss at each of these intermediate points, often called through return loss, is high enough, the point is transparent regarding echo. Echo, then, can be considered only as a function of the return loss at the terminating points of transformation, often called terminating return loss. If a two-wire toll switch has an echo return loss of 27 dB on at least 50% through-connections, it meets through return loss objectives, and as far as VNL design is concerned, it may be considered transparent regarding echo.

However, a number of two-wire toll exchanges do not meet this minimum criterion. Therefore, a penalty is assigned to each in the design of VNL. This penalty is called the *B* factor. *B* factor is an amount of loss which must be added to the VNL value of each trunk (incoming as well as outgoing) to compensate for the excessive amount of echo and singing current reflection which this office will create in an intertoll connection. The following table provides median office ERL and corresponding *B* factors:

Median Office ERL (dB)	*B* Factor (dB)
27	0
21	0.3
18	0.6
16	0.9
15	1.2
14	1.5

2.7 NOTES ON NETWORK HIERARCHY

Telephone networks require some form of hierarchy regarding switches to route traffic effectively and economically. Earlier in the chapter we discussed the subscriber as being connected to a local office, a switch. Once the subscriber lifts a telephone "off-hook," he seizes a line at the switch, gets a dial tone in return, and by dialing gives himself access to all other subscribers in the common user network to which he is connected. Figure 2.8 shows the hierarchy of the North American network. Figure 2.9 shows the CCITT hierarchy.

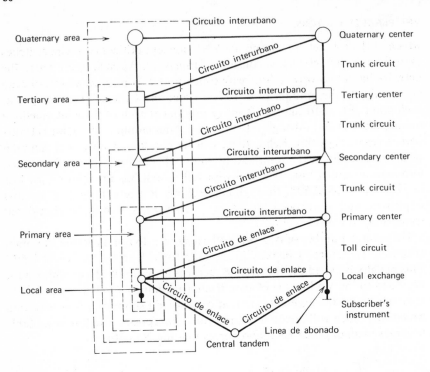

Figure 2.9 CCITT routing plan (network hierarchy). Linea de abonado = subscriber line; central tandem = tandem switch; circuito de enlace = junction (trunk); circuito interurbano = long distance (trunk) circuit. Courtesy ITU—CCITT.

Networks have evolved and so developed in accordance with the principal requirements of traffic volume, geographic coverage, and community of interest of the population as spread across that geographic area. A simple network is one where all switches are mutually interconnected. This, by definition, is a mesh connection, and is a common form of interconnection in many European metropolitan areas, where traffic volumes between all switches are high. In today's complex society where the demand for telephone service is very high with resulting higher traffic volumes which are nonuniform, such a simple configuration is inefficient and cannot be supported economically.

A natural development from the simple network is the star network. A simple star configured network is shown below.

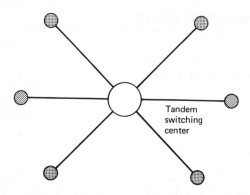

A star network allows efficient utilization of interconnecting facilities when traffic between switches is relatively small. This done by connecting each switch to a central point where tandem switching may be carried out to interconnect outlying switches by trunk groups. A trunk group is a grouping of telephone circuits representing a concentration of traffic.

As a network grows in extension and in traffic volume carried, successive stages of concentration are required to switch relatively small traffic volumes greater distances. The hierarchical network naturally evolves from a series of simple star configurations. Figures 2.8 and 2.9 are good examples. Both the North American and CCITT networks comply with the following criteria:

- Any switch, anywhere, can be connected with any other switch with no more than nine links connected in tandem for the worst connection (national).
- It is possible for calls to bypass one or more intermediate offices on the final route path.
- Direct trunk groups available between switching centers on the final route may be employed to advance a call toward its destination.

Table 2.3 compares the CCITT hierarchy switching/routing nomenclature (Figure 2.9) with that of North America (Figure 2.8).

2.8 DESIGN OF SUBSCRIBER LOOP

2.8.1 Introduction

The subscriber loop connects a subscriber telephone subset with a local switching center (Class 5 Office in Figure 2.8). A subscriber loop in nearly

Table 2.3   Hierarchy Nomenclature

CCITT	North American
Quaternary center (highest ranking)	Class 1 office
Tertiary center	Class 2 office
Secondary center	Class 3 office
Primary center	Class 4 office
Local exchange	Class 5 office

all cases is two-wire with simultaneous transmission in both directions. The simplified drawing below will help to illustrate the problem.

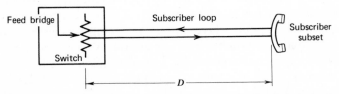

Distance D, the loop length, is most important. By Section 2.4 D must be limited in length owing to (1) *attenuation* of the voice signal and (2) dc resistance for signaling.

The attenuation is taken from the national transmission plan covered in Section 2.6.4. For our discussion we shall assign 6 dB as the loop attenuation limit (referred to 800 Hz). For the loop resistance limit, many crossbar switches will accept up to 1300 Ω.* From this figure we subtract 50 Ω, nominal resistance for the telephone subset in series with the loop, leaving us with a 1250-Ω limit for the wire pair (bridge resistance disregarded). Therefore in the paragraphs that follow we shall use the following figures:

6 dB (attenuation limit for a loop)†

1250 Ω (resistance limit)

2.8.2   Basic Resistance Design

To calculate the dc loop resistance for copper conductors the following formula is applicable:

$$R_{dc} = \frac{0.1095}{d^2}$$

---

* Many semielectronic switches will accept 1800-Ω loops, and with special line equipment, 2400-Ω.
† In the United States this value may be as high as 9 db.

where R_{dc} = loop resistance Ω/mi

d = diameter of the conductor (in.)

If we want a 10-mi loop and allow 125 Ω/mi of loop (for the 1250-Ω limit) what diameter of copper wire would we need?

$$125 = \frac{0.1095}{d^2}$$

$$d^2 = \frac{0.1095}{125}$$

$$d^2 = 0.0089$$

$$d = 0.03 \text{ in. or } 0.76 \text{ mm (round off to } 0.80 \text{ mm)}$$

Using Table 2.4, we can compute maximum loop lengths for 1250-Ω signaling resistance. As an example, for a 26-gauge loop:

$$\frac{1250}{83.5} = 14.97, \text{ or } 14,970 \text{ ft}$$

This, then, is the signaling limit and not the loss (attenuation) limit, or what some call the "transmission limit' referred to in "transmission design."

To assist relating American Wire Gauge (AWG) to cable diameter in millimeters, Table 2.5 is presented.

2.8.3 Transmission Design

The second design consideration mentioned above was attenuation or loss. The attenuation of a wire pair used on a subscriber loop varies with frequency, resistance, inductance capacitance and leakage conductance. Resistance of the line will depend on temperature. For open-wire lines

Table 2.4

Gauge of Conductor	$\Omega/1000$ ft of loop	Ω/mi of loop	Ω/km of loop
26	83.5	440	237
24	51.9	274	148
22	32.4	171	92.4
19	16.1	85	45.9

Table 2.5   American Wire Gauge (B & S) versus Wire Diameter and Resistance

American Wire Gauge	Diameter (mm)	Resistance (Ω/km) [a] at 20°C
11	2.305	4.134
12	2.053	5.210
13	1.828	6.571
14	1.628	8.284
15	1.450	10.45
16	1.291	13.18
17	1.150	16.61
18	1.024	20.95
19	0.9116	26.39
20	0.8118	33.30
21	0.7229	41.99
22	0.6439	52.95
23	0.5733	66.80
24	0.5105	84.22
25	0.4547	106.20
26	0.4049	133.9
27	0.3607	168.9
28	0.3211	212.9
29	0.2859	268.6
30	0.2547	338.6
31	0.2268	426.8
32	0.2019	538.4

[a] These figures must be doubled for loop/km. Remember it has a "go" and "return" path.

attenuation may vary ±12% between winter and summer conditions For buried cable, which we are more concerned with, loss variations due to temperature are much less.

Table 2.6 gives losses of some common subscriber cable per 1000 ft. If we are limited to a 6-dB (loss) subscriber loop, then by simple division, we can derive the maximum loop length permissible for transmission design considerations for the wire gauges shown

$$26 \qquad \frac{6}{0.51} = 11.7 \text{ kft}$$

$$24 \qquad \frac{6}{0.41} = 14.6 \text{ kft}$$

$$22 \qquad \frac{6}{0.32} = 19.0 \text{ kft}$$

$$19 \qquad \frac{6}{0.21} = 28.5 \text{ kft}$$

$$16 \qquad \frac{6}{0.14} = 42.8 \text{ kft}$$

Table 2.6 Loss/1000 ft of Subscriber Cable[a]

Cable Gauge	Loss/1000 ft (dB)
26	0.51
24	0.41
22	0.32
19	0.21
16	0.14

[a] Cable is low capacitance type (i.e., under 0.075 nF/mi).

2.8.4 Loading

In many situations it is desirable to extend subscriber loop lengths beyond the limits described in Section 2.8.3. Common methods to attain longer loops without exceeding loss limits are the following:

1. Increase conductor diameter.
2. Use amplifiers and/or loop extenders.*
3. Use inductive loading

Loading tends to reduce transmission loss on subscriber loops and other types of voice pairs at the expense of good atteunation-frequency response beyond 3000–3400 Hz. Loading a particular voice pair loop consists of inserting series inductances (loading coils) into the loop at fixed intervals. Adding load coils tends to

- Decrease the velocity of propagation.
- Increase impedance.

* A loop extender is a device that increases battery voltage on a loop extending signaling range. It may also contain an amplifier, thereby extending transmission loss limits.

Loaded cables are coded according to the spacing of the load coils. The standard code for load coils regarding spacing is shown in Table 2.7.

Table 2.7   Code for Load Coil Spacing

Code Letter	Spacing (ft)	Spacing (m)
A	700	213.5
B	3000	915
C	929	283.3
D	4500	1372.5
E	5575	1700.4
F	2787	850
H	6000	1830
X	680	207.4
Y	2130	649.6

Loaded cables typically are designated 19-H-44, 24-B-88, and so forth. The first number indicates the wire gauge, the letter is taken from Table 2.7 and is indicative of the spacing, and the third item is the inductance of the coil in millihenries (mH). 19-H-66 is a cable commonly used for long-distance operation in Europe. Thus the cable has 19-gauge voice pairs loaded at 1830-m intervals with coils of 66-mH inductance. The most commonly used spacings are B, D and H.

Table 2.8 will be useful to calculate attenuation of loaded loops for a given length. For example, for 19-H-88 (last entry in table) cable, the attenuation per kilometer is 0.26 dB (0.42 dB/statute mi). Thus for our 6-dB loop loss limit, we have 6/0.26, limiting the loop to 23 km in length (14.3 statute mi).

When determining signaling limits in loop design, add about 15 Ω per load coil as series resistors.

The tendency in many administrations is to use a new loading technique. This has been taken from "unigauge design" discussed in the next Section. With this technique no loading is required on any loop less than 5000 m long (15,000 ft). For loops longer than 5000 m, loading starts at the 4200-m point and load coils are installed at 1830-m intervals thereon. The loading intervals should not vary by more than 2%.

Table 2.8 Some Properties of Cable Conductors

Diameter (mm)	AWG No.	Mutual Capacitance (nF/km)	Type of Loading	Loop Resistance (Ω/km)	Attenuation at 1000 Hz (dB/km)
0.32	28	40	None	433	2.03
		50	None		2.27
0.40		40	None	277	1.62
		50	H-66		1.42
		50	H-88		1.24
0.405	26	40	None	270	1.61
		50	None		1.79
		40	H-66	273	1.25
		50	H-66		1.39
		40	H-88	274	1.09
		50	H-88		1.21
0.50		40	None	177	1.30
		50	H-66	180	0.92
		50	H-88	181	0.80
0.511	24	40	None	170	1.27
		50	None		1.42
		40	H-66	173	0.79
		50	H-66		0.88
		40	H-88	174	0.69
		50	H-88		0.77
0.60		40	None	123	1.08
		50	None		1.21
		40	H-66	126	0.58
		50	H-88	127	0.56
0.644	22	40	None	107	1.01
		50	None		1.12
		40	H-66	110	0.50
		50	H-66		0.56
		40	H-88	111	0.44
0.70		40	None	90	0.92
		50	H-66		0.48
		40	H-88	94	0.37
0.80		40	None	69	0.81
		50	H-66	72	0.38
		40	H-88	73	0.29
0.90		40	None	55	0.72
0.91	19	40	None	53	0.71
		50	None		0.79
		40	H-44	55	0.31
		50	H-66	56	0.29
		50	H-88	57	0.26

Source: ITT, *Telecommunication Planning Documents—Outside Plant.*

2.8.5   Other Approaches to Subscriber Loop Design

INTRODUCTION

Between 30 and 50% of a telephone company's investment is tied up in what is generally referred to as "outside plant." Outside plant, for this discussion, can be defined as that part of the telephone plant that takes the signal from the local switch and delivers it to the subscriber. Much of this expense is attributable to copper in the subscriber cable. Another important expense is cable installation, such as that incurred in tearing up city streets to augment present installation or install new plant. Much work today is being done to devise methods to reduce these expenses. Among these methods are unigauge design, dedicated plant, and fine gauge design. All have a direct impact on the outside plant transmission engineer.

UNIGAUGE DESIGN

Unigauge is a concept developed by the Bell System of North America to save on the expense of copper in the subscriber loop plant. This is done by reducing the gauge (diameter) of wire pairs to a minimum, retaining specific resistance and transmission limits. The description that follows is an attempt at applying Unigauge to the more general case

To start with, review the basic rules set down up to this point for subscriber loop design. These are as foollows:

- Maximum loop resistance 1250 Ω (*Note* 1300 Ω is used in North America).
- Maximum loop loss, 6 dB (or a figure taken from the National Transmission Plan).

To these, let us add two more, namely:

- The use of modern, equalized telephone subsets [or at least on all loops longer than 10,000 ft (3000 m)].
- Cable gauges limited to AWG 19, 22, 24, and 26 while maintaining the 1250-Ω limit or whichever limit that needs to be established inside which signaling can be effected.

All further argument will be based on minimizing the amount of copper used (e.g., using the smallest diameter of wire possible within the limits prescribed above).

The reader should keep in mind that many telephone companies must install subscriber loops in excess of 15,000 (5000 m) and thus must use a conductor size (or sizes) larger than 26 gauge. The Bell System, for example, found that 20% of its subscriber loops exceeded 15,000 ft in 1964. To meet transmission and signaling objectives, the cost of such loops tends to mount excessively for those loops over 15,000 ft in length.

The concept of unigauge design basically is one which takes advantage of relatively inexpensive voice frequency gain devices and range extenders to permit the use of 26-gauge conductors on the greater percentage of its longer loops. Also, unigauge design makes it mandatory that a unified design approach be used. In other words, the design of subscriber loop plant is an integrated process and not one of piece-by-piece engineering. Unigauge allows the use of 26-gauge cable on subscriber loops up to 30,000 ft in length. This results in a very significant savings in copper and a general overall economy in present-day outside plant installation

APPLICATION OF UNIGAUGE DESIGN

A typical layout of a subscriber plant based on unigauge design is shown in Figure 2.10. The example is taken from the Bell System (Ref. 14).

In the figure it will be seen that subscribers within 15,000 ft of the switch are connected over loops made up of 26-gauge nonloaded cable with standard 48-V battery Their connection at the switch is conventional. It is also seen that 80% of Bell System subscribers are within this radius. Loops 15,000–30,000 ft long are called unigauge loops. Subscribers in the range 15,000–24,000 ft from the switch are connected by 26-gauge, nonloaded cable as well but require a range extender to provide sufficient voltage for signaling and supervision In the drawing a 72-V range extender is shown equipped with an amplifier that gives a midband gain of 5 dB. The output of the amplifier "emphasizes" the higher frequencies This offsets that additional loss suffered at the higher frequencies of the voice channel on the long, nonloaded loops. To extend the loops to the full 30,000 ft, the Bell System adds 88 mH loading coils at the 15,000 and 21,000-ft points.

For long loops (more than 15,000 ft long) the range extender/amplifier combinations are not connected on a line-for-line basis. It is standard practice to equip four or five subscriber loops with only one range extender/amplifier used on a shared basis. When the subscriber goes "off-hook" on a long, unigauge loop, a line is seized and the range extender is switched in. This concentration is another point in favor of unigauge because of the economics involved. It should be noted that the long, 15,000-ft, nonloaded sections, which the switch faces into, provide a fairly uniform impedance for all conditions when an active amplifier is switched in. This is a positive factor with regard to stability.

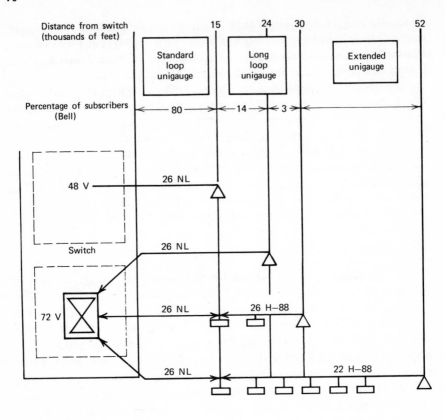

Figure 2.10 Layout of a unigauge subscriber plant.

Loops more than 30,000 ft long may also use the unigauge principle and often are referred to as "extended unigauge." Such a loop is equipped with 26-gauge nonloaded cable from the switch out to the 15,000-ft point. Beyond 15,000 ft 22-gauge cable is used with H-88 loading. As with all loops more than 15,000 ft long, a range extender-amplifier is switched in when the loop is in use. The loop length limit for this combination is 52,000 ft.

Loops more than 52,000 ft long may also be installed by using a gauge of diameter larger than 19.

It should be noted that the Bell System replaces its line relays with ones that are sensitive up to 2500 Ω of loop resistance with 48-V battery. Such a modification to the switch is done on long loops only. The 72 V supplied by the range extender is for pulsing, and the ringing voltage (to ring the distant telephone) is superimposed on the line with the subscriber's subset being in an "on-hook" condition.

Besides the savings in the expenditure on copper, unigauge displays some small improvements in transmission characteristics over older design methods of subscriber loops. These are:

- Unigauge has a slightly lower average loss (when we look at a statistical distribution of subscribers).
- There is 15-dB average return loss on the switch side of an amplifier, compared with an average of 11 dB for older design methods.

These two points echo Bell System experience with unigauge design (Ref. 10).

DEDICATED PLANT

Dedicated plant assigns wire pairs to subscribers or would-be subscriber locations. In the past bridged taps were used so that subscriber loops were more versatile regarding assignment. The idea of the bridged tap is shown below:

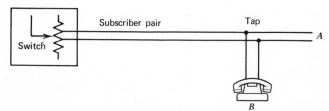

In this case station *A* is not connected; it is not in use but available in case station *B* disconnects telephone service.

The primary impact of dedicated plant is that it eliminates the use of bridged taps. Bridged taps deteriorate the quality of transmission by notable increasing the capacitance of a loop.

OTHER LOOP DESIGN TECHNIQUES

Fine gauge and "minigauge" techniques essentially are refinements of the uniguage concept. In each case the principal object is to reduce the amount of copper Obviously, one method is to use still smaller gauge pairs on shorter loops. Consideration has been given to the use of gauges as small as 32. Another approach is to use aluminum as the conductor. When aluminum is used, a handy rule of thumb to follow is that ohmic and attenuation losses of aluminum may be equated to copper in that aluminum wire always be the next "standard gauge" larger than its copper counterpart if copper were to be used.

Copper	Aluminum
19	17
22	20
24	22
26	24

Aluminum has some drawbacks as well; the major ones are summed up as follows:

- Not to be used on the first 500 yd of cable where the cable has a large diameter (i.e., more loops before branching).
- More difficult to splice than copper.
- More brittle.
- Because the equivalent conductor is larger than its copper counterpart, an equivalent aluminum cable with the same conductivity/loss characteristics as copper will have a smaller pair count in the same sheath.

2.9 DESIGN OF LOCAL AREA TRUNKS (JUNCTIONS)

Exchanges in a common local area often are connected on a full mesh basis (see Section 2.7 for definition of mesh connection). Depending on distance and certain other economic factors, these trunk circuits use VF transmission over cable. In view of the relatively small number of these trunk circuits when compared to the number of subscriber lines,* it is generally economically desirable to minimize attenuation in this portion of the network.

One approach used by some telephone companies (administrations) is to allot $\frac{1}{3}$ of the total end-to-end reference equivalent to each subscriber's loop and $\frac{1}{3}$ to the trunk network. Figure 2.11 illustrates this concept.

---

* Due to the inherent concentration in local switches (e.g., 1 trunk for 8–25 subscribers, depending on design).

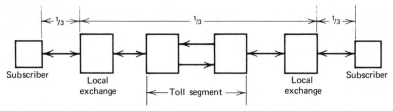

Figure 2.11 One approach to loss assignment.

For instance, if the transmission plan called for a 24-dB ORE, then $\frac{1}{3}$ of 24 dB, or 8 dB, would be assigned to the trunk plant. Of this we may assign 4 dB to the four-wire portion or toll segment of the network, leaving 4 dB for local VF trunks or 2 dB at each end. The example has been highly simplified of course.

For the toll connecting trunks (e.g., those trunks which connect the local network to the toll network), if a good return loss cannot be maintained on all or nearly all connections, losses on two-wire toll connecting trunks may have to be increased to reduce possibilities of echo and singing. Sometimes the range of loss for these two-wire circuits must be extended to 5 or 6 dB. It is just these circuits that the four-wire toll network looks into directly.

From this it can be seen that the approach to the design of VF trunks varies considerably from that used for subscriber loop design. Although we must ensure that signaling limits are not exceeded, most always the transmission limit will be exceeded well before the signaling limit. The tendency to use larger diameter cable on long routes is also evident.

One major difference from the subscriber loop approach is in the loading. If loading is to be used, the first load coil is installed at distance $D/2$, where D is the normal separation distance between load points. Take the case of H loading, for instance. The distance between load points is 1830 m, but the first load coil from the exchange is placed at $D/2$ or 915 m from the exchange. Then, if an exchange is bypassed, a full load section exists. This concept is illustrated in Figure 2.12.

Now consider this example. A loaded 500-pair VF trunk cable extends across town. A new switching center is to be installed along the route where 50 pairs are to be dropped and 50 inserted. It would be desirable to establish the new switch midway between load points. At the switch 450 circuits will bypass the office (switch). Using this $D/2$ technique, these circuits need no conditioning; they are full-load sections (i.e., $D/2 + D/2 = 1D$, a full-load section). Meanwhile, the 50 circuits entering from each direction are terminated for switching and need conditioning so each looks electri-

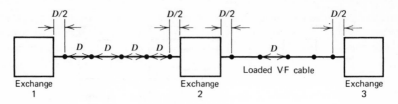

Figure 2.12 Loading of VF trunks (junctions).

cally like a full-load section. However, the physical distance from the switch out to the first load point is $D/2$ or, in the case of H loading, 915 m. To make the load distance electrically equivalent to 1830 m, line build out (LBO) is used. This is done simply by adding capacity to the line.

Suppose the location of the new switching center was such that it was not half-way, but some other fractional distance. For the section consisting the shorter distance, LBO is used. For the other, longer run, often a half-load coil is installed at the switching center and LBO is added to trim up the remaining electrical distance.

2.10 VF REPEATERS

VF repeaters in telephone terminology imply the use of *uni*directional amplifiers at voice frequency on VF trunks. On a two-wire trunk two amplifiers must be used on each pair with a hybrid in and a hybrid out. A simplified block diagram is shown in Figure 2.13.

The gain of a VF repeater can be run up as high as 20 or 25 dB, and originally they were used on 50-mi, 19-gauge loaded cable in the long-distance (toll) plant. Today they are seldom found on long-distance circuits, but do have application on local trunk circuits, where the gain require-

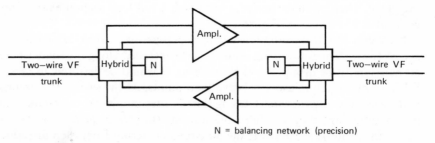

Figure 2.13 Simplified block diagram of a VF repeater.

ments are considerably less. Trunks using VF repeaters have the repeater's gain adjusted to the equivalent loss of the circuit minus the 4-dB loss to provide the necessary singing margin. In practice a repeater is installed at each end of the trunk circuit to simplify maintenance and power feeding. Gains may be as high as 6–8 dB.

An important consideration with VF repeaters is the balance at the hybrids. Here precision balancing networks may be used instead of the compromise networks employed at the two-wire, four-wire interface (Section 2.6.3) It is common to achieve a 21-dB return loss, 27 dB is also possible, and theoretically, 35 dB can be reached.

Another repeater commonly used on two-wire trunks is the negative-impedance repeater. This repeater can provide a gain as high as 12 dB, but 7 or 8 dB is more common in practice. The negative-impedance repeater requires a line build out (LBO) at each port and is a true two-way, two-wire repeaters The repeater action is based on regenerative feedback of two amplifiers. The advantage of negative-impedance repeaters is that they are transparent to dc signaling. On the other hand, VF repeaters require a composite arrangement to pass dc signaling. This consists of a transformer bypass.

2.11 TRANSMISSION CONSIDERATIONS OF TELEPHONE SWITCHES IN THE LONG-DISTANCE NETWORK (FOUR-WIRE)

Any device placed in an analog transmission circuit tends to degrade the quality of transmission. Telephone switches, unless properly designed, well selected from a systems engineering point of view, and properly installed, may seriously degrade a transmission network. Transmission specifications of a switch must be set forth clearly when switches are to be purchased. One reference which sets forth a series of specifications for toll switches is CCITT Rec. Q.45. The reader should also consult the G.100 series of CCITT recommendations (see Appendix A).

In practice it has been found that much of the criterion specified in CCITT Rec. Q.45 is rather loose. Table 2.9 compares Q.45 recommendations with a stricter criteria (based on Ref. 19).

2.12 CCITT INTERFACE

2.12.1 Introduction

To facilitate satisfactory communications between telephone subscribers in different countries, the CCITT has established certain transmission

**Table 2.9   Transmission Characteristics of Switches
(Four-Wire Switching)**

Item	1 (Q.45)[c]	2 (Ref. 19 Criterion)[c]
Loss	0.5 dB	0.5 dB
Loss, dispersion [a]	<0.2 dB	<0.2 dB
Attenuation/	300–400 Hz: −0.2/+0.5 dB	−0.1/+0.2 dB
frequency response	400–2400 Hz: −0.2/+0.3 dB	−0.1/+0.2 dB
	2400–3400 Hz: −0.2/+0.5 dB	−0.1/+0.3 dB
Impulse noise	5 in 5 min above −35 dBm0	5 in 5 min, 12 dB above floor of random noise [d]
Noise		
Weighted	200 pWp	25 pWp
Unweighted	1000,000 pW	3000 pW
Unbalance against	300–600 Hz: 40 dB	300–3000 Hz: 55 dB
ground	600–3400 Hz: 46 dB	3000–3400 Hz: 53 dB
Crosstalk		
Between go and	60 dB	65 dB ⎫ in the band
return paths		
Between any two	70 dB	80 dB ⎭ 200–3200 Hz
paths		
Harmonic distortion	—	50 db down with a −10 dBm signal for 2nd harmonic, 60 dB down for 3rd harmonic at 200 Hz: 15 dB
Impedance variation	300–600 Hz: 15 dB	300 Hz: 18 dB
with frequency [b]	600–3400 Hz: 20 dB	500–2500 Hz: 20 dB
		3000 Hz: 18 dB
		3400 Hz: 15 dB

[a] Dispersion loss is the variation in loss from calls with the highest loss to those with the lowest loss. This important parameter affects circuit stability.
[b] Expressed as return loss.
[c] Reference frequency, where required, is 800 Hz for column 1, and 1000 Hz for column 2.
[d] Taken from standard measurement techniques for impulse noise.

criteria in the form of recommendations. According to CCITT a connection is satisfactory if it meets certain criteria for the following:

- Reference equivalent.
- Noise.
- Echo.
- Singing.

The following paragraphs outline and highlight these criteria from the point of view of a telephone company or administration's international interface. This is the point where the national network meets the international system.

2.12.2 Maximum Number of Circuits in Tandem

If we consult Figure 2.14, we will count six links in the portion of a connection that we call international. The interface for the country of origin is a CT3 (CT—Central Transito, ranked 3 in the hierarchy) and the country of destination, the other CT3. A call must be limited from local exchange (origin) to local exchange (destination) to have no more than 14 links (circuits) in tandem, of which four may be in the country of origin and four in the country of destination. The number of links are limited to assure maintenance of limits of noise, stability, and reference equivalent.

The number of links in a national chain is further limited) CCITT Rec. G.101). If the average distance to a subscriber is 600 mi (1000 km) from the international interface, at most three national four-wire circuits can be connected on a four-wire basis between each other. In countries that have average distances to subscribers in excess of 600 mi, a fourth or possibly a fifth national circuit may be added on a four-wire basis.

2.12.3 Noise

CCITT Rec. G.103 treats noise. It equates noise to length of a circuit. This is equated at the rate of 4 pWp/km. Assume a national connection is 2500 km long, or 2500 km to reach the international interface. Then we would expect to measure no more than 10,000 pWp at that point. The reader should consult CCITT Recs. G.152 and G.153 for further information. Actual design objectives should improve on these specifications by proper choice of equipment and system layout and engineering.

Absolute noise maximums on the receive side of an international connection should not exceed 50,000 pWp referred to a zero relative level point of the first circuit in the chain (Rec. G.143). This maximum noise level is permissible if there are six international circuits in tandem.

2.12.4 Variation of Transmission Loss with Time

This important parameter affects stability. CCITT Rec. G.151C states that "The standard deviation of the variation in transmission loss of a circuit

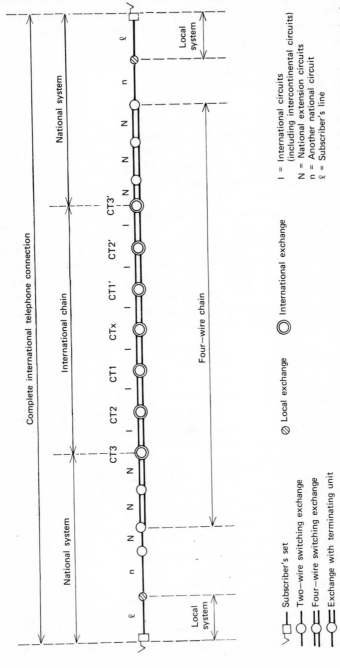

Figure 2.14 A CCITT international telephone connection with national extensions. Courtesy International Telecommunications Union—CCITT.

should not exceed 1 dB . . . . The difference between the mean value and the nominal value of the transmission loss for each circuit should not exceed 0.5 dB."

2.12.5 Crosstalk (CCITT Rec. G.151D)

The near-end and far-end crosstalk (intelligible crosstalk only) measured at audio frequencies at trunk exchanges between two complete circuits in terminal service position should not numerically be less than 58 dB.

Between "go" and "return" paths of the same circuit in a four-wire long-distance (toll) exchange intelligible crosstalk should be at least 43 dB down.

2.12.6 Attenuation Distortion

The worst condition for attenuation distortion is shown in Figure 2.15 (taken from CCITT Rec. G.131). It assumes that the nominal 4-kHz voice channel is used straight through. The attenuation distortion as shown in the figure is a result of 12 circuits of a four-wire chain in tandem. Note that the slope from 300 to 600 Hz and from 2400–3400 Hz is approximately 6.5 dB.

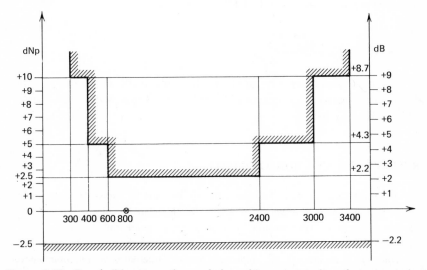

Figure 2.15 Permissible attenuation variation with respect to its value measured at 800 Hz (objective for worldwide four-wire chain of 12 circuits in terminal service) (CCITT Rec. G.132). Courtesy International Telecommunications Union—CCITT.

The slope for one link, therefore, would be 6.5/12 or approximately 0.5 dB. From this we can derive the attenuation distortion permissible for each link to the international interface. For further information, consult Chapter 3 for a discussion of attenuation distortion of FDM carrier equipment back-to-back.

2.12.7 Reference Equivalent

"For 97 percent of the connections made in a country of average size (1000 km average distance to the CT), the nominal reference equivalent between a subscriber and the 4-wire terminals of the international circuit (the international interface) . . . should not exceed 20.8 dB sending and 12.2 dB receiving" (CCITT Rec. G.111). This gives an ORE of 33 dB. It is recommended that telephone companies and administrations attempt to lower this figure as much as possible to improve subscriber satisfaction. A good target would be an ORE from a low of 6 dB to a maximum of 20 dB.

2.12.8 Propagation Time

The propagation time at 800 Hz of the national sending or receiving system should not exceed 50 ms (CCITT Rec. G.114).

2.12.9 Echo Suppressors

Echo suppressors should be used on all connections where the *round-trip* delay (propagation time) exceeds 50 ms. Echo suppressors are covered in CCITT Rec. G.161.

REFERENCES AND BIBLIOGRAPHY

1. *Reference Data for Radio Engineers*, 5th ed., Howard W. Sams & Co., Indianapolis.
2. *U.S. Army Technical Manual*, TM-486-5, "Outside plant."
3. M. A. Clement, "Transmission," *Telephony* (Magazine) reprint.
4. CCITT White Books, Mar del Plata, 1968, Vol. III, G. Recommendations.
5. *National Networks for the Automatic Service*, International Telecommunications Union, Geneva, Chap. V.
6. *Overall Communications System Planning*, Vols. I–III, IEEE New Jersey Section Seminar, 1964.

7. *Transmission Systems for Communications*, rev. 3rd ed., rev. 4th ed., Bell Telephone Laboratories.

8. F. T. Andrew and R. W. Hatch, "National Telephone Network Planning in the ATT," *IEEE ComTech J.*, June 1971.

9. "Terminal Balance, Description and Test Methods," *Autom. Electr. Tech. Bull.*305–351.

10. P. A. Gresh et al., "A Unigauge Design Concept for Telephone Customer Loop Plant," *IEEE ComTech J.*, April 1968.

11. *Principles of Electricity as Applied to Telephone and Telegraph Work*, ATT Long Lines Department, New York, 1961.

12. *Rural Electrification Administration Telephone Engineering and Construction Manual*, Section 400 series.

13. *ITT Pentaconta Manual PC-1000A/B/B1*.

14. H. R. Huntley, "Transmission Design of Intertoll Telephone Trunks," *Bell Syst. Tech. J.*, Sept. 1953.

15. Military Standard 188B with Notice 1, U.S. Department of Defense, Washington, D.C.

16. *DCA Circ.* 330-175-1, through Change 9, U.S. Department of Defense, Washington, D.C.

17. *Lenkurt Demodulator*, Lenkurt Electric Corporation, San Carlos, Calif.: July 1960, Oct. 1962, March 1964, Jan. 1964, Aug. 1966, June 1968, July 1966, Aug. 1973.

18. *Local Telephone Networks*, CCITT, International Telecommunications Union, Geneva, 1968.

19. USITA Symposium, April 1970, Open Questions 18–37.

20. Telecommunication Planning Documents, *Outside Plant*, ITT Laboratories Spain, 1973.

3 | FREQUENCY DIVISION MULTIPLEX

3.1 INTRODUCTION

Multiplex deals with the transmission of two or more signals simultaneously over a single transmission facility. Multiplexing may be accomplished either in the domain of frequency or in the time domain. The first is called frequency division multiplex (FDM), and the second, time division multiplex (TDM). This chapter covers the former. The latter is discussed in Chapter 11.

Before we launch into multiplexing, keep in mind that all multiplex systems work on a four-wire basis. The transmit and receive paths are separate. Two-wire and four-wire transmission and conversion from two-wire to four-wire systems are covered in Section 2.6.3.

Carrier is a word that has become associated with FDM systems, probably owing to the fact that FDM employs "carrier" waves. Unfortunately its use has spread to TDM systems as well. As used in this text, however, carrier refers exclusively to FDM systems.

It is assumed that the reader has some background in how an SSB (single sideband) signal is developed as well as how a carrier is suppressed in SSBSC systems.

A simplified diagram of an FDM link is shown in Figure 3.1.

3.2 MIXING

The heterodyning or mixing of signals of frequencies A and B is shown below. What frequencies may be found at the output of the mixer?

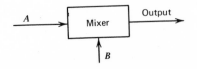

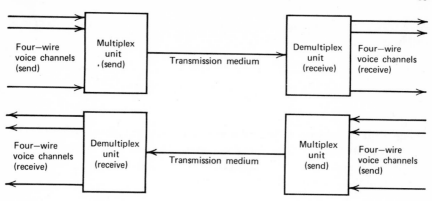

Figure 3.1 Simplified block diagram of a frequency division multiplex link.

Both the original signals will be present, as well as signals representing their sum and their difference in the frequency domain. Thus we will have present at the output of the mixer above signals of frequency A, B, $A + B$, and $A - B$. Such a mixing process is repeated many times in frequency division multiplex (FDM) equipment.

Let us look at the boundaries of the nominal 4-kHz voice channel. These are 300 and 3400 Hz. Let us further consider these frequencies as simple tones of 300 and 3400 Hz. Now consider the mixer below and examine the possibilities at its output.

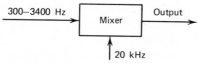

First, the output may be the sum or

$$
\begin{array}{r}
20{,}000 \text{ Hz} \\
+ \quad 300 \text{ Hz} \\
\hline
20{,}300 \text{ Hz}
\end{array}
\qquad
\begin{array}{r}
20{,}000 \text{ Hz} \\
+ \ 3{,}400 \text{ Hz} \\
\hline
23{,}400 \text{ Hz}
\end{array}
$$

A simple low-pass filter could filter out all frequencies below 20,300 Hz.

Now imagine that instead of two frequencies, we have a continuous spectrum of frequencies between 300 and 3400 Hz (i.e., we have the voice channel). We represent the spectrum as a triangle.

300   3400 Hz

As a result of the mixing process (translation) we have another triangle as follows:

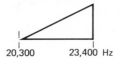

When we take the sum, as we did above, and filter out all other frequencies, we say we have selected the upper sideband. Thus we have a triangle facing to the right, and we call this an upright or erect sideband.

We can also take the difference, such that

$$
\begin{array}{rr}
20{,}000 \text{ Hz} & 20{,}000 \text{ Hz} \\
-\quad 300 \text{ Hz} & -\ 3{,}400 \text{ Hz} \\
\hline
19{,}700 \text{ Hz} & 16{,}600 \text{ Hz}
\end{array}
$$

and we see that in the translation (mixing process) we have had an inversion of frequencies. The higher frequencies of the voice channel become the lower frequencies of the translated spectrum, and the lower frequencies of the voice channel become the higher when the difference is taken. We represent this by a right triangle facing the other direction (left):

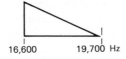

This is called an inverted sideband. To review, when we take the sum, we get an erect sideband. When we take the difference, frequencies invert and we have an inverted sideband represented by a triangle facing left.

Now let us complicate the process a little by translating three voice channels into the radio electric spectrum for simultaneous transmission on a specific medium, a pair of wire lines, for example. Let the local oscillator (mixing) frequency in each case be 20, 16, and 12 kHz. The mixing process is shown in Figure 3.2.

From Figure 3.2 the difference frequencies are selected in each case as follows:

$$
\begin{array}{lcc}
\text{For channel No. 1} & 20{,}000 \text{ Hz} & 20{,}000 \text{ Hz} \\
 & -\quad 300 \text{ Hz} & -\ 3{,}400 \text{ Hz} \\
\cline{2-3}
 & 19{,}700 \text{ Hz} & 16{,}600 \text{ Ha} \\
\\
\text{For channel No. 2} & 16{,}000 \text{ Hz} & 16{,}000 \text{ Hz} \\
 & -\quad 300 \text{ Hz} & -\ 3{,}400 \text{ Hz} \\
\cline{2-3}
 & 15{,}700 \text{ Hz} & 12{,}600 \text{ Hz}
\end{array}
$$

For channel No. 3   12,000 Hz   12,000 Hz
                        —   300 Hz   — 3,400 Hz
                        _____   _____
                          11,700 Hz   8,600 Hz

In each case the lower sidebands have been selected as mentioned above and all frequencies above 19,700 Hz have been filtered from the output as well as the local oscillator carriers themselves. The outputs from the modulators terminate on a common bus. The common output appearing on this bus is a band of frequencies between 8.6 and 19.7 kHz containing the three voice channels which have been translated in frequency. They now

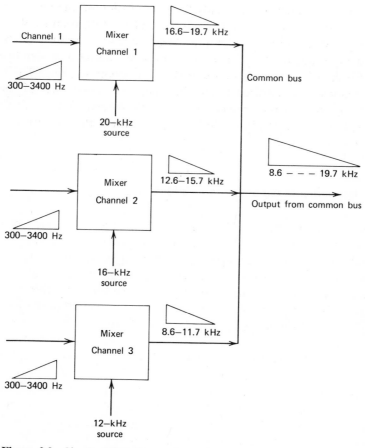

Figure 3.2 Simple FDM (transmit portion only shown).

appear on one two-wire circuit ready for transmission. They may be represented by a single inverted triangle as shown:

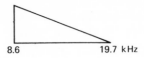

8.6   19.7 kHz

3.3 THE CCITT MODULATION PLAN

3.3.1 Introduction

A modulation plan sets forth the development of a band of frequencies called the line frequency (i.e., ready for transmission on the line or transmission medium). The modulation plan usually is a diagram showing the necessary mixing, local oscillator mixing frequencies, and the sidebands selected by means of the triangles described previously in a step-by-step process from voice channel input to line frequency output. The CCITT has recommended a standardized modulation plan with a common terminology. This allows large telephone networks, on both national and multinational systems, to interconnect. In the following paragraphs the reader is advised to be careful with terminology.

3.3.2 Formation of the Standard CCITT Group

The standard *group* as defined by the CCITT occupies the frequency band 60–108 kHz and contains 12 voice channels. Each voice channel is the nominal 4-kHz channel occupying the 300–3400 Hz spectrum. The group is formed by mixing each of the 12 voice channels with a particular carrier frequency associated with the channel. Lower sidebands are then selected. Figure 3.3 shows the preferrred approach to the formation of the standard CCITT group. It should be noted that in the 60–108 kHz band voice channel 1 occupies the highest frequency segment by convention, between 104 and 108 kHz. The layout of the standard group is shown in Figure 3.4. The applicable CCITT recommendation is G.232.

Single sideband suppressed carrier modulation techniques are recommended except under special circumstances discussed later in this chapter. CCITT recommends that carrier leak be down to at least -26 dBm0 referred to a zero relative level point (see Section 1.9.5).

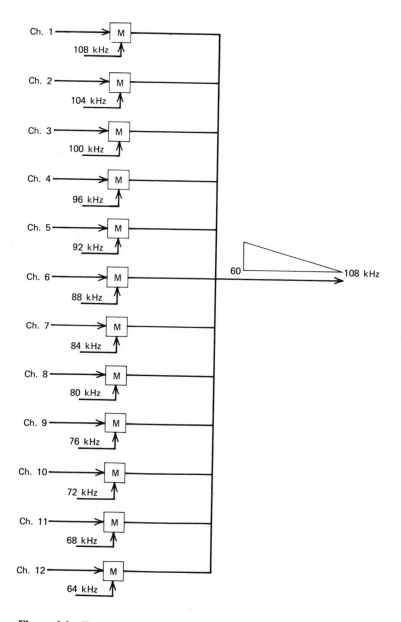

Figure 3.3 Formation of the standard CCITT group.

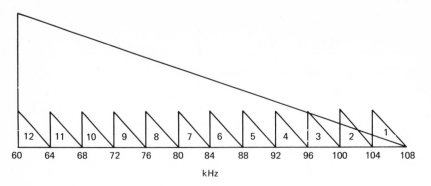

Figure 3.4   The standard CCITT group.

3.3.3   Alternative Method of Formation of the Standard CCITT Group

Economy of filter design has caused some manufacturers to use an alternate method to form a group. This is done by an intermediate modulation step forming four pregroups. Each pregroup translates three voice channels in the intermediate modulation step. The translation process for this alternative method is shown in Figure 3.5. For each pregroup the first voice channel modulates a 12-kHz carrier, the second a 16-kHz carrier, and the third a 20-kHz carrier. The upper sidebands are selected in this case and carriers suppressed.

The result is a subgroup occupying a band of frequencies from 12 to 24 kHz. The second modulation step is to take four of these pregroups so formed and translate them, each to their own frequency segment, in the band 60–108 kHz. To achieve this the pregroups are modulated by carrier frequencies of 84, 96, 108, and 120 kHz and the lower sidebands are selected, properly inverting the voice channels. This dual modulation process is shown in Figure 3.5.

The choice of the one-step or two-step modulation is an economic trade-off. Adding a modulation stage adds noise to the system.

3.3.4   Formation of the Standard CCITT Supergroup

A supergroup contains five standard CCITT groups, equivalent to 60 voice channels. The standard supergroup before translation occupies the frequency band 312–552 kHz. Each of the five groups making up the supergroup is translated in frequency to the supergroup band by mixing with the proper carrier frequencies. The carrier frequencies are 420 kHz for Group

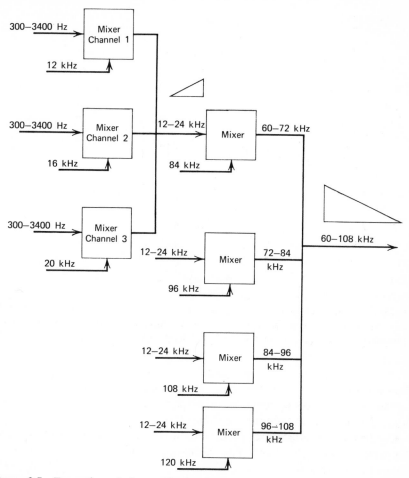

Figure 3.5 Formation of the standard CCITT group by two steps of modulation (mixing).

1, 468 kHz for Group 2, 516 kHz for Group 3, 564 kHz for Group 4, and 612 kHz for Group 5. In the mixing process the difference is taken (lower sidebands are selected). This translation process is shown in Figure 3.6.

3.3.5 Formation of the Standard CCITT Basic Mastergroup and Supermastergroup

The basic mastergroup contains five supergroups, 300 voice channels. It occupies the spectrum 812–2044 kHz. It is formed by translating the five

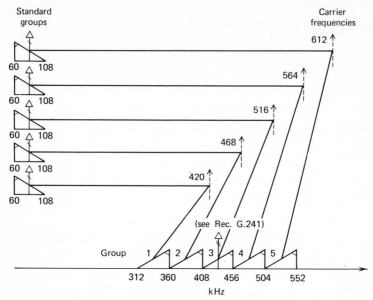

Figure 3.6 The formation of the standard CCITT supergroup (CCITT Rec. G.233).
Note. Vertical arrows show group level regulating pilot tones (see Section 3.5).
Courtesy International Telecommunications Union—CCITT.

standard supergroups, each occupying the 312–552 kHz band, by a process similar to that used to form the supergroup from five standard CCITT groups. This process is shown in Figure 3.7.

The basic supermastergroup contains three mastergroups and occupies the band 8516–12388 kHz. The formation of the supermastergroup is shown in Figure 3.8.

3.3.6   The "Line" Frequency

The band of frequencies that the multiplex applies to the line, whether the line is a radiolink, coaxial cable, wire pair or open-wire line, is called the line frequency. Another expression often used is HF (or high frequency), not to be confused with high frequency radio, discussed in Chapter 4.

The line frequency in this case may be the direct application of a group or supergroup to the line. However, more commonly a final translation stage occurs, particularly on high density systems. Several of these line configurations are shown below.

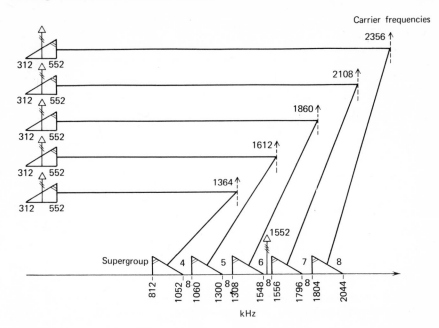

Figure 3.7 Formation of the standard CCITT mastergroup (CCITT Rec. G.233). Courtesy International Telecommunications Union—CCITT.

Figure 3.9 shows the makeup of the basic 15-supergroup assembly. Figure 3.10 shows the makeup of the standard 15-supergroup assembly No. 3 as derived from the basic 15-supergroup assembly shown in Figure 3.9. Figure 3.11 shows the development of a 600-channel standard CCITT line frequency.

3.4 LOADING OF MULTICHANNEL FDM SYSTEMS

3.4.1 Introduction

Most of the FDM (carrier) equipment in use today carries speech traffic, sometimes misnamed in North America "message traffic." In this context we refer to full-duplex conversations by telephone between two "talkers." However, the reader should not lose sight of the fact that there is a marked increase in the use of these same intervening talker facilities for data transmission.

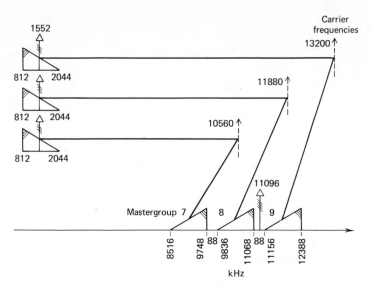

Figure 3.8 Formation of the standard CCITT supermastergroup (CCITT Rec. G.233). Courtesy International Telecommunications Union—CCITT.

For this discussion the problem essentially boils down to that of human speech and how multiple telephone users may load a carrier system. If we load a carrier system too heavily, meaning here that the input levels are too high, intermodulation noise and crosstalk will become intolerable. If we do not load the system sufficiently, the signal-to-noise ratio will suffer. The problem is fairly complex because speech amplitude varies:

1. With talker volume.
2. At a syllabic rate.
3. At an audio rate.
4. With varying circuit losses as different loops and trunks are switched into the same channel bank voice channel input.

Also the loading of a particular system varies with the busy hour.*

3.4.2 Speech Measurement

The average power measured in dBm of a typical single talker is:

$$P_{\mathrm{dBm}} = V_{\mathrm{VU}} - 1.4 \qquad (3.1)$$

---

* The *busy hour* is a term used in traffic engineering and is defined by the CCITT as "The uninterrupted period of 60 minutes during which the average traffic flow is maximum."

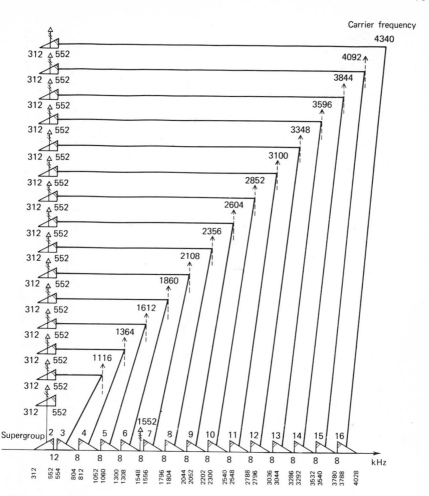

Figure 3.9 Makeup of the basic CCITT 15-supergroup assembly. Courtesy International Telecommunications Union—CCITT.

where V_{VU} is the reading of a VU meter (see Section 1.9.6). In other words a 0 VU talker has an average power of -1.4 dBm.

Empirically for a typical talker the peak power is about 18.6 dB higher than average power. The peakiness of speech level means that carrier equipment must be operated at a low average power to withstand voice peaks so as not to overload and cause distortion. Thus the primary concern is that of voice peaks or spurts. These can be related to an activity factor

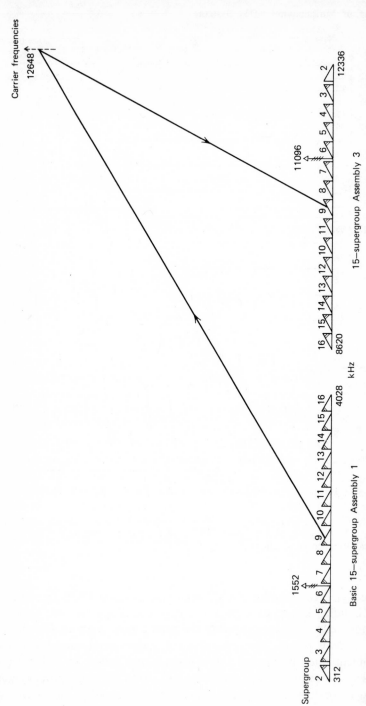

Figure 3.10 Makeup of standard 15-supergroup Assembly No. 3 as derived from basic 15-supergroup assembly. Courtesy International Telecommunications Union—CCITT.

T_a. T_a is defined as that proportion of the time that the rectified speech envelope exceeds some threshold. If the threshold is about 20 dB below the average power, the activity dependence on threshold is fairly weak. We can now rewrite our equation for average talker power in dBm relating it to the activity factor T_a as follows:

$$P_{dBm} = V_{VU} + 10 \log T_a$$

If $T_a = 0.725$, the results will be the same as for equation 3.1.

Consider now adding a second talker operating on a different frequency segment on the same equipment, but independent of the first talker. Translations to separate frequency segments are described earlier in the chapter. With the second talker added the system average power will increase 3 dB. If we have N talkers, each on a different frequency segment, the average power developed will be

$$P_{dBm} = V_{VU} - 1.4 + 10 \log N \tag{3.2}$$

where P_{dBm} is the power developed across the frequency band occupied by all the talkers.

Empirically we have found that the peakiness or peak factor of multi-talkers over a multichannel analog system reaches the characteristics of random noise peaks when the number of talkers, N, exceeds 64. When $N = 2$, the peaking factor is 18 dB; for 10 talkers, 16 dB; for 50, 14 dB; and so forth. Above 64 talkers the peak factor is 12.5 dB.

An activity factor of 1, which we have been using above, is unrealistic. This means that someone is talking all the time on the circuit. The traditional figure for activity factor accepted by CCITT and used in North American practice is 0.25. Let us see how we reach this lower figure.

For one thing, the multichannel equipment cannot be designed for N callers and no more. If this were true, a new call would have to be initiated every time a call terminated or calls would have to be turned away for an "all trunks busy" condition. In the real life situation, particularly for automatic service, carrier equipment, like switches, must have a certain margin by being overdimensioned for busy hour service. For this over dimensioning we drop the activity factor to 0.70. Other causes reduce the figure even more. For instance, circuits are essentially inactive during call setup as well as during pauses for thinking during a conversation. The 0.70 of before now reduces to 0.50. This latter figure is divided in half owing to the talk-listen effect. If we disregard isolated cases of "double-talking," it is obvious that on a full-duplex telephone circuit, while one end is talking, the other is listening. Thus a circuit (in one direction) is idle half the time during "listen" period. The resulting activity factor is 0.25.

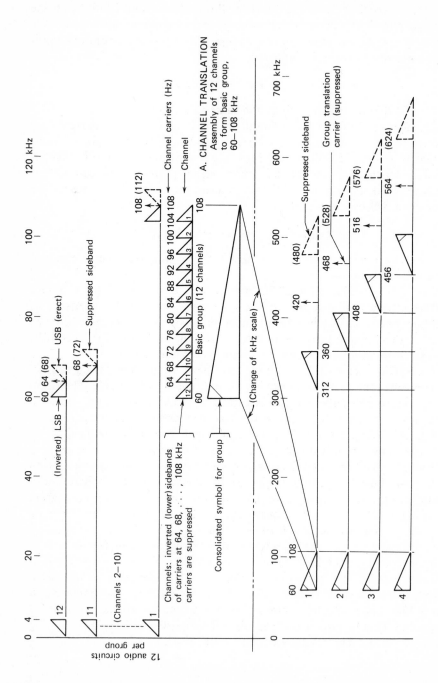

120 kHz

100

80

60

40

20

4

0

12 audio circuits per group

(Inverted) LSB →

USB (erect)

Suppressed sideband

60 64 (68)

68 (72)

108 (112)

Channel carriers (Hz)

Channel

108 104 100 96 92 88 84 80 76 72 68 64 60

1 2 3 4 5 6 7 8 9 10 11 12

Basic group (12 channels)

Channels: inverted (lower) sidebands of carriers at 64, 68, . . . , 108 kHz carriers are suppressed

Consolidated symbol for group

(Channels 2–10)

12

11

1

A. CHANNEL TRANSLATION
Assembly of 12 channels to form basic group, 60–108 kHz

0   100   200   300   400   500   600   700 kHz

(Change of kHz scale)

60   108

312   360

408   456   (480)   468   516   564   (576)   624

(528)

Suppressed sideband

Group translation carrier (suppressed)

1

2

3

4

420

96

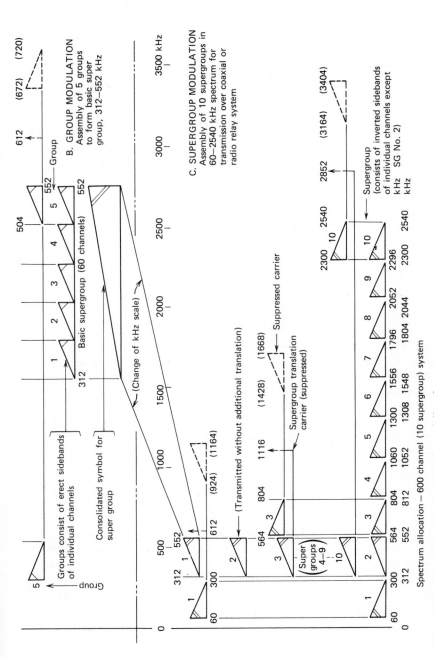

Figure 3.11 Development of a 600-channel line frequency.

3.4.3   Overload

In Section 1.9.7, where we discussed intermodulation (IM) products, we showed that one cause of IM products is overload. One definition of overload (Ref. 4) is, "The overload point, or overload level, of a telephone transmission system is 6 dB higher than the average power in dBm0 of each of two applied sinusoids of equal amplitude and of frequencies a and b, when these input levels are so adjusted that an increase of 1 dB in both their separate levels causes an increase, at the output, of 20 dB in the intermodulation product of frequency $2a - b$."

Up to this point we have been talking about average power. Overload usually occurs when instantaneous signal peaks exceed some threshold. Consider that peak instantaneous power exceeds average power of a simple sinusoid by 3 dB. For multichannel systems the peak factor may exceed that of a sinusoid by 10 dB.

White noise is often used to simulate multitalker situations for systems with more than 64 operative channels.

3.4.4   Loading

For loading of multichannel FDM systems, CCITT recommends:

It will be assumed for the calculation of intermodulation below the overload point that the multiplex signal during the busy hour can be represented by a uniform spectrum [of] random noise signal, the mean absolute power level of which, at a zero relative level point—(in dBm0)

$$P_{av} = -15 + 10 \log N \qquad (3.3)$$
$$\text{when } N \geq 240$$

and

$$P_{av} = -1 + 4 \log N \qquad (3.4)$$
$$\text{when } 12 \leq N < 240 \ldots.$$

(CCITT Rec. G.223) (all logs to the base 10). *Note.* These equations apply only to systems without preemphasis and using independent amplifiers in both directions. Preemphasis is discussed in Section 5.5.2. An activity factor of 0.25 is assumed. See Figure 5.28 and the discussion therewith. Examples of the application of these formulas are discussed in paragraph 5.6.3. It should also be noted that the formulas above include a small margin for loads caused by signaling tones, pilot tones, and carrier leaks.

Example 1. What is the average power of the composite signal for a 600-voice channel system using CCITT loading? N is greater than 240; thus equation 3.3 is valid.

$$P_{av} = -15 + 10 \log 600$$
$$= -15 + 10 \times 2.7782$$
$$= +12.782 \text{ dBm0}$$

Example 2. What is the average power of the composite signal for a 24-voice channel system using CCITT loading? N is less than 240; thus equation 3.4 is valid.

$$P_{av} = -1 + 4 \log 24$$
$$= -1 + 4 \times 1.3802$$
$$= +4.5208 \text{ dBm0}$$

3.4.5 Single Channel Loading

A number of telephone administrations have attempted to standardize on -16 dBm0 for single channel speech input to multichannel FDM equipment. With this input, peaks in speech level may reach -3 dBm0. Tests indicated that such peaks will not be exceeded more than 1% of the time. However, the conventional value of average power per voice channel allowed by the CCITT is -15 dBm0. (Refer to formula 3.3 and CCITT Rec. G.223.) This assumes a standard deviation of 5.8 dB and the traditional activity factor of 0.25. Average talker level is assumed to be at -11.5 VU. We must turn to the use of standard deviation because we are dealing with talker levels that vary with each talker, and thus with the mean or average.

3.4.6 Loading with Constant Amplitude Signals

Speech on multichannel systems has a low duty cycle or activity factor. We established the traditional figure of 0.25. Certain other types of signals transmitted over the multichannel equipment have an activity factor of 1. This means that they are transmitted continuously, or continuously over fixed time frames. They are also characterized by constant amplitude. Examples of these types of signals follow:

- Telegraph tone or tones.
- Signaling tone or tones.
- Pilot tones.
- Data signals (particularly FSK and PSK; see Chapter 8).

Here again, if we reduce the level too much to ensure against overload, the signal-to-noise ratio will suffer, and hence the error rate will suffer.

For typical constant amplitude signals, traditional* transmit levels (input to the channel modulator on the carrier (FDM) equipment) are as follows:

- Data: -13 dBm0.
- Signaling (SF supervision), tone-on when idle: -20 dBm0.
- Composite telegraph: -8.7 dBm0.

For one FDM system now on the market with 75% speech loading, 25% data/telegraph loading with more than 240 voice channels, the manufacturer recommends the following:

$$P_{rms} = -11 + 10 \log N$$

using -5 dBm0 per channel for the data input levels† and -8 dBm0 for the composite telegraph level.

Table 3.1 shows some of the standard practice for data/telegraph loading on a per-channel basis. Data and telegraph should be loaded uniformly. For instance, if equipment is designed for 25% data and telegraph loading, then voice channel assignment should, whenever possible, load each group and supergroup uniformly. 75% of one group should not be loaded with data and none of another.

Table 3.1   Voice Channel Loading of Data/Telegraph Signals

Signal Type	CCITT	North American
High speed data	-10 dBm0 simplex -13 dbm0 duplex	-10 dBm0 switched network -8 dBm0 leased line -5 dBm0 occasionally
Medium speed	—	-8 dBm0 total power
Telegraph (multichannel)		
≤ 12 channels	-19.5 dBm0/channel -8.7 dBm0 total	
≤ 18 channels	-21.25 dBm0/channel -8.7 dBm0 total	
≤ 24 channels	-22.25 dBm0/channel -8.7 dBm0 total	

Source: *Lenkurt Demodulator*, July 1968.

* As taken from CCITT.
† All VF channels may be loaded at -8 dBm0 level, whether data or telegraph with this equipment, but for -5 dBm0 level data, only 2 channels/group may be assigned this level, the remainder voice, or the group must be "deloaded" (i.e., idle channels).

Data should also be assigned to voice channels that will not be near group band edge. Avoid Channels 1 and 12 on each group for the transmission of data, particularly medium and high speed data. It is precisely these channels that display the poorest attenuation distortion and group delay due to the sharp roll-off of group filters. See Chapter 8.

3.5 PILOT TONES

3.5.1 Introduction

Pilot tones in FDM carrier equipment have essentially two purposes:

• Control of level.
• Frequency synchronization.

Separate tones are used for each application. However, it should be noted that on a number of systems frequency synchronizing pilots are not standard design features, owing to the improved stabilities now available in master oscillators.

Secondarily, pilots are used for alarms.

3.5.2 Level Regulating Pilots

The nature of speech, particularly its varying amplitude, makes it a poor prospect as a reference for level control. Ideally, simple, single sinusoid, constant amplitude signals with 100% duty cycles provide simple control information for level regulating equipment. Multiplex level regulators operate in the same manner as automatic gain control circuits on radio systems, except that their dynamic range is considerably smaller.

Modern carrier systems initiate a level regulating pilot tone on each group at the transmit end. Individual level regulating pilots are also initiated on all supergroups and mastergroups. The intent is to regulate system level within ±0.5 dB.

Pilots are assigned frequencies that are part of the transmitted spectrum yet do not interfere with voice channel operation. They usually are assigned a frequency appearing in the guard band between voice channels or are residual carriers (i.e., partially suppressed carriers). CCITT has assigned the following as group regulation pilots:

84.080 kHz (at a level of −20 dBm0)
84.140 kHz (at a level of −25 dBm0)

Table 3.2   Frequency and Level of CCITT Recommended Pilots

Pilot for	Frequency (kHz)	Absolute Power Level at a Zero Relative Level Point (dB) (Np)
Basic group B	84.080	−20 (−2.3)
	84.140	−25 (−2.9)
	104.080	−20 (−2.3)
Basic supergroup	411.860	−25 (−2.9)
	411.920	−20 (−2.3)
	547.920	−20 (−2.3)
Basic mastergroup	1552	−20 (−2.3)
Basic supermastergroup	11096	−20 (−2.3)
Basic 15-supergroup assembly (No. 1)	1552	−20 (−2.3)

The Defense Communications Agency of the U.S. Department of Defense recommends 104.08 kHz ± 1 Hz for group regulation and alarm.

For CCITT group pilots, the maximum level of interference permissible in the voice channel is −73 dBm0p. CCITT pilot filters have essentially a bandwidth at the 3-dB points of 50 Hz. Refer to CCITT Rec. G.232.

Table 3.2 presents other CCITT pilot tone frequencies as well as those standard for group regulation. Respective levels are also shown. This table was taken from CCITT Rec. G.241. The operating range of level control equipment activated by pilot tones is usually about ±4 or 5 dB. If the incoming level of a pilot tone in the multiplex receive equipment drops outside the level regulating range, then an alarm will be indicated (if such an alarm is included in the system design). CCITT recommends such an alarm when the incoming level varies 4 dB up or down from the nominal (Rec. G.241).

3.5.3   Frequency Synchronization Pilots

End-to-end frequency tolerance on international circuits should be better than 2 Hz. To maintain this accuracy carrier frequencies used in FDM equipment must be very accurate or a frequency synchronizing pilot must be used.

The basis of all carrier frequency generation for modern FDM equipment is a master frequency source. On the transmit side, called the master station, the frequency synchronizing pilot is derived from this source. It is thence transmitted to the receive side, called a slave station. The receive master oscillator is phase locked to the incoming pilot tone. Thus for any variation in the transmit master frequency source, the receive master frequency source at the other end of the link is also varied. The Defense Communication Agency recommends 96 kHz as a frequency synchronizing pilot on Group 5 of Supergroup 1 (*DCA Circ. 330-175-1*); other systems use 60 kHz. Transmit level is at -16 dBm0.

CCITT Rec. G.225 does not recommend a frequency synchronizing pilot. Individual master frequency sources should have sufficient stability and accuracy to meet the following:

Virtual channel carrier frequency, $\pm 10^{-6}$
Group and supergroup carrier frequencies, $\pm 10^{-7}$
Mastergroup and supermastergroup carrier frequencies
    for 12 MHz (line frequency), $\pm 5 \times 10^{-8}$
    for 60 MHz (above 12 MHz), $\pm 10^{-8}$

3.6 FREQUENCY GENERATION

In modern FDM carrier equipment a redundant master frequency generator serves as the prime frequency source from which all carriers are derived or to which they are phase locked. Providing redundant oscillators with fail-safe circuitry gives markedly improved reliability figures.

One equipment on the market has a master frequency generator with three outputs: 4, 12, and 124 kHz. Automatic frequency synchronization is available as an option. This enables the slave terminal to stay in exact frequency synchronization with the master terminal providing drop-to-drop frequency stability (See Section 3.5.3).

The 4-kHz output of the master supply drives a harmonic generator in the channel-group carrier supply. Harmonics of the 4-kHz signal falling between 64 and 108 kHz are selected for use as channel carrier frequencies. The 12-kHz output is used in a similar manner to derive translation frequencies to form the basic CCITT supergroup (420, 468, 516, 564, and 612 kHz). The 124-kHz output drives a similar harmonic generator providing the necessary carriers to translate standard supergroups to the line frequency.

These same carrier frequencies are also used on demultiplex at a slave terminal, or at the demultiplex at a master terminal if that demultiplex is not slaved to a distant terminal.

A simplified block diagram of a typical single sideband suppressed carrier multiplex-demultiplex terminal is shown in Figure 3.12.

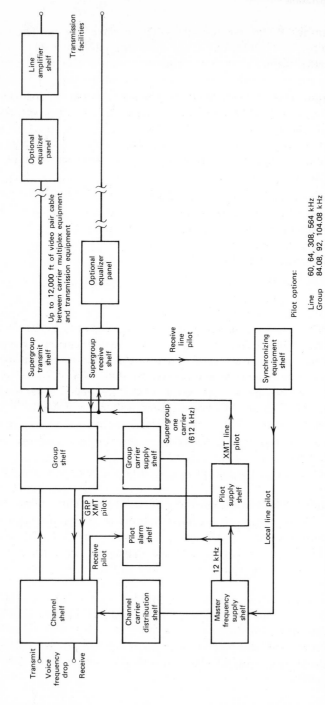

Figure 3.12 Simplified block diagram of typical 120-channel terminal arrangement. Courtesy GTE Lenkurt Incorporated, San Carlos, Calif.

104

3.7 NOISE AND NOISE CALCULATIONS

3.7.1 General

Carrier equipment is the principal contributor of noise on coaxial cable systems and other metallic transmission media. On radiolinks it makes up about one-quarter of the total noise. The traditional approach is to consider noise from the point of view of a hypothetical reference circuit. Two methods are possible, depending on the application. The first is the CCITT method, which is based on a 2500-km hypothetical reference circuit. The second is used by the U.S. Department of Defense in specifying communication systems. Such military systems are based on a 6000-nautical mi reference circuit with 1000-mi links and 333-mi sections.

3.7.2 CCITT Approach

. . . The mean psophometric power, which corresponds to the noise produced by all modulating (multiplex) equipment . . . shall not exceed 2500 pW at a zero relative level point. This value of power refers to the whole of the noise due to various causes (thermal, intermodulation, crosstalk, power supplies, etc.). Its allocation between various equipments can be to a certain extent left to the discretion of design engineers. However, to ensure a measurement agreement in the allocation chosen by different administrations, the following values are given as a guide to the target values:

for 1 pair of channel modulators	200 to 400 pW
for 1 pair of group modulators	60 to 100 pW
for 1 pair of supergroup modulators	60 to 100 pW

The following values are recommended on a provisional basis:

for 1 pair of mastergroup modulators	40 to 60 pW
for 1 pair of supermastergroup modulators	40 to 60 pW
for 1 pair of 15-supergroup assembly modulators	40 to 60 pW

(CCITT Rec. G.222).

Experience has shown that often these target figures can be improved upon considerably. The CCITT notes that they purposely loosened the value for channel modulators. This permits the use of the subgroup modulation scheme shown above as the alternate method for forming the basic 12-channel group.

For instance, one solid-state equipment now on the market, when operated with CCITT loading has the following characteristics:

1 pair of channel modulators	224 pWp
1 pair of group modulators	62 pWp
1 pair of supergroup modulators	25 pWp
1 (single) line amplifier	30 pWp

If out-of-band signaling is used, the noise in a pair of channel modulators reduces to 75 pWp. Out-of-band signaling is discussed in Section 3.13.

Using the same solid-state equipment mentioned above and increasing the loading to 75% voice, 17% telegraph tones, and 8% data, the following noise information is applicable:

1 pair of channel modulators	322 pWp
1 pair of group modulators	100 pWp
1 pair of supergroup modulators	63 pWp
1 (single) line amplifier	51 pWp

If this equipment is used on a real circuit with the heavier loading, the sum for noise for channel modulators, group modulators, and supergroup modulator pairs is 485 pWp. Thus a system would be permitted to demodulate to voice only five times over a 2500-km route (i.e., $5 \times 485 = 2425$ pWp). This leads to the use of through-group and through-supergroup techniques discussed in Section 3.9. Figure 3.13 shows a typical application of this same equipment using CCITT loading.

3.7.3   U.S. Military Approach

As set forth in Military Standard 188B with Notice 1, 5000 pWp0 of the total noise is assigned to multiplex equipment over the 6000 nautical mi reference circuit. The noise in this case covers both thermal and intermodulation. Mean noise allowances may be allocated as follows:

1 pair of channel modulators	345 pWp0
1 pair of group modulators	70 pWp0
1 pair of supergroup modulators	50 pWp0

Loading is not stated.

On the other hand, *DCA Circ. 330-175-1* issued by the Defense Communication Agency offers the following. Idle channel noise for looped-

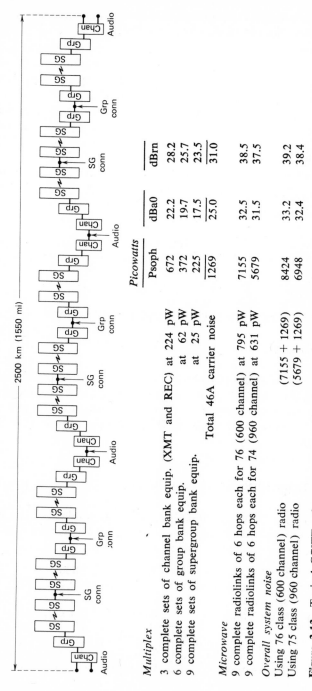

Figure 3.13 Typical CCITT reference system noise calculations. *Note.* Calculations are for a typical equipment with CCITT loading. Courtesy GTE Lenkurt Incorporated, San Carlos, Calif.

The data accompanying the figure:

2500 km (1550 mi)

Multiplex

		Picowatts		
		Psoph	dBa0	dBrn
3 complete sets of channel bank equip. (XMT and REC)	at 224 pW	672	22.2	28.2
6 complete sets of group bank equip.	at 62 pW	372	19.7	25.7
9 complete sets of supergroup bank equip.	at 25 pW	225	17.5	23.5
Total 46A carrier noise		1269	25.0	31.0

Microwave

9 complete radiolinks of 6 hops each for 76 (600 channel) at 795 pW		7155	32.5	38.5
9 complete radiolinks of 6 hops each for 74 (960 channel) at 631 pW		5679	31.5	37.5

Overall system noise

Using 76 class (600 channel) radio	(7155 + 1269)	8424	33.2	39.2
Using 75 class (960 channel) radio	(5679 + 1269)	6948	32.4	38.4

107

back channels (i.e., a channel modulator properly terminated in an equivalent channel demodulator) is 35 pWp0.

Loading all channels but one with white noise and using the same loopback arrangement, the total thermal plus intermodulation noise in the idle channel shall not exceed 200 pWp with input level to each of the loaded channels at an input level equivalent to −5.0 dBm0.

With a similar approach, loading all channels but one with white noise at −5.0 dBm0:

1 pair of group modulators	100 pWp0
1 pair of supergroup modulators	100 pWp0
1 through-supergroup equipment with regulator	50 pWp0

This last example using −5 dBm0 white noise loading gives the reader some idea of what to expect from well-designed carrier equipment that is heavily loaded.

3.8   OTHER CHARACTERISTICS OF CARRIER EQUIPMENT

3.8.1   Attenuation Distortion

Our interest in this discussion is centered on the attenuation distortion of the voice channel (i.e., not the group, supergroup, etc.) (see CCITT Rec. G.232A). Figure 3.14 shows limits of attenuation distortion (amplitude-frequency response) as set forth in this CCITT recommendation as well as those found in *DCA Circ. 330-175-1*.

3.8.2   Envelope Delay Distortion (Group Delay)

The cause and results of envelope and group delay are similar. CCITT Rec. G.232C provides guidance on group delay for a pair of channel modulators (back-to-back) as follows: typical values are

300 Hz, 4.2 ms; 400 Hz, 2.9 ms; 2000 Hz, 1.2 ms; 3000 Hz,
            1.8 ms; and 3400 Hz, 3.4 ms (reference 800 Hz).

On U.S. military circuits under DCA specifications, *DCA Circ. 330-175-1* states:

"The envelope delay distortion of a channel from audio to [the] group distribution frame (GDF) in the transmit direction and from the GDF to audio in

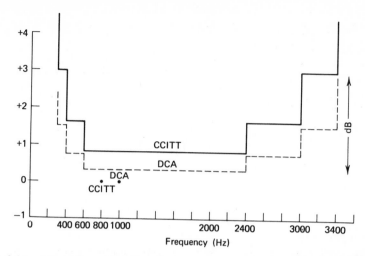

Figure 3.14 Attenuation distortion limits for channel modulators. For CCITT, "Limits for any pair of channel transmitting and receiving equipment." Reference frequency, 800 Hz. For DCA, "The individual transmitting and receiving branch insertion loss-frequency characteristic audio to GDF (group distribution frame) *or* GDF to audio." Reference frequency, 1000 Hz.

the receive direction, taken individually, shall not exceed 90 microseconds over the 600 to 3200 Hz portion of a channel and 55 microseconds over the 1000 to 2500 Hz portion of a channel."

3.9 THROUGH-GROUP, THROUGH-SUPERGROUP TECHNIQUES

Section 3.7 showed where modulation/translation steps in long carrier equipment systems must be limited to avoid excessive noise accumulation. One method widely used is that of employing group connectors and through-supergroup devices. A simple application of supergroup connectors (through-supergroup devices) is shown in Figure 3.15. Here Supergroup 1 is passed directly from point *A* to *B* while Supergroup 2 is dropped at *C*, a new supergroup is inserted for onward transmission to *E*, and so forth. At the same time Supergroups 3–15 are passed directly from *A* to *E* on the same line frequency (baseband).

The expression "drop and insert" is terminology used in carrier systems to indicate that at some way point a number of channels are "dropped" to voice (if you will) and an equal number are "inserted" for transmission back in the opposite direction. If channels are dropped at *B* from *A*, *B* necessarily must insert channels going back to *A* again.

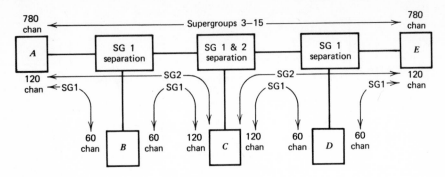

Figure 3.15 Typical drop and insert of supergroups.

Through-group and through-supergroup.techniques are much more used on long trunk routes where excessive noise accumulation can be a problem. Such route plans can be very complex. However, the savings on equipment and reduction of noise accumulation is obvious.

When through-supergroup techniques are used, the supergroup pilot may be picked off and used for level regulation. Nearly all carrier equipment manufacturers include level regulators as an option on through-supergroup equipment, whereas through-group equipment does not usually have the option. Figure 3.16 is a simplified block diagram showing how a supergroup may be dropped (separated and inserted).

CCITT Recs. G.242 and G.243 apply.

3.10 LINE FREQUENCY CONFIGURATIONS

3.10.1 General

When applying carrier techniques to a specific medium such as open-wire, radiolink, etc., consideration must be given to some of the following:

- Type of medium, metallic or radio?
- If metallic, is it pair, quad or coaxial, or open-wire?
- If metallic, what are amplitude distortion limitations, envelope delay, loss?
- If radio, what are the bandwidth limitations?
- What are considerations of international regulations, CCITT recommendations?
- What are we looking into at the other end or ends?

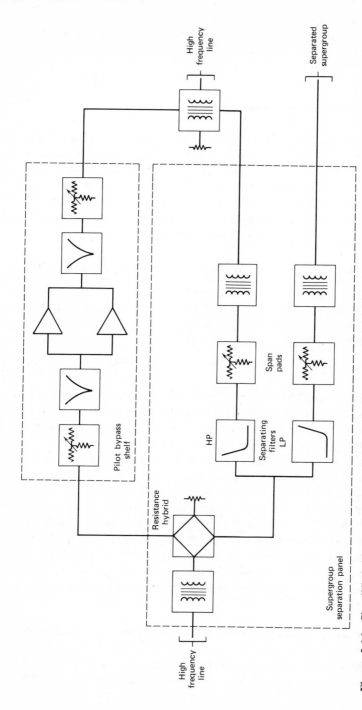

Figure 3.16 Simplified block diagram of typical supergroup drop equipment with pilot tone bypass. Courtesy GTE Lenkurt Incorporated, San Carlos, Calif.

The following paragraphs review some of the more standard line frequency configurations and their applications. The idea is to answer the questions posed above.

3.10.2   Twelve-Channel Open-Wire Carrier

This is a line frequency configuration that permits transmission of 12 full-duplex voice channels on open-wire pole lines using single sideband suppressed carrier multiplex techniques. The industry usually refers to the "go" channels as west-east and the "return" channels as east-west. The standard modulation approach is to develop the standard CCITT 12-channel basic group and through an intermediate modulation process translate the group to one of the following (CCITT terminology):

System Type	West-East (kHz)	East-West (kHz)
SOJ-A-12	26–84	92–140
SOJ-B-12*	36–84	95–143
SOJ-C-12	36–84	93–141
SOJ-D-12	36–84	94–142

* Recommended configuration in *DCA Circ. 330-175-1*.

It should be noted that this type of carrier arrangement replaces a single voice pair on open wire with 12 full-duplex channels. The go and return segments on the same pair are separated by directional filters. Here CCITT Rec. G.311 applies with regard to modulation plans.

Baseband or line frequency repeaters have gains on the order of 43–64 dB. Loss in an open-wire section should not exceed 34 dB under worst conditions at the highest transmitted frequency. This follows the intent of CCITT Rec. G.313. The Bell System uses a similar configuration called J-carrier. Here average repeater spacing is on the order of 50 mi. Two pilot tones are used. One is for flat gain at 80 kHz for W-E and 92 kHz for E-W directions. The other pilot is called a slope pilot. Such slope pilots regulate the lossier part of the transmitted band as weather conditions change (i.e., dry to wet conditions or the reverse) as these losses increase and decrease.

3.10.3   Carrier Transmission over Nonloaded Cable Pairs (K-Carrier)

This technique increases the capacity of nonloaded cable facilities. Typical is the North American cable carrier called K-carrier. Separate cable pairs are used in each direction. Each pair carries 12 voice channels in the band 12–60 kHz. This is the standard CCITT Subgroup A, which is the standard CCITT basic group, 60–108 kHz, translated in frequency. On 19-gauge cable repeaters are required about every 17 mi. Pilots are used on one of the following frequencies: 12, 28, 56, or 60 kHz for automatic regulation.

For U.S. military application *DCA Circ. 330-175-1* specifies one pair for both "go" and "return" channels: 6–54 kHz for the east-west path, 60–108 for the west-east. Intermediate repeaters, when the go and return paths are on the same pair, are of the "frogging" type. For an explanation of the word "frogging," see the following section.

3.10.4 Type N-Carrier for Transmission over Nonloaded Cable Pairs

N-carrier is designed to provide 12 full-duplex voice channels on nonloaded cable pairs over distances of 20–200 mi. The modulation plan is nonstandard. N-carrier uses double sideband emitted carrier with carrier spacings every 8 kHz. Nominal voice channel bandwidth is 250–3100 Hz. The 12 channels for one direction of transmission are contained in a band of frequencies from 44 to 140 kHz called the low band. The other direction of transmission is in a high band, 164–260 kHz. The emitted carriers serve as pilot tones do in other systems, as level references in level regulating equipment.

A technique known as "frogging" or "frequency frogging" is used with N-carrier whereby the frequency groups in each direction of transmission are transposed and reversed at each repeater so that all repeater outputs are always in one frequency band and all repeater inputs are always in the other. This minimizes the possibility of "interaction crosstalk" around the repeaters through paralleling VF (voice frequency) cables. This reversal or transposition of channel groups at each repeater provides automatic self-equalization. A 304-kHz oscillator is basic to every repeater.

3.10.5 Carrier Transmission on Star or Quad Type Cables

Standard carrier systems recommended by the CCITT for transmission over star or quad cables allow transmission of 12, 24, 36, 48, 60, or 120 full-duplex VF channels. The cables in all cases are nonloaded or deloaded. Repeater sections CCITT Rec. G.321 recommends that repeater sections have no more than a 41-dB loss at the highest modulating frequency for systems with one, two, or three groups, and 36 dB for those with four or five groups

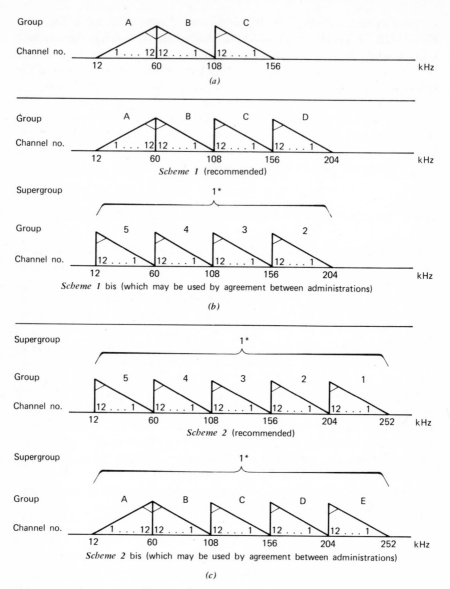

Figure 3.17 Line frequency configurations for star or quad type cables. (*a*) Systems providing one, two, or three groups; (*b*) systems providing four groups; (*c*) systems providing five groups. Courtesy International Telecommunications Union.

and up to two supergroups. We refer here to transistor type repeaters. Figure 3.17 shows CCITT line frequency configurations recommended for star or quad cables.

DCA Circ. 330-175-1 specifies up to a 60-channel configuration on quadded cables from 12 to 252 kHz in 12 channel increments of 48 kHz.

3.10.6 Coaxial Cable Carrier Transmission Systems

Coaxial cable transmission systems using FDM carrier configurations are among the highest density transmission media in common use today. Nominal cable impedance is 75 Ω. Repeater spacing is a function of the highest modulating frequency.

A number of line frequency arrangements are recommended by the CCITT. Only several of these are discussed here. Figures 3.18 and 3.19 show several configurations for 12-MHz. cables Here repeater spacing is on the order of 3 mi (4.5 km). CCITT Rec. G.332 applies.

CCITT recommends 12,435 kHz to be used for the main line pilot on 12-MHz line frequency configurations. This is the main level regulating pilot and should maintain a frequency accuracy of $\pm 1 \times 10^{-5}$. For auxiliary line pilots, CCITT recommends 308 and/or 4287 kHz.

The CCITT 12-MHz hypothetical reference circuit for coaxial cable systems is shown in Figure 3.20. The circuit is 2500 km long (1550 mi) consisting of nine homogeneous sections. Such an imaginary circuit is used as guidance in real circuit design for the allocation of noise. This particular circuit has in each direction of transmission

> 3 pairs of channel modulators
> 3 pairs of group modulators
> 6 pairs of supergroup modulators
> 9 pairs of mastergroup demodulators

Noise allocation for each of these may be found in Section 3.7. When referring to pairs of modulators, the intent is that a "pair" consists of one modulator and a companion demodulator.

Each telephone channel at the end of the 12-MHz (2500-km) coaxial cable carrier system should not exceed 10,000 pWp of noise during any period of 1 hr. 10,000 pWp is the sum of intermodulation and thermal noise. No specific recommendation regarding allocation to either type of noise is made.

Of the 10,000 pWp total noise, 2500 pWp is assigned to terminal equipment and 7500 to line equipment. Refer to CCITT Rec. G.332.

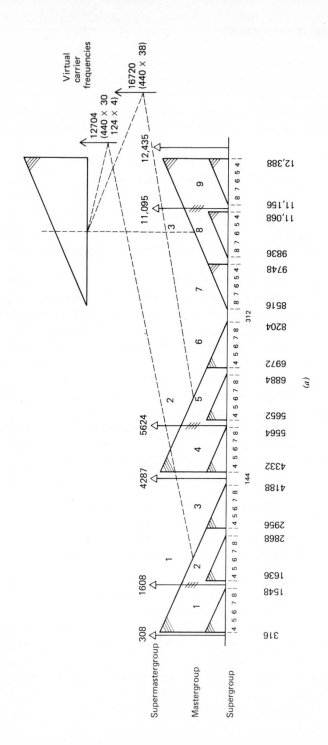

(a)

116

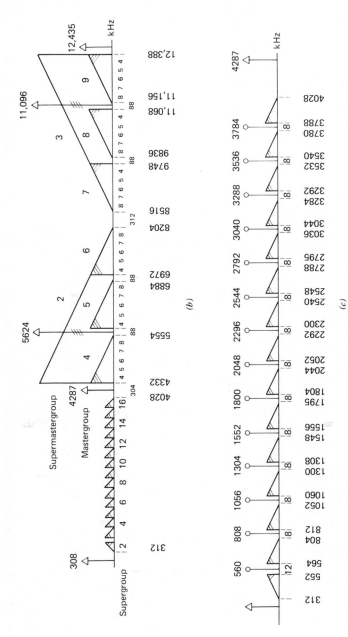

Figure 3.18 (*a*) Plan 1A frequency arrangement for 12-MHz systems; (*b*) Plan 1B frequency arrangement for 12-MHz systems; (*c*) Plan 1B frequency arrangement for 12-MHz system, frequencies below 4287 kHz. Courtesy International Telecommunication Union.

117

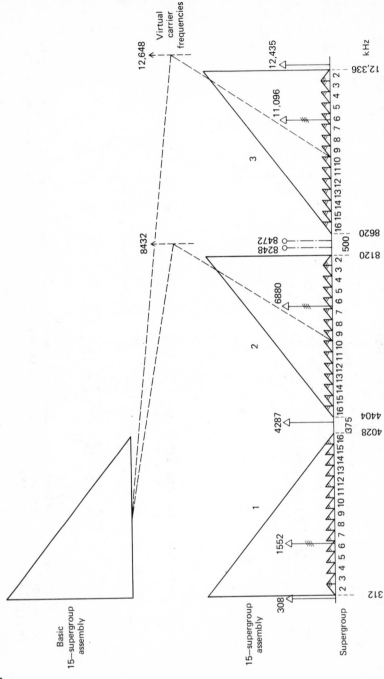

Figure 3.19 Plan 2 frequency arrangement for 12-MHz systems. Courtesy International Telecommunications Union

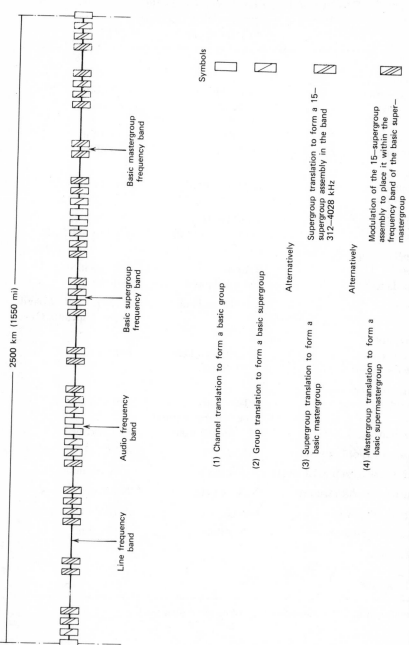

2500 km (1550 mi)

Line frequency band

Audio frequency band

Basic supergroup frequency band

Basic mastergroup frequency band

Symbols

(1) Channel translation to form a basic group

(2) Group translation to form a basic supergroup

Alternatively

(3) Supergroup translation to form a basic mastergroup

Supergroup translation to form a 15—supergroup assembly in the band 312—4028 kHz

Alternatively

(4) Mastergroup translation to form a basic supermastergroup

Modulation of the 15—supergroup assembly to place it within the frequency band of the basic super—mastergroup

Figure 3.20 CCITT 12-MHz hypothetical reference circuit for coaxial cable systems (CCITT Rec. G.332). Courtesy International Telecommunications Union.

Table 3.3 L-Carrier and CCITT Comparison Table

Item	ATT L-Carrier	CCITT
Level		
Group		
Transmit	−42 dBm	−37 dBm
Receive	−5 dBm	−8 dBm
Supergroup		
Transmit	−25 dBm	−35 dBm
Receive	−28 dBm	−30 dBm
Impedance		
Group	130 Ω balanced	75 Ω unbalanced
Supergroup	75 Ω unbalanced	75 Ω unbalanced
VF channel	200–3350 Hz	300–3400 Hz
Response	+1.0 to −1.0 dB	+0.9 to −3.5 dB
Channel carrier		
Levels	0 dBm	Not specified
Impedances	130 Ω balanced	Not specified
Signaling	2600 Hz in band	3825 Hz [a] out of band (see Appendix C)
Group pilot		
Frequencies	92 or 104.08 kHz	84.08 kHz
Relative levels	−20 dBm0	−20 dBm0
Supergroup carrier		
Levels	+19.0 dBm per mod or demod	Not specified
Impedances	75 Ω unbalanced	Not specified
Supergroup pilot frequency	315.92 kHz	411.92 kHz
Relative supergroup pilot levels	−20 dBm0	−20 dBm0
Frequency synchronization	Yes, 64 kHz	Not specified
Line pilot frequency	64 kHz	60/308 kHz
Relative line pilot level	−14 dBm0	−10 dBm0
Regulation		
Group	Yes	Yes
Supergroup	Yes	Yes

[a] Recommended, but depends on system; see Appendix C.

For line frequency allocations other than 12-MHz, consult the following:

2.6 MHz: repeater spacing 6 mi/9 km, line frequency
60–2540 kHz

4 MHz: repeater spacing 6 mi/9 km, 15 supergroups
with the line frequency 60–4028 kHz

DCA Circ. 330-175-1 recommends for 12-MHz systems what is essentially contained in CCITT Plans 1A and B. Refer to Figure 3.18.

3.10.7 The L-Carrier Configuration

L-carrier is the generic name given by the Bell System of North America to their long-haul SSB carrier system. Its development of the basic group and supergroup assemblies is essentially the same as in the CCITT recommended modulation plan (Section 3.3).

The basic mastergroup differs, however. It consists of 600 VF channels (i.e., 10 standard supergroups). The L600 configuration occupies the 60–2788 kHz band. The U600 configuration occupies the 564–2084 kHz band. The mastergroup assemblies are shown in Figure 3.21. Bell identifies specific long-haul line frequency configurations by adding a simple number after the letter L. For example, L3-carrier, which is used on coaxial cable and the TH microwave,* has three mastergroups plus one supergroup comprising 1860 VF channels occupying the band 312–8284 kHz. L4 consists of six U600 mastergroups in a 3600 VF channel configuration. L5 is discussed in Chapter 11.

Table 3.3 compares some basic L-carrier and CCITT system parameters.

3.11 SUBSCRIBER CARRIER–STATION CARRIER

In the subscriber distribution plant, when an additional subscriber line is required, it is often more economical to superimpose a carrier signal above the voice frequency signal on a particular subscriber pair. This is particularly true when a cable has reached or is near exhaustion (i.e., all the subscriber voice pairs are assigned). This form of subscriber carrier is often referred to as a 1 + 1 system and is the most commonly encountered today.

The terms subscriber carrier and station carrier often are used synonymously because both systems perform the same function in much the same

---

* A Bell System type of microwave.

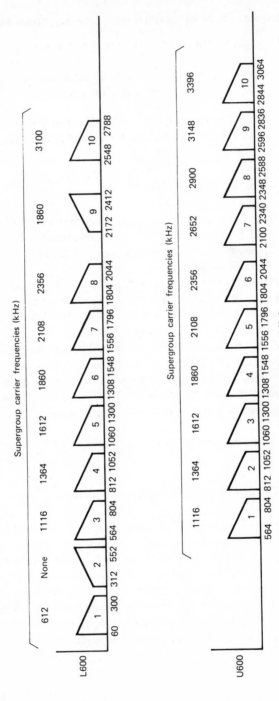

Figure 3.21 L-carrier mastergroup assemblies. All frequencies in kilohertz.

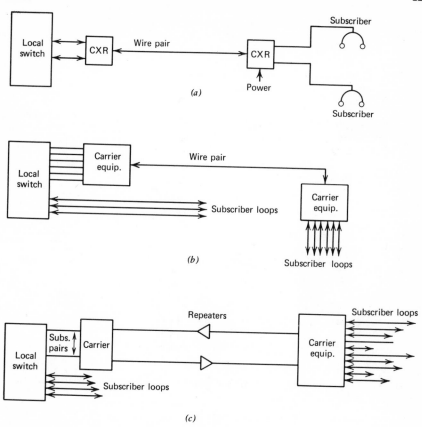

Figure 3.22 (a) A typical 1 + 1 subscriber carrier system. (b) A typical station carrier system serving six subscribers from a single wire pair. (c) A typical station carrier system using a wire pair for each direction of transmission to the remote carrier equipment.

way. There is a difference, however, By convention we say that subscriber carrier systems are powered locally at remote distribution points. Station carrier systems are powered at the serving local switch (central office). Figures 3.22 and 3.23 outline general applications of these types of systems.

Figure 3-22c shows one type of station carrier system. Several are available on the market today. Some allow a local switch to serve subscribers more than 30 mi (50 km) away. This particular system carries 20 voice channels on two pairs of cable to a distant distribution point. From the distribution point standard voice pair subscriber loops can be installed with loop resistance up to 1000 Ω (see Section 2.8.2). Another system pro-

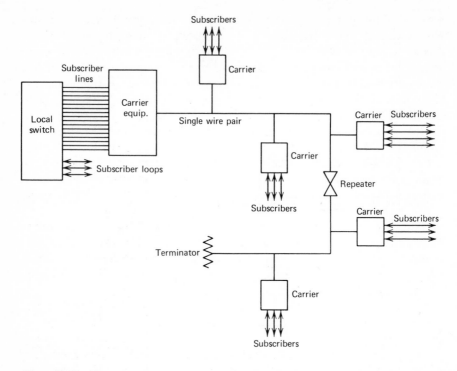

Figure 3.23   An arrangement of station carrier equipment with distributed subscribers.

vides for the addition of six subscribers to a single wire pair (Figure 3.22*b*). In this case transmission from the serving switch to the subscriber is in the band 72–140 kHz. From the subscribers to the switch, frequency assignment is in the band 8–56 kHz.

An important point when applying subscriber carrier to an existing subscriber loop is that load coils and bridged taps must be removed. The taps may act as tuned stubs and the load coils produce a sharp cutoff, usually right in the band of interest. The load points are candidate locations for line repeaters if required.

3.12   ECONOMICS OF CARRIER TRANSMISSION

We often hear that a particular brand of FDM carrier equipment costs so much per channel end. A channel end is taken to be a full-duplex voice channel including the cost of the channel modulator-demodulator pair, its portion of the group modulation equipment, supergroup, and so forth. However, such figures are illusory. It is illusory in that much meaning is lost or hidden in what is called common equipment expense. Common equipment is that equipment that serves more than one module. For instance, a 108-kHz carrier frequency source will be used for both modulation and demodulation of voice channel number one in each and every channel bank. A 24-channel system must amortize this frequency source for four modules. A 600-channel system amortizes the same source across 240 modules.

For channel modulation and demodulation equipment (channel banks) when using the standard CCITT modulation plan, 12 frequency sources are required. These are carriers starting at 64 kHz and spaced at 4-kHz intervals extending to 108 kHz. Let us say, for example, that the cost of these frequency sources totaled $1200 including the master frequency generator. Thus the cost of frequency sources for a 24-channel system would total $50/channel added to another channel bank costs. The cost of a 600-channel system would be only $2/channel. Thus per-channel costs tend to drop as we increase the number of channels in a system.

A telecommunication planner must decide: is it more economical to use wire pairs for a particular number of voice channels to be transmitted or carrier on one wire pair? It may cost $10,000 to lay a cable for 60 voice pairs for a distance of 10 mi, $20,000 for 20 mi, etc. At some point a decision is made that to extend the system further, we will abandon wire pairs and use carrier systems because of the economy involved.

There are many inputs to this decision-making process. Today more than ever, we usually deal with service expansion. Here wire pair systems are already in existence, such as interexchange trunks, toll connecting trunks, and tandem office trunks. One case that arises is that the exhaustion date is approaching (the date when nearly all the cable pairs have been assigned). Here a telecommunication planner must decide: is it more economical to rip up the streets and add pair cable, or to add carrier systems to the existing plant?

One term we use in this regard is "prove-in distance." For one type of carrier system we say that on junctions (trunks) the prove-in distance is 8–12 km (5-7.5 mi). Here we mean that if a trunk route is less than 8 km in length and we wish to expand it, we add pairs. If it is more than 12 km long, our decision is to use a specific carrier system. For intermediate ranges a careful economic study is necessary to decide whether pair cable or carrier should be used.

Prove-in distances can be reduced by reducing the costs of carrier systems. It is here that double sideband emitted carrier systems show feasibility from an economic standpoint. One such double sideband emitted carrier

(DSBEC) system has been discussed in Section 3.10.4 (N-carrier). Here, by reducing the cost of equipment, prove-in distance is reduced as well, usually at the expense of deteriorating some of the desirable features. For instance, DSBEC systems have increased intermodulation noise over their SSBSC counterparts. One major source of the noise is the fact that carriers are always present. More power is consumed because of the presence of the carriers and loading problems exist.

3.13 COMPANDORS

The word compandor is derived from two words which describe its functions: Compressor-expandor, to compress and to expand. A compandor does just that. It compresses a signal on one end of a circuit and expands it on the other. See the simplified functional diagram and its analogy, Figure 3.24.

The compressor compresses the intensity range of speech signals at the input circuit of a communication channel by imparting more gain to weak signals than to strong signals. At the far-end output of the communication circuit the expandor performs the reverse function. It restores the intensity of the signal to its original dynamic range. We cover only syllabic compandors in this discussion.

The advantages of compandors are three:

1. Tend to improve the signal-to-noise ratio on noisy speech circuits.
2. Limit the dynamic power range of voice signals reducing the chances of overload of carrier systems.
3. Reduce the possibility of crosstalk.

A basic problem in telephony stems from the dynamic range of talker levels. This intensity range can vary 70 dB for the weakest syllables of the weakest

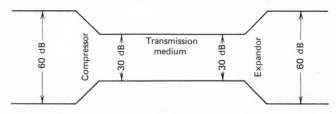

Figure 3.24 Functional analogy of a compandor.

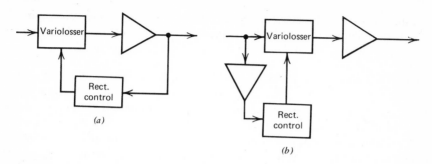

Figure 3.25 Simplified functional block diagrams of (a) compressor and (b) expandor.

talker to the loudest syllables of the loudest talker. The compandor brings this range down to more manageable proportions.

An important parameter of a compandor is the compression-expansion ratio. It is the degree to which speech energy is compressed and expanded. It is expressed by the ratio of the *input* to the *output* power (dB) in the compressor and expandor, respectively. Compression ratios are always greater than 1 and expansion ratios are less than 1. The most common compression ratio is 2 (2:1). The corresponding expansion ratio is thus $\frac{1}{2}$. The meaning of a compression ratio of 2 is that the dynamic range of the speech volume has been cut in half from the input of the compressor to its output. Figure 3.25 is a simplified functional block diagram of a compressor and expandor.

Another important criterion for a compandor is its companding range. This is the range of intensity levels a compressor can handle at its input. Usually 50–60 dB is sufficient to provide the expected signal-to-noise ratio and reduce the possibility of distortion. High level signals appearing outside this range are limited without markedly affecting intelligibility.

But just what are the high and the low levels? Such a high or low is referred to as an "unaffected level" or focal point. CCITT Rec. G.162 defines the unaffected level as follows:

> ... the absolute level, at a point of zero relative level on the line between the compressor and expandor of a signal at 800 Hz, which remains unchanged whether the circuit is operated with the compressor or not.

CCITT goes on to comment:

> The unaffected level should be, in principle, 0 dBm0. Nevertheless, to make allowances for the increase in mean power introduced by the compressor, and to avoid the risk of increasing the intermodulation noise and overload which might result, the unaffected level may, in some cases, be reduced as much as

5 dB. However, this reduction of unaffected level entails a diminution of improvement in signal-to-noise ratio provided by the compandor . . . . No reduction is necessary, in general, for systems with less than 60 channels.

CCITT recommends a range of level from $+5$ to -45 dBm0 at the compressor input and $+5$ to -50 dBm0 at the nominal output of the expandor. Figure 3.26 shows diagrammatically a typical compandor range of $+10$ to -50 dBm.

A syllabic compandor operates much in the fashion of any level control device where output level acts as a source for controlling input to the device (see Figure 3.25). AGC (automatic gain control) on radio receivers operates in the same manner. This brings up the third important design parameter of syllabic compandors: "attack" and "recovery" times. These are the response to suddenly applied signals such as a loud speech syllable or burst of syllables. Attack and recovery times are a function of design time constants and are adjusted by the designer to operate as a function of the speech envelope (syllabic variations) and *not* with instantaneous amplitude changes (such as used in PCM). If the operation time is too fast, wide bandwidths would be required for faithful transmission. When attack times are too slow, the system may be prone to overload.

CCITT Rec. G.162 specifies an attack time equal to or less than 5 ms and a recovery time equal to or less than 22.5 ms.

The signal-to-noise ratio advantage of a compandor varies with the multichannel loading factor of FDM equipment and thus depends on the

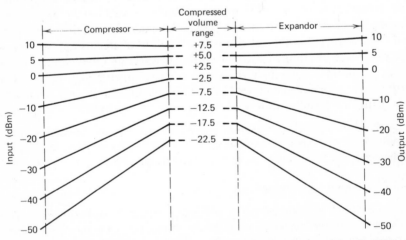

Figure 3.26 Input-output characteristics of a compandor. Copyright © 1970, by Bell Telephone Laboratories.

voice level into the FDM channel modulation equipment. At best an advantage of 20 dB may be attained on low level signals.

3.14 SIGNALING ON CARRIER SYSTEMS

Appendix C provides an overview of telephone signaling. Nevertheless, in any discussion of carrier equipment, some space must be dedicated to signaling over carrier channels.

For this discussion signaling may be broken down into two categories, supervisory and address. Supervisory signaling conveys information regarding on-hook and off-hook conditions. Address signaling routes the calls through the switching equipment. It is that signaling containing the dialing information. For this discussion we are concerned with only the former, "supervisory."

The problem stems from the fact that it would be desirable to have continuous supervisory information being exchanged during an entire telephone conversation. This may be done by one of two methods. These are in-band signaling and out-of-band signaling.

In-band signaling accomplishes both supervisory and address signaling inside the operative voice band spectrum (i.e., in the band 300–3400 Hz). The supervisory function is carried out when a call is set up and when it is terminated. Thus it is not continuous. The most common type in use today is called SF signaling. SF means single frequency and is a tone, usually at 2600 Hz. Other frequencies also may be used, and most often are selected purposely in the higher end of the voice band.

A major problem with in-band signaling is the possibility of "talk-down." Talk-down refers to the activation or deactivation of supervisory equipment by an inadvertent sequence of voice tones through normal speech usage of the channel. One approach is to use slot filters to bypass the tones as well as a time-delay protection circuit to avoid the possibility of talk-down.

With out-of-band signaling supervisory information is transmitted out-of-band. Here we mean above 3400 Hz (i.e., outside the speech channel).

Supervisory information is binary, either "on-hook" or "off-hook." Some systems use "tone-on" to indicate "on-hook" and others use tone-off. One expression used in the industry is "tone-on when idle." When a circuit is idle, it is "on-hook." The advantage of "out-of-band" signaling is that either system may be used, tone-on or tone-off when idle. There is no possibility of talk-down occurring because all supervisory information is passed out-of-band, away from the voice.

The most common out-of-band signaling frequencies are 3700 and 3825 Hz. 3700 Hz finds more application in North America; 3825 Hz is that recommended by CCITT (Rec. Q.21).

In the short run, out-of-band signaling is attractive from an economic and design standpoint. One drawback is that when patching is required, signaling leads have to be patched as well. In the long run, the signaling equipment required may indeed make out-of-band signaling even more costly owing to the extra supervisory signaling equipment and signaling lead extensions required at each end and at each time the FDM equipment demodulates to voice. The advantage is that continuous supervision is provided, whether "tone-on" or "tone-off," during the entire telephone conversation.

REFERENCES AND BIBLIOGRAPHY

1. *Reference Data for Radio Engineers*, 5th ed., Howard W. Sams & Co., Indianapolis.
2. *Lenkurt Demodulator*, Lenkurt Electric Corp., San Carlos, Calif.: Dec. 1965, May 1966, March 1965, Oct. 1964, Oct. 1972, Oct. 1964, March 1970, July 1968, Sept. 1967, Sept. 1973.
3. CCITT White Books, Mar del Plata, Vol. III, G. Recommendations.
4. *Transmission Systems for Communications*, 4th ed., Bell Telephone Laboratories.
5. Military Standard 188B with Notice 1.
6. *Principles of Electricity Applied to Telephone and Telegraph Work*, American Telephone and Telegraph Co., 1961.
7. D. Hamsher, *Communication System Engineering Handbook*, McGraw-Hill, 1967.
8. *DCA Circ. 330-175-1*, through Change 8, Defense Communication Agency, U.S. Department of Defense, Washington, D.C.
9. R. L. Marks et al., *Some Aspects of Design for FM Line-of-Sight Microwave and Troposcatter Systems*, Rome Air Development Center NY 1965, U.S. Technical Information Service AD 617-686, Springfield, Va.
10. B. D. Holbrook and J. T. Dixon, *Load Rating Theory for Multichannel Amplifiers*, Bell Syst. Tech. J., Monograph B-1183, Feb. 1963.
11. David Talley, *Basic Carrier Telephony*, John R. Rider, New York, 1960 (No. 268).
12. Philip F. Panter, *Communication System Design—Line-of-Sight and Troposcatter Systems*, McGraw-Hill, 1972.
13. *Engineering and Equipment Considerations—46A*, Issue 1, Lenkurt Electric Corp., San Carlos, Calif.

4 | HIGH FREQUENCY RADIO

4.1 GENERAL

Radio frequency transmission between 3 and 30 MHz by convention is called high frequency radio, or simply HF. HF radio is in a class of its own because of certain characteristics of propagation; the phenomenon is such that many radio amateurs at certain times carry out satisfactory communication halfway around the world with 1-2 W of radiated power.

4.2 BASIC HF PROPAGATION

HF propagation is characterized by a groundwave component and a skywave component. The groundwave follows the surface of the earth and can provide useful communications up to about 400 mi (640 km) from the transmitter location, particularly over water, in the lower part of the band. It is the skywave, however, that gives HF an advantage. Transmission engineers design systems to take advantage of skywave propagation which permits reliable communication (90% path reliability) for distances up to 4000 mi (6400 km). On some links somewhat greater than 90% reliability has been reported. The same reliability has been obtained on even longer paths, except in years with low sunspot number, using oblique ionospheric sounding in parallel with the link to accurately determine optimum transmitting frequency. Ionospheric sounding is discussed in Section 4.14.

4.2.1 Skywave Transmission

The skywave transmission phenomenon of HF depends on ionospheric refraction. Transmitted radio waves hitting the ionosphere are bent or refracted. When they are bent sufficiently, the waves are returned to earth at a distant location. Often at the distant location they are reflected back to the sky again, only to be returned to earth still again, even further from the transmitter.

The ionosphere is the key to HF skywave communication. Look at the ionosphere as a layered region of ionized gas above the earth. The amount of refraction varies with the degree of ionization. The degree of ionization is primarily a function of the sun's ultraviolet radiation. Depending on the intensity of the ultraviolet radiation, more than one ionized layer may form (see Figure 4.1). The existence of more than one ionized layer in the atmosphere is explained by the existence of different ultraviolet frequencies in the sun's radiation. The lower frequencies produce the upper ionospheric layers, expending all their energy at high altitude. The higher frequency ultraviolet waves penetrate the atmosphere more deeply before producing appreciable ionization. Inonization of the atmosphere may also be caused by particle radiation from sunspots, cosmic rays, and meteor activity.

For all practical purposes four layers of the ionosphere have been identified and labeled as follows:

D region. Not always present, but when it does exist, it is a daytime phenomenon. It is the lowest of the four layers. When it exists, it occupies an area between 50 and 90 km above the earth. The D region is usually highly absorptive.

E layer. A daylight phenomenon, existing between 90 and 140 km above the earth. It depends directly on the sun's ultraviolet radiation and thus it is most dense directly under the sun. The layer disappears shortly after sunset. Layer density varies with seasons owing to variation in sun's zenith angle with seasons.

F_1 layer. A daylight phenomenon existing between 140 and 250 km above the earth. Its behavior is similar to that of the E layer in that it tends to follow the sun (i.e., most dense under the sun). At sunset the F_1 layer rises, merging with the next higher layer, the F_2 layer.

F_2 layer. This layer exists day and night between 150–250 km (night) and 250–300 km above the earth (day). During winter in the daytime, it extends from 150 to 300 km above the earth. Variations in height are due to solar heat. It is believed that the F_2 layer is also strongly influenced by the earth's mag-

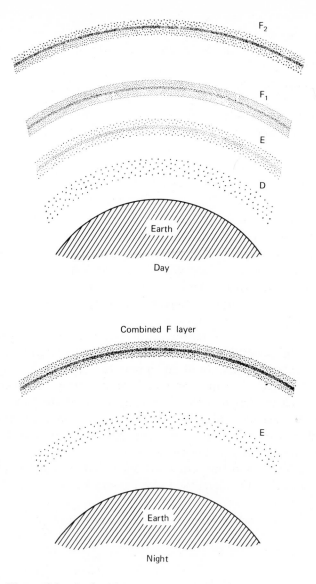

Figure 4.1 Ionized layers of the atmosphere.

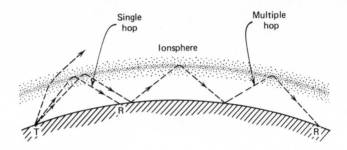

Figure 4.2   Single and multihop HF skywave transmission. T, transmitter; R, receiver.

netic field. The earth is divided into three magnetic zones representing different degrees of magnetic intensity called east, west, and intermediate. Monthly F_2 propagation predictions are made for each zone.* The north and south auroral zones are also important for F_2 propagation, particularly during high sunspot activity.

Consider these layers as mirrors or partial mirrors depending on the amount of ionization present. Thus transmitted waves striking an ionospheric layer, particularly the F layer, may be refracted directly back to earth and received after their first hop, or they may be reflected from the earth back to the ionosphere again and repeat the process several times before reaching the distant receiver. The latter phenomenon is called multihop transmission. Single and multihop transmission are illustrated diagrammatically in Figure 4.2.

To give some idea of the estimated least possible number of F layer hops as related to path length, the following may be used as a guide:

Number of Hops	Path Length (km)
1	<4,000
2	4,000– 8,000
3	8,000–12,000

An important concept at this point in the discussion is that the higher and more strongly ionized layers refract progressively higher frequencies. At some point as the frequency is increased, the wave will not be refracted

* Monthly propagation forecasts are made by the CRPL, Central Radio Propagation Laboratory, U.S. National Bureau of Standards.

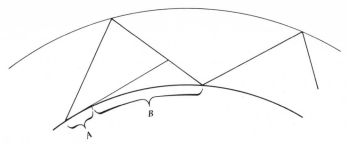

Figure 4.3 The skip zone. *A*, limit of groundwave communications; *B*, skip zone.

but will pierce the ionosphere and continue out into space. This point is called the "critical frequency" and varies with time of day, geographical position, season of the year, and the time position in the sunspot cycle.

HF propagation above about 8 MHz encounters what is called a skip zone. This is an "area of silence" or a zone of no reception extending from the outer limit of groundwave communication to the inner limit of skywave communication (first hop). The skip zone is shown graphically in Figure 4.3.

4.3 OPERATING FREQUENCY

One of the elements most important to the successful operation of an HF system using the skywave phenomenon is an operating frequency that will assure 100% path reliability day and night, year-round. Seldom, if ever, can this be achieved. To arrive as close as possible to this ideal, we must deal with two factors: (*a*) the design of an efficient HF installation,* and (*b*) the determination of an assigned frequency that will provide optimum communication.

The latter portion of this section deals with item *a*. Our concern here is the proper determination of operating frequency. From the point of view of an HF installation operator on long-haul, point-to-point service, HF propagation varies with time of day, from day to day, with season of the year, and on a cyclic yearly basis. He will tell you that propagation variations are most notable around local sunrise and sunset. As midday approaches, conditions seem to become more stable and higher frequencies prove more attractive. After local sunset, depending on the location of the distant end (i.e., whether it is in darkness, too), lower

---

* The reader should not lose sight of the interaction of item *b* on item *a*.

frequencies become more desirable. These are only generalities, which might be set forth from the point of view of a serious shortwave listener (SWL). To the communicator, the propagation problem of HF is complex. The vagaries (as it seems) of the ionosphere are beyond comprehension. Long circuits, particularly those crossing high latitude regions, are particularly troublesome.

The HF operator, to maintain good communication, is required to make frequency changes during any 24-h period on long-haul circuits. At this juncture we can make a generality. Well designed HF point-to-point installations not operating in or crossing the auroral zones, during the greater portion of the sunspot cycle will require operating frequency changes (QSY's) no more than two to four times a day.

There are three methods to determine operating frequency:

1. By experience.
2. By use of CRPL predictions.
3. By oblique ionospheric sounder information. Sounders are covered in Section 4.14.

Many old-time operators rely on experience. First they listen to their receivers, then judge if an operating frequency change is required. An operator may well feel that "yesterday I had to QSY at 7 PM to 13.7 MHz, so today I must do the same." Listening to his receiver will confirm or deny this belief. He hears his present operating frequency (from the distant end) start to take deep fades. He checks a new frequency on a spare receiver, listening for other identifiable signals nearby to determine conditions on the new frequency. If he finds conditions to his liking, he orders the transmitter operator to change frequency (QSY at the distant end). If not, he may check other assigned frequencies and then order a move to one of these. This is the "experience method."

The next approach is to carefully study the CRPL predictions issued by the U.S. Bureau of Standards. These are published monthly, 3 months in advance of their effective date.

We choose an optimum operating frequency to minimize communication outrages—to keep the circuit operating at optimum. On skywave HF communication circuits fading is endemic. Fading must be reduced to some acceptable minimum for good communications. Fading on HF is due to the variations in the ionosphere, with skywave received signal intensities varying from minute to minute, day to day, month to month, and year to year.

Consider an HF system designed for a 95% path (propagation) reliability, a high design goal on long circuits. The median received signal level intensity must be increased on the order of 8 dB to overcome slow variations of skywave field intensity, 11 dB to overcome rapid variations of skywave field intensity, and 13 dB to overcome variations in atmospheric noise. Therefore, a good operational figure is that we need at least 32 dB above median signal level—that level which will provide good communication 50% of the time—to give us margin over fades for good communication 95% of the time. We shall discuss other approaches to meet this figure further on in our discussion. Our concern now is the primary approach—choice of the right frequency.

When dealing with HF propagation and the choice of operating frequency, the terms MUF, FOT, and LUF come into play. It is these terms that are used on the CRPL predictions.

MUF and LUF are the upper and lower limiting frequencies for skywave communications between points X and Y. MUF is the median maximum usable frequency and LUF the lowest usable frequency. Suppose we chose the MUF or LUF as an operating frequency. We would be operating near boundary limits, regions of heavy fade. Depending on conditions, one or the other or both may drop out (i.e., communications would be lost entirely). Therefore, there is some other frequency above the LUF, usually just somewhat below the MUF, that is an optimum frequency. Use of this frequency, the FOT (fréquence optimum de travail), OWF in English (optimum working frequency), will provide a received signal with less fading and minimize frequency changes. CRPL predictions give MUF, LUF, and FOT. The OWF is equal to 85% of the maximum usable frequency for the F_2 layer.

For the second approach in the determination of operating frequency, the operator consults the CRPL prediction and chooses an assigned frequency as close as is available to the OWF.

Another concept that comes into play at this point will be used later in system design. This is the angle of incidence, or the angle at which the transmitted ray or beam strikes the ionosphere. Here we want to determine the maximum usable frequency and from it we can determine the incidence angle. This can be derived from the following formula:

$$f_0 = f_n \sec i \qquad (4.1)$$

where $f_0 =$ maximum usable frequency at oblique angle i

$f_n =$ maximum frequency that will be reflected back at vertical incidence

$i \ =$ angle of incidence (i.e., that angle between the direction of propagation and a line perpendicular to the earth)

As sec i increases, the angle of incidence, i, is increased. For most effective HF transmission, it is desirable to strive for the maximum concentration of energy (i.e., the transmitted ray) at low angles with respect to the horizon because higher frequencies can be used which are less affected by absorption than lower frequencies. Thus we design for low takeoff angles (i.e., high angles of incidence), particularly on longer paths.

Important natural phenomena that affect the ionosphere and, likewise, skywave propagation are as follows:

> Sunspots
> Magnetic storms
> Sudden ionospheric disturbances (SID)
> Sporadic E (layer)

Sunspot activity greatly influences skywave HF propagation. It can cause the loss of long-haul HF communication for extended periods or permit phenomenal worldwide communication with relatively low power and homemade installations for weeks at a time. Sunspot activity vastly increases ultraviolet radiation. The number and intensity of sunspots are indicative of ion density in the ionosphere. Thus, in turn, sunspot activity is a measure of the probability of skywave communication. Sunspot activity is cyclic, with a full cycle being on the order of 11.1 years. The activity is measured by the Wolf *sunspot number*. High numbers indicate high activity, and low numbers, low activity.

Magnetic storms are associated with solar activity and are most likely to occur during periods of high sunspot numbers and reoccur in 27-day cycles, the rotation period of the sun. Magnetic storms are so named because the solar radiation seems to be deflected by the earth's magnetic field. Therefore, their effects are experienced most severely in the regions of the magnetic poles or in the northern and southern auroral zones. Magnetic storms tend to come on suddenly, last for several days, and slowly subside. During this period skywave communication may be spotty or cut entirely. Magnetic storms often occur from 18 to 36 h after a sudden ionospheric disturbance.

A sudden ionospheric disturbance (SID) is an occasional daytime phenomena rendering HF skywave communication impossible or nearly so. It is believed that a SID is caused by a chromospheric eruption on the sun resulting in marked increase in ion density in the D region with a moderate increase in the E region, increasing their absorptive properties.

Sporadic E propagation occurs in the E region but cannot be accounted for by normal E propagation theory. One theory is that it is caused by particle radiation caused by meteor bursts (i.e., meteors entering the atmosphere). Sporadic E has been found to permit skywave propagation from 25 to 50% of the time on frequencies up to 15 MHz.

4.4 HF RADIO SYSTEMS

The most elementary HF installation operating on a point-to-point basis requires as a minimum a transmitter at one end and a receiver at the other. Each must have an antenna and each must employ some sort of signal transducer, say, a microphone and headsets.

This configuration is valid for one-way transmission such as that of broadcasting.

The radio amateur requires a half-duplex arrangement where he can talk to a distant acquaintance and listen as well. His equipment arrangement will appear as follows:

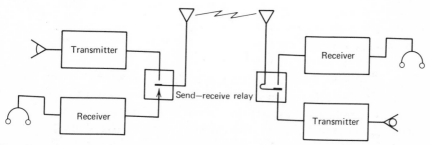

In this case, while one transmits the other receives. When invited to do so, the transmitting station ceases and switches to receive (i.e., he switches his antenna to receive and stops transmitting), and the other station goes into a transmit condition.

For commercial or military point-to-point service a full-duplex operation is necessary. This means that both ends transmit and receive simultaneously. With sufficient frequency separation between transmit and receive frequencies and some good filters, this can be done with only one antenna on each end. However, most land installations use at least two antennas.

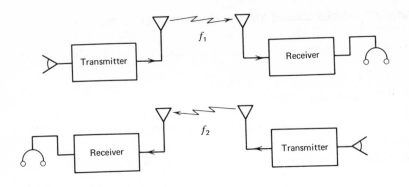

4.5 PRACTICAL HF POINT-TO-POINT COMMUNICATION

4.5.1 General

Reliable HF point-to-point facilities are complex, often occupying more than one site. In such cases the transmitters are separated physically from their receiving counterparts by 3–45 km so that noise from high power transmitters will not affect reception.

4.5.2 Emission Types

Table 4.1 provides a list of emission types used on HF. For many years the backbone of HF transmission was what we familiarly call CW; CW stands for continuous wave. Information is transmitted by turning on and off (keying) the carrier. The carrier is keyed using short and long pulses called dots and dashes. An alphanumeric code is used, the International Morse Code (see Table 4.2). CW transmission is listed as A1 in Table 4.1.

Standard teletypewriter service is also used widely. For the most part for single channel service wide shift FSK is employed. FSK stands for frequency shift keying and is discussed in Chapter 8. The standard frequency shift is ±425 Hz. This is listed in Table 4.1 as F1 emission. Table 4.3 shows comparative emissions and their assigned bandwidths.

HF is a highly congested band. Conservation of bandwidth is extremely important. The goal of HF transmission is to consume minimum bandwidth, and achieve maximum energy transfer in a heavily fading environment.

Table 4.1 Designation of Specific Emission Types

Desig- nation	Modulation Type	Type of Transmission	Supplementary Characteristics
A0	Amplitude modulation	With no modulation	None
A1	"	Telegraphy without the use of a modu- lating audio frequency (by on-off keying)	None
A2	"	Telegraphy by the on-off keying of an amplitude-modulating audio fre- quency or audio frequencies, or by the on-off keying of the modulated emission (special case: an unkeyed emission amplitude modulated)	None
A3	"	Telephony	Double sideband
A3A	"	Telephony	Single sideband, reduced carrier
A3J	"	Telephony	Single sideband suppressed carrier
A3B	"	Telephony	Two independent sidebands
A4	"	Facsimile (with modulation of main carrier either directly or by a fre- quency modulated subcarrier)	None
A4A	"	Facsimile (with modulation of main carrier either directly or by a fre- quency modulated subcarrier)	Single sideband, reduced carrier
A5C	"	Television	Vestigial sideband
A7A	"	Multichannel voice frequency teleg- raphy	Single sideband, reduced carrier
A9B	"	Cases not covered by the above, e.g., a combination of telephony and telegraphy	Two independent sidebands
F1	Frequency (or phase) modulation	Telegraphy by frequency shift keying without the use of a modulating audio frequency: one of two fre- quencies being emitted at any in- stant	None
F2	"	Telegraphy by the on-off keying of a frequency modulating audio fre- quency or by the on-off keying of a frequency modulated emission (spe- cial case: an unkeyed emission, fre- quency modulated)	None

(Continued)

Table 4.1 (*Continued*)

Desig-nation	Modulation Type	Type of Transmission	Supplementary Characteristics
F3	"	Telephony	None
F4	"	Facsimile by direct frequency modulation of the carrier	None
F5	"	Television	None
F6	"	Four-frequency duplex telegraphy	None
F9	Frequency (or phase) modulation	Cases not covered by the above, in which the main carrier is frequency modulated	None
P0	Pulse modulation	A pulsed carrier without any modulation intended to carry information (e.g., radar)	None
P1D	"	Telegraphy by the on-off keying of a pulsed carrier without the use of a modulating audio frequency	None
P2D	"	Telegraphy by the on-off keying of a modulating audio frequency or audio frequencies, or by the on-off keying of a modulated pulsed carrier (special case: an unkeyed modulated pulsed carrier)	Audio frequency or audio frequencies modulating the amplitude of the pulses
P2E	"	"	Audio frequency or audio frequencies modulating the width (or duration) of the pulses
P2F	"	"	Audio frequency or audio frequencies modulating the phase (or position) of the pulses
P3D	"	Telephony	Amplitude modulated pulses
P3E	"	Telephony	Width (or duration) modulated pulses
P3F	"	Telephony	Phase (or position) modulated pulses
P3G	"	Telephony	Code modulated pulses (after sampling and quantization)

Table 4.1   (*Continued*)

Desig- nation	Modulation Type	Type of Transmission	Supplementary Characteristics
P9	"	Cases not covered by the above in which the main carrier is pulse modulated	None
		Bandwidths Whenever the full designation of an emission is necessary, the symbol for that emission, as given above, shall be preceded by a number indicating in kilohertz the necessary bandwidth of the emission. Bandwidths shall generally be expressed to a maximum of three significant figures, the third figure being almost always a zero or a five.	

Source: *Radio Regulations*, Geneva, 1968, Article 2 (RR2-1).

4.5.3   Recommended Modulation (Emission Type)

Single sideband suppressed carrier (SSBSC) systems [including independent sideband (ISB) systems] for multichannel application are recommended. This recommendation is made according to the following criteria when considering the transmission of voice or other complex analog signals.

- Spectrum conservative.
- More efficient from the point of view of power.
- Superior to other emission types during periods of poor propagation conditions.

After a short description of single sideband modulation, the traditional comparison with double sideband (DSB) emitted carrier will be made.

4.5.4   Single Sideband Suppressed Carrier

A single sideband (SSB) signal is simply an audio frequency signal translated to the radio frequency (RF) spectrum. If we could mix

Table 4.2 The International Morse Code

Character	International Morse	Character	International Morse
A	.—	.	.—.—.—
B	—...	;	—.—.—.
C	—.—.	,	.—.——
D	—..	:	———...
E		?	..——..
F	..—.	!	——.——
G	——.	'	.————.
H	....	-	—....—
I	..	/	—..—.
J	.———	Ā	.—.—
K	—.—	A or Å	.——.—
L	.—..	E	..—..
M	——	CH	————
N	—.	Ñ	——.——
O	———	Ö	———.
P	.——.	Ü	..——
Q	——.—	(OR)	—.——.—
R	.—.	”	.—..—.
S	...	—	..——.—
T	—	SOS	...———...
U	..—	Attention	—.—.—
V	...—	CQ	—.—.——.—
W	.——	DE	—..
X	—..—	Go ahead	—.—
Y	—.——	Wait	.—...
Z	——..	Break	—...—.——
1	.————	Understand	...—.
2	..———	Error	....
3	...——	OK	.—.
4	....—	Separator, heading—	
5	.....	text and end of	
6	—....	message	—...—
7	——...	End of work	...—.—
8	———..		
9	————.		
0	—————		

Table 4.3 Emission Bandwidths (kHz)

Type of Service	Remarks	Bandwidth (kHz)
Double sideband radio-telephony	Speech grade quality at 100% modulation	6
Standard broadcast (AM)	High quality service	10
Single sideband radio-telephony	Speech grade quality, carrier suppressed, 10, 20, or 40 dB, single channel	3
Single sideband radio-telephony	Speech grade quality, carrier suppressed, 10, 20, or 40 dB, two channel	6
Independent sideband radio-telephony	Speech grade quality, carrier suppressed, 10, 20, or 40 dB, four 3-kHz voice channels	12
Manual continuous wave radio telegraphy	30 words per minute	<2
Modulated manual continuous wave telegraphy	30 words per minute	2
Radio teleprinter frequency shift	Up to 150 words per minute	1.7
Radio teleprinter audio frequency shift, SSB	100 words per minute, single channel	<1
Radio teleprinter audio frequency shift, SSB	16 channels, 100 words per minute each channel, carrier suppressed, 20 or 40 dB	3
Frequency modulation broadcast service	Broadcast quality	150
Facsimile	AM subcarrier modulated	6
Facsimile-SSB	NBFM [a] modulation, carrier suppressed, 20 or 40 dB	3

[a] NBFM, narrow band frequency modulation.

(modulate) two signals A and B in a linear device we have the following appearing at the output:

Signal A
Signal B
Sum of signals, $A + B$
Difference of signals, $A - B$

As an example, modulate (mix) a 4000-kHz RF carrier with a 1.000-kHz audio signal. Following the above rules, we would have:

4,000 kHz (= 4,000,000 Hz)   Carrier
1.000 kHz (= 1,000 Hz)   Modulating signal
4,000,000 Hz + 1,000 Hz   Upper sideband
4,000,000 Hz − 1,000 Hz   Lower sideband

By the use of filters one or the other sideband is suppressed. The modulating signal as a single entity disappears because the bandpass of the modulator filter is far above the modulating frequency. The carrier itself is suppressed by using a special balanced modulator.

Let us go through the exercise again, this time considering a voice channel occupying the audio spectrum 300–3000 Hz. Here we must define boundaries, the highest modulating frequency, 3000 Hz, and the lowest, 300 Hz, as though they were simple sinusoidal tones. We use the same RF carrier, 4000 kHz. Modulation again is mixing or heterodyning. Thus we have the following:

4,000 kHz = 4,000,000 Hz   RF carrier
4,000,000 Hz − 300 Hz ⎫
4,000,000 Hz − 3000 Hz ⎭   Difference or lower sideband
4,000,000 Hz + 300 Hz ⎫
4,000,000 Hz + 3000 Hz ⎭   Sum or upper sideband

The resulting lower sideband will then be a frequency of 3,999,700 Hz which represents the 300 Hz modulating tone and a frequency of 3,997,000 Hz representing the 3,000 Hz modulating tone. Note that for the lower sideband an inversion has taken place. This means that the higher modulating frequency became the lower RF frequency, and the lower became the higher. Such a sideband is called an inverted sideband.

The upper sideband will be the frequencies 4,0003,000 Hz and 4,000,300 Hz. This is called an erect or upright sideband. The lowest modulating frequency is the lowest RF frequency, and the highest is the highest RF frequency.

As mentioned earlier, in most modern SSB systems the carrier is not transmitted. It is suppressed by 50 or 60 dB at the transmitter. At the SSB receiver the carrier is reinserted locally as closely as possible to the original suppressed carrier frequency. Another approach is to suppress the carrier only about 20 dB, and use the remaining reduced carrier (pilot carrier) as a frequency reference at the receiver. This pilot carrier technique is discussed below.

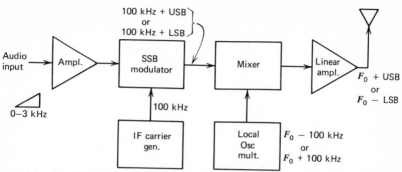

Figure 4.4 Simplified functional block diagram of a typical SSB transmitter (100 kHz IF). USB, upper sideband; LSB, lower sideband; F_o, operating frequency; IF, intermediate frequency.

4.5.5 SSB Operation and its Comparison with Conventional AM

Before proceeding further, the reader should consult Figures 4.4 and 4.5, simplified functional block diagram of a typical SSB transmitter and receiver respectively.

TRANSMITTER OPERATION

A nominal 3-KHz voice channel amplitude modulates a 100-kHz* stable IF carrier; one sideband is filtered and the carrier is suppressed. The resultant sideband occupies the RF spectrum of 100 + 3 kHz or 100 − 3 kHz, depending on whether the upper or lower sideband is to be transmitted. This signal is then mixed with a local oscillator to provide output at frequency F_0. Thus the local oscillator output to the mixer is $F_0 + 100$ or $F_0 − 100$ kHz. Note that frequency inversion takes place when lower sidebands are selected (see Section 3.2, which discusses inverted sidebands). The output of the mixer is fed to the linear power amplifier (LPA). The output of the LPA is radiated by the antenna.

RECEIVER OPERATION

The incoming signal, consisting of a suppressed carrier plus a sideband, is amplified by one or several RF amplifiers, mixed with a stable local oscillator to produce an IF (assume again that the IF is 100 kHz*). Several IF amplifiers will then increase the signal level. Demodulation

* Other common IF frequencies are 455 and 1750 kHz.

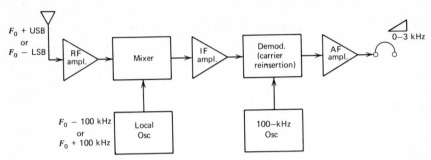

Figure 4.5 Simplified functional block diagram of a typical SSB receiver (100 kHz IF). USB, upper sideband; LSB, lower sideband; F_o, operating frequency; IF, intermediate frequency.

takes place by reinserting the carrier at IF. In this case it is from a stable 100-kHz oscillator, and detection is usually via a product detector. The output of the receiver is the nominal 3-kHz voice channel.

As we see, SSB is a special form of conventional AM, which has been around for a long time. Conventional AM transmits a carrier plus two sidebands. The transmitted intelligence is in the sidebands. The intelligence is identical in both sidebands. If the intelligence is identical, why not just transmit only one sideband? The carrier, which consumes two-thirds of our power on a system that is 100% modulated, is needed only in the demodulation process. Why not suppress the carrier and reinsert it at the receiver at a low level?

Consider an AM transmitter operating with 100% modulation, modulated with a 1000-Hz sine wave with 1500 W output. This 1500 W is allocated as follows: 1000 W in the carrier and 250 W in each sideband. It is the sidebands that contain the useful information and only one is necessary (leaving aside phase relationships). It is apparent, therefore, that we are transmitting at a cost of 1500 W to communicate 250 W of useful energy. In the case of SSB only the one sideband is transmitted; the carrier is suppressed and the unwanted sideband eliminated. At the receiver the carrier is reinserted and its phase in relation to the received SSB signal is irrelevant.

Because only one sideband is being received the effective receiver bandwidth is reduced to half in the case of SSB when compared to an equivalent AM system. This conserves spectrum and gives SSB a comparable 3-dB noise advantage over AM.

It can be shown using voice tests that under fading conditions SSB may give up to a 9-dB advantage over AM for equal power outputs.

4.5.6 Key to SSB Transmission

One of the most important considerations in the development of the SSB signal at the near end (transmitter) and its demodulation at the far end (receiver), other than linearity, carrier suppression, and unwanted sideband suppression, is that of accurate and stable frequency generation and reinsertion. It is equally important (frequency generation) at the transmitter as at the receiver (generation and reinsertion).

Most of us have heard the "monkey talk," or high or low pitched speech, just barely intelligible on some poor telephone circuits or while "shortwave" listening on an HF receiver. This is the characteristic of SSB systems that are not operating on the same frequency and/or are not properly "frequency synchronized." The quotes are purposeful, as we shall see later.

Generally SSB circuits can maintain tolerable intelligibility when the transmitter or receiver are no more than 50 Hz out of "synchronization." Telegraph (VFTG) and low speed data require "synchronization" of 2 Hz or better.

There are two approaches to achieve "synchronization": (1) pilot carrier operation, and (2) synthesized operation. In the first the carrier frequency is suppressed only 20 dB (10 dB during poor propagation conditions or through interference). The remaining carrier is called the pilot carrier and is used to actuate an AFC (automatic frequency control) circuit such that the tuning local oscillator centers the receiver's converted incoming signal on the center of the first IF (intermediate frequency). The pilot carrier may also be regenerated and used after conversion for reinsertion; this is seldom the case, however. It is more usual to use a fully independent locally generated reinsertion carrier. Its frequency is usually so low that even without any frequency control (oven) of the carrier oscillator, there will not be more than a fraction of a hertz of error in the received demodulated signal due to carrier reinsertion.

Synthesized transmission and reception are used widely today in HF. A synthesizer is no more than a highly stable frequency generator. It gives one or several simultaneous sinusoidal RF outputs on discrete frequencies in the HF range. In most cases it provides the frequency supply for all RF carrier needs in SSB applications. For example, it will supply:

Transmitter IF carrier
Transmitter local oscillator supply
Receiver local oscillator supply (supplies)
IF carrier reinsertion supply

The following are some of the demands we must place upon an HF synthesizer:

Frequency stability
Frequency accuracy
A number of frequency increments
Supplementary outputs
Capability of being slaved to a frequency standard
Spectral purity of RF output(s)

For VFTG* operation, as noted before, end-to-end frequency error may not exceed 2 Hz ($\pm$ 1 Hz) to maintain a satisfactory error rate. If the HF transmitter and receiver were the only devices in a system to inject or cause frequency error, we could then allow $\pm$ 0.5 Hz error for each. Thus the frequency stability and accuracy must be maintained to $\pm$ 0.5 Hz under all conditions for synthesizers used at both ends of the circuit. First consider a system operating at 10 MHz (1×10^7 Hz). It would require a transmitter and receiver stability of $\pm 0.5 \times 10^{-7}$ or $\pm 5 \times 10^{-8}$. If the maximum operating frequency of the system were 30 MHz, then the stability would be three times more stringent than the 10-MHz system, or $\pm 5/3 \times 10^{-8}$, or 1.66 parts in 10^8. Synthesizers are usually based on oven-controlled crystal oscillators. These crystals drift in frequency as they age. Thus the specification must state stability requirements over a time limit. Usually the time frame is the month. A requirement should also be stated for the periodic readjustment of the synthesizer to assure an accuracy with a known standard.

A synthesizer provides outputs on discrete RF frequencies. One requirement may be that outputs shall be from 3 to 30 MHz in 100-kHz increments. Another system may need a tighter requirement with frequency steps of 1 kHz, or even 1 Hz (or less) from 3 to 30 MHz.

Tight requirements increase the price of the equipment.

4.5.7 Synthesizer Application Block Diagrams

Synthesizers, when used for HF SSB operation, provide stable frequency sources. Figure 4.6 shows a typical application for an SSB transmitter; Figure 4.7 for an SSB receiver.

To simplify operation of the synthesizer the frequency readout on the front panel is "offset" by the IF and is not the true local oscillator injection

---

*Voice frequency carrier telegraph.

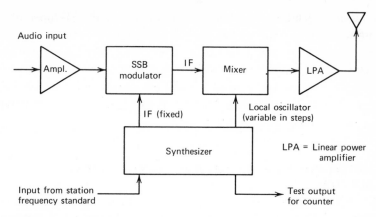

Figure 4.6 Transmitter.

frequency (applicable for both transmitters and receivers), but the RF operating frequency. However the "output for counter" will read the local oscillator frequency, and the operator must subtract or add the IF (first IF) (depending on whether the local oscillator is offset above or below the operating frequency).

If the synthesizer is phase locked to a station frequency standard, it will take on the stability of the standard. Some station frequency standards, such as a cesium atomic standard, have frequency stabilities better than 1×10^{-11}/month.

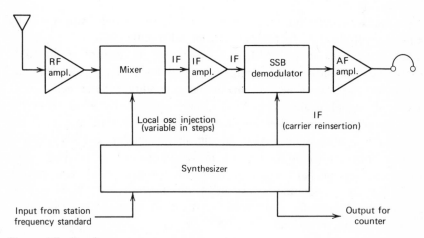

Figure 4.7 Receiver.

The use of synthesizers allows more efficient HF SSB operation by permitting equivalent full carrier suppression (-50 to -60 dB) for some additional cost. If an installation is to operate on a very limited number of RF frequencies, simple oven-controlled crystals may give a more economic solution. Such crystals now display stabilities on the order of 1×10^{-7}/ month or better. This certainly is sufficient stability for voice operation of SSB.

4.6 LINEAR POWER AMPLIFIERS IN POINT-TO-POINT SSB SERVICE

The power amplifier in an SSB installation raises the power of a low level signal input without distorting the signal. That is, the envelope of the signal output must be as nearly as possible an exact replica of the signal input. By definition the power amplifier that will perform this function is a linear power amplifier (LPA). For HF application peak envelope power outputs of LPA's are on the order of 1 or 10 kW, and in some installations 40- or 50-kW amplifiers are used. Air cooling is almost universal.

Desirable characteristics of an LPA are the following:

1. High efficiency.
2. High gain.
3. Low grid-to-plate capacitance.
4. Good linearity at all operating frequencies.

The most common tube used in the LPA is the tetrode, which usually requires neutralization to prevent it from self-oscillating.

Automatic load control (ALC) is an important feature of an LPA. It provides the means of keeping the signal level adjusted so that the power amplifier works near its maximum capability without being over-driven on signal peaks. The ALC circuit receives its input from the envelope peaks in the power amplifier and uses its output to control the gain of the exciting signal. ALC circuits in transmitters are very similar in their function to AGC circuits in receivers. As in receiver AGC circuits, important parameters of ALC are proper attack and release times. For voice operation a fast attack on the order of 10 ms is used to remove gain rapidly and avoid overload. A release of about $\frac{1}{10}$ s returns the gain to normal.

4.7 INTERMODULATION DISTORTION

Intermodulation (IM) distortion must be rigidly monitored and controlled, particularly on SSB/ISB transmitting installations. Measuring IM distortion is one quick method of determining linearity. IM distortion may be measured in two different ways:

Two-tone test.
Tests using white noise loading.

The two-tone test is carried out by applying two tones at the audio input of the SSB or ISB* transmitter. A 3:5 frequency ratio between the two test tones is desirable to identify the intermodulation products easily. In a 6-kHz input we could apply tones of 3 and 5 kHz, and to a 3-kHz channel, tones of 1500 and 2500 Hz.

The test tones are applied at equal amplitudes and gains are increased to drive the transmitter to full power output. Exciter or transmitter output is sampled and observed in a spectrum analyzer. The amplitudes of all undesired products (see Section 1.9.6) and the carrier are measured in terms of decibels below either of the two equal amplitude test tones. The decibel difference is the signal-to-distortion ratio. This should be at least 40 dB or better on HF SSB (ISB) systems. Normally the highest level IM product is the third order product. This product is two times the frequency of one tone minus the frequency of the second tone. For example, two test tones are 1500 and 2500 Hz; thus

$$2 \times 1500 - 2500 = 500 \text{ Hz or } 2 \times 2500 - 1500 = 3500 \text{ Hz}$$

and consequently, the third order products will be 500 and 3500 Hz. The presence of IM products numerically lower than 40 dB indicates maladjustment or deterioration of one or several stages, or overdrive.

The white noise test for IM distortion more nearly simulates operating conditions of a complex signal such as voice. One simple approach is to load three of the four voice channels* in an ISB* system with white noise and measure the level of the noise in the idle channel. Next measure the noise level on the same channel with no noise loading on the other channels. This is the idle noise. The IM noise level is the difference in levels in the idle channel when noise loading is applied and the idle noise.

* Section 4.11 introduces the reader to ISB and three- and four-channel operation.

4.8 HF ANTENNAS

4.8.1 General

Perhaps the most important element in an HF telecommunication system is the antenna. To achieve the 90% or better path reliability, great care must be taken in the selection and design of antennas.

For HF point-to-point operation two basic antennas types are available:

Rhombic
Log periodic. (LP)

Both the rhombic and the LP should be considered for selection on any path over 2500 km in length.

Precision-designed rhombics outperform any other antennas on point-to-point paths more than 4000 km long.

4.8.2 Basic Antenna Considerations

To design an HF antenna system for point-to-point service, we would want to consider the following:

1. Broadband nature of the antenna (VSWR characteristics).
2. Its efficiency and dissipation losses on transmission.
3. Gain.
4. Fixed or variable takeoff angle.
5. Polarization (in the case of LP's).
6. Height above ground.
7. Land area required.
8. Need of counterpoise (ground screen).
9. Transmission line requirements.
10. Side lobe suppression.
11. Transmit VSWR.
12. Cost.

The rhombic and log periodic antennas are compared regarding the 12 points listed above in Table 4.4.

4.8.3 Rhombic Antennas

The rhombic is a horizontally polarized antenna and in its basic configuration is composed of four horizontal long-wire radiators arranged in the

Table 4.4   Comparison Chart—Rhombic and Log Periodic Antennas

Consideration	Rhombic	Log Periodic
1. Broadband	1–2 octaves	3 octaves
2. Efficiency	50–70%	95%
3. Gain over an isotropic	8–20 dB	6–12 dB (vertical LP)
		10–17 dB (horizontal LP)
4. Takeoff angle	Variable with frequency, down to 4° over poor earth	Fixed; horizontal LP down to 12°, vertical LP down to 8°
5. Polarization	Horizontal	Horizontal and vertical
6. Height	Fixed	Sloping
6. Land area	5–15 acres	2–5 acres
8. Need of counterpoise	No	Yes on many installations
9. Transmission line	Open wire	Coaxial
	Usually uses 600-Ω balanced line	50-Ω coaxial /300-Ω balanced
10. Side lobe suppression	6 dB	12 dB
11. Transmit VSWR	2:1	2:1
12. Cost	$7000–14,000 exclusive of labor and land	$15,000–30,000 exclusive of labor and land

shape of a rhombus. One apex of the rhombus is terminated with a resistance which normally is slightly greater in value than the characteristic impedance of the antenna. The slightly larger value compensates for the loss of energy by variation as the wave travels the length of the antenna. An unterminated rhombic antenna is bidirectional. The termination makes the antenna essentially unidirectional. Figure 4.8 is a drawing of a typical rhombic antenna.

The characteristic impedance Z_0 of a rhombic antenna has a mean value of 800 Ω with some small variation with respect to frequency. Z_0 may be lowered to approximately 600 Ω by using properly spaced multiple conductors for the individual radiators. The spacing between these conductors (usually 3) varies from zero at the apexes to several feet at the side masts or towers. The resulting 600-Ω characteristic impedance is convenient for the common varieties of open-wire transmission line.

The basic limiting factor on the amount of power that a rhombic can handle is the termination resistance. Nominally one-third to one-half of this power is dissipated by the terminator. For reception the terminator is

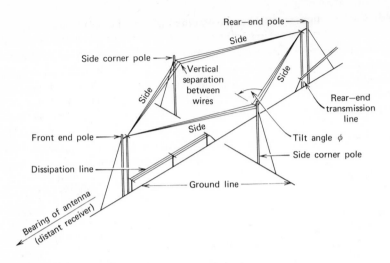

Figure 4.8 A rhombic antenna (transmitting).

normally a carbon resistor. For transmission the terminator is an iron wire dissipation line.

The physical size of the rhombic antenna is comparatively great because the antenna is composed of long-wire radiators and depends on wave interference for gain. The rhombic generally requires more land area than any other commonly used HF antenna; 5–15 acres of level ground is a typical requirement for each antenna. Table 4.5 gives dimensions for standard compromise rhombic antennas.

Because of the variation of transmission properties with frequency, a single rhombic antenna is not likely to maintain a satisfactory lobe orientation or power gain over the HF range of frequencies. Therefore, at many installations, two or even three rhombics are used on a transmission path, each being designed for a particular range of frequencies. This optimizes frequency and lobe orientation (in both planes) for the path involved.

4.8.4 Basic Rhombic Design

Considering the distant point Y in a path between X and Y and the propagation condition of the path, a wave angle W can be determined. This angle can be determined from the incidence angle formula (4.1).

Table 4.5 Dimensions of Standard Compromise Rhombics

Rhombic Type	Range (mi) (km)	Side Length (ft) (m)	Tilt Angle (°)	Ht (ft) (m)	Length End-to-End (ft) (m)	Width Side-to-Side (ft) (m)
A	>3000 (4800)	375 (115)	70	65 (20)	723 (222)	258 (82.4)
B	2000–3000 (3200–4800)	350 (107.7)	70	60 (18.5)	676 (208)	251 (77.2)
C	1500–2000 (2400–3200)	315 (96.8)	70	57 (17.5)	611 (188)	228 (70.2)
D	1000–1500 (1600–2400)	290 (89.2)	67.5	55 (16.9)	553 (170)	234 (72.0)
E	600–1000 (960–1600)	270 (83)	65	53 (16.3)	506 (155)	240 (73.8)
F	400–600 (640–960)	245 (75.3)	62.5	51 (15.7)	453 (139.3)	238 (73.2)
G	200–400 (320–640)	225 (69.2)	60	50 (15.3)	407 (125.2)	237 (73.0)

Source: Ref. 2.

For given wave angle W, the height H, tilt angle ϕ, and leg length L can be determined from the following formulas:

$$H = \frac{1}{4 \sin W} \tag{4.2}$$

$$L = \frac{1}{2 \sin W} \tag{4.3}$$

$$\sin \phi = \cos W \tag{4.4}$$

H and L are given in wavelengths, the tilt angle in degrees. This is the "maximum output design method" where

H = height of the antenna above ground
λ = wave length
L = leg length (the length of a side of the rhombus)
W = wave angle (angle of maximum radiation) in degrees
ϕ = one-half of the side apex angle, which is the *tilt angle*

Such a method leads to a maximum output at a given angle of radiation, although the lobe maximum may not fall along the desired wave angle.

Table 4.6 gives dimensions for rhombic antennas for several selected wave angles using "maximum output design" criteria. To convert values from Table 4.6 to more applicable parameters, the following formulas may be used:

Table 4.6 Rhombic Antenna Maximum Output Design Dimensions

Wave Angle (°)	Height (Wavelength)	Leg Length (Wavelength)	Tilt Angle (°)
10	1.42	16.3	80
11	1.30	13.8	79
12	1.20	11.5	78
13	1.10	9.9	77
14	1.04	8.5	76
15	0.96	7.5	75
16	0.90	6.7	74
17	0.86	5.8	73
18	0.81	5.2	72
19	0.77	4.7	71
20	0.73	4.3	70
21	0.70	3.9	69
22	0.67	3.6	68
23	0.64	3.3	67
24	0.615	3.0	66
25	0.59	2.8	65
26	0.57	2.7	64
27	0.55	2.5	63
28	0.54	2.4	62
29	0.52	2.2	61
30	0.50	2.0	60

Source: Ref. 11.

$$\lambda_{ft} = \frac{984}{f_{MHz}} \tag{4.5}$$

$$L' = \frac{492(N - 0.05)}{f_{MHz}} \tag{4.6}$$

where f is the frequency in megahertz and N is the number of half-wavelengths and may be taken from the table. Formula 4.5 may be used to convert the height into feet when the frequency in megahertz is given. This is the traditional free-space formula to determine wavelength when frequency in megahertz is given. It is sufficiently accurate to aid in calculation height from the table.

For leg length, use formula 4.6, where L' is the length in feet of a leg taking end effects into account; N is the number of half-wavelengths (double the value in Table 4.6) and f again is in megahertz.

The "alignment design" method uses the following formulas to find the three variable dimensions:

$$H = \frac{\lambda}{4 \sin W} \tag{4.7}$$

$$L = \frac{0.371}{\sin^2 W} \tag{4.8}$$

$$\sin \phi = \cos W \tag{4.4}$$

Note. Lengths are in wavelengths, angles in degrees. The alignment design method provides for exact alignment of the major lobe of the directive pattern with the wave angle and gives the antenna a better discrimination factor. The major difference in dimensions of the alignment method over the maximum output method involves a decrease in element length of approximately 25% in the alignment design.

The adjusted design method is one in which one of the three basic parameters is limited. For this method refer to Tables 4.7a and 4.7b where tilt and wave angles may be determined when height and leg length of rhombic antennas are given.

Rhombic antennas may be fed with coaxial lines by the use of a balun (balance-to-unbalance transformer). Nearly all receiving rhombics use coaxial transmission lines. For the case of transmission, the balun must handle the peak power to be transmitted. However, baluns are now available handling up to 200-kW peak power (50-kW average) with an insertion VSWR of 1.2:1 and operating into a load VSWR of 2.5:1 or less. Balun efficiencies may reach 98%.

4.8.5 Terminating Resistances

For receiving applications, terminating resistors are used. For single-wire rhombics, two low-wattage, series-connected carbon resistors of 400 Ωs each are employed in a terminating network. For the three-wire rhombic

**Table 4.7a   Determination of Rhombic Antenna Height and
Tilt Angle When Wave Angle and Leg Length Are Known**

Wave Angle (°)	2-Wavelength Leg		3-Wavelength Leg		4-Wavelength Leg	
	Height (Wavelengths)	Tilt Angle (°)	Height (Wavelengths)	Tilt Angle (°)	Height (Wavelengths)	Tilt Angle (°)
10	1.45	56.0	1.45	63.0	1.45	67.2
11	1.30	56.5	1.30	63.5	1.30	67.8
12	1.20	57.0	1.20	64.0	1.20	68.4
13	1.10	57.4	1.10	64.5	1.10	69.0
14	1.04	57.8	1.04	65.0	1.04	69.6
15	.97	58.2	.97	65.5	.97	70.2
16	.90	58.6	.90	66.0	.90	70.9
17	.86	59.0	.86	66.5	.86	71.7
18	.81	59.4	.81	67.2	.81	72.8
19	.77	59.8	.77	68.0	.77	74.0
20	.73	60.2	.73	69.0	.73	75.3
21	.70	60.9	.70	70.0	.70	77.0
22	.67	61.7	.67	71.0	.67	78.8
23	.65	62.5	.65	72.0	.65	80.8
24	.62	63.4	.62	73.5	.62	83.1
25	.60	64.3	.60	75.4	.60	85.0
26	.58	65.3	.58	77.2	.58	—
27	.56	66.4	.56	79.0	.56	—
28	.54	67.6	.54	81.8	.54	—
29	.52	68.8	.52	85.0	.52	—
30	.50	70.0	.50	—	.50	—

Source: Ref. 11.

the value of each resistor drops to 300 Ω. Each resistor is connected to a leg of the antenna, and the common connection between the resistors is well grounded. It should be noted that the impedance of single-wire rhombics, as the frequency is increased from 4 to 20 MHz, drops from about 900 to 600 Ω, and for a three-wire rhombic the impedance varies around 600 ± 50 Ω.

For transmitting applications the terminating resistance should be capable of dissipating 50% of the available power. For a 50-kW average power output from the transmitter, the dissipation line should be able to dissipate 25 kW. In applications for systems transmitting less than 5 kW, resistors are available. These are usually mounted in a box at the base

Table 4.7b   Determination of Tilt and Wave Angles When Height and Leg Length of Rhombic Antennas Are Known

Height (Wavelength)	2-Wavelength Leg		3-Wavelength Leg		4-Wavelength Leg	
	Wave Angle (°)	Tilt Angle (°)	Wave Angle (°)	Tilt Angle (°)	Wave Angle (°)	Tilt Angle (°)
0.5	30	70	—	—	—	—
0.6	24.8	64	24.8	75	24.8	85
0.7	21	61	21	69.8	21	77
0.8	18.2	59.3	18.2	67.2	18.2	73
0.9	16	58.5	16	65.8	16	71
1.0	14.5	57.9	14.5	65.0	14.5	70
1.1	13	57.2	13	64.2	13	69
1.2	12	56.9	12	63.8	12	68.2
1.3	10.9	56.4	10.9	63.5	10.9	67.6
1.4	10.2	56.0	10.2	63.2	10.2	67.2

Source: Ref. 11.

of the terminating end pole. The resistor is fed with a standard 600-Ω (or 800) line. The center of the resistor must be well grounded and the box must be weather proof.

For applications with higher power, iron, steel, or magnetic material alloy resistive lines are used. A wire loop is constructed with a 650-Ω resistance. A 640-Ω impedance is maintained by using proper spacing. The wires are often folded back and forth between relatively few poles. The electrical resistive center of the line must be grounded. The resistance of the lines may be adjusted for peak performance of the antenna.

4.8.6   Transmission Lines

With an air dielectric, the characteristic impedance of an open-wire transmission line is given by the formula:

$$Z_0 = 276 \log \frac{b}{a} \qquad (4.9)$$

where Z_0 = characteristic impedance (Ω)

b = center-to-center spacing between the two feed lines

a = radius of each conductor

The most common open-wire line uses No. 6 AWG copperweld wire with 12-in. spacing. Line spacing must be measured carefully and kept uniform to avoid impedance discontinuities. Supporting poles must be placed at varying distances and with no adjacent pole spans alike. Sharp turns in the line should be avoided. If they must be made, they must be made gradual using jumpers.

Turns should be made by twisting the line to a vertical position and returned to horizontal after the turn. The line should be afforded lightning protection. This is usually done by using horn gaps to ground. Care must be taken against sag and letting the line blow and sway in the wind. Insulators must be sufficient for the voltage on the line.

Coaxial lines may be attractive for transmitting feed lines using the balun transformer mentioned above. They may be buried, which is an advantage. If they are of the air dielectric type, they must be pressurized. Coaxial lines adapt to modern antenna switching matrices and to direct VSWR and power measurement. On the other hand, long runs are lossy at higher frequencies when compared to open-wire lines. Transmitting coaxial lines are much more expensive than open wire.

Information on typical coaxial cable is contained in Chapter 5, Figures 5.24A and 5.24B.

4.8.7 Log Periodic Antennas

A logarithmically periodic (LP) antenna has electrical properties such that both its pattern and impedance characteristics repeat periodically with the natural logarithm of frequency. By making the variation of pattern and input impedance small within any period, operating characteristics are obtained which are nearly independent of frequency over the entire working frequency range. Although the LP antenna is a traveling wave type of antenna like the rhombic, it can be designed to have substantially the same pattern of radiation over a wide frequency range with a fairly constant impedance.

There are three types of LP antennas:

1. Trapezoidal.
2. Horizontal dipole.
3. Vertical monopole.

Figure 4.9 shows the three types of antennas.

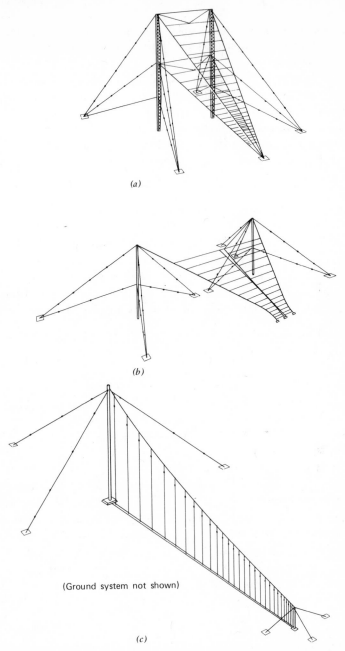

(a)

(b)

(Ground system not shown)

(c)

Figure 4.9 The three basic types of LP antennas used in HF point- to-point service. (a) Trapezoidal outline LP antenna; (b) horizontal dipole LP antenna; (c) vertical monopole LP antenna.

The horizontal dipole and vertical monopole antennas require ground planes. The first two antennas are horizontally polarized; the third is vertically polarized.

The dipole type of LP antenna is fed by a balanced transmission line with 180° transpositions at alternate dipole elements (see Figure 4.10). The dipole elements range approximately from those which are a half-wavelength at the highest frequency of operation to be covered, to those which are a half wavelength at the lowest frequency of operation. As shown in Figure 4.11, the antenna feed is at the short element end and the direction of transmission is from the large element end toward the short element end.

Beginning at the largest element, each successive element is reduced in size by the same scale factor τ. The values for τ range from approximately 0.75 to about 0.975. As mentioned previously, the antenna is periodic. This is accomplished by having the antenna elements spaced from each other in a periodic manner defined by σ, which is the ratio of adjacent dipole spacing to twice the length of the longer element. The ideal LP antenna should have all elements display the same length-to-

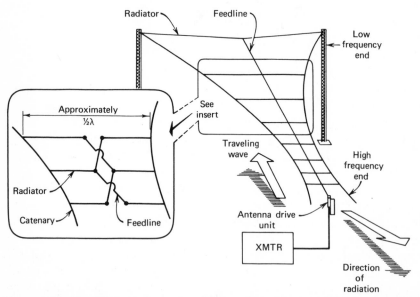

Figure 4.10 Simplified diagram of a LP antenna. Courtesy Granger Associates, Menlo Park, Calif.

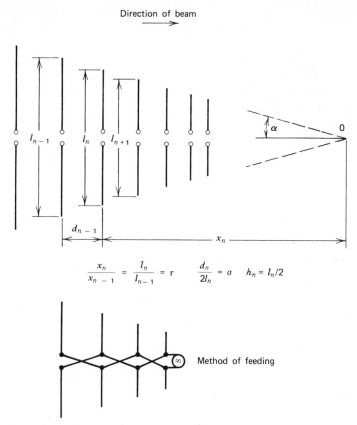

$$\frac{x_n}{x_{n-1}} = \frac{l_n}{l_{n-1}} = \tau \qquad \frac{d_n}{2l_n} = \sigma \qquad h_n = l_n/2$$

Figure 4.11 The dipole type of LP antenna.

radius ratio. In practice, however, the ratio is maintained on an average basis for a number of elements. Figure 4.11 illustrates the basic geometric considerations of the dipole type of LP antenna.

Consider a traveling wave launched from the balanced feed of an LP antenna of the dipole type. The wave first encounters the highest frequency radiators, which represent a very high impedance across the line and essentially do not affect the wave as it passes up the antenna. As the wave continues up the antenna toward the lower frequency end, it will meet radiators (elements) which will nearly match the characteristic impedance of the line at the frequency of transmission, and an active region is developed. Here most of the energy of the incident wave is radiated into space. Beyond the resonant region the lower frequency antenna elements again offer a high impedance to the excitation remaining.

The monopole LP array (Figure 4.9c) cannot be considered as one-half of a dipole array fed against the ground. This is because the transmission line cannot be transposed from one element to the next as accomplished in the dipole type of LP antenna shown in Figures 4.9b and 4.10. To get around this problem the monopole array provides proper phase shift between elements, involving lumped constant sections between elements that use series inductance and shunt capacitance. These sections also act as filters near cutoff to confine array currents to those elements that are a quarter-wavelength long.

4.8.8 Horizontal and Vertical Polarization

The rhombic antenna is horizontally polarized. The LP antenna offers versions either horizontally or vertically polarized. What will determine the choice of polarization?

Vertically polarized antennas are preferred:

- Where long distances are to be covered by skywave, and a horizontal antenna cannot be raised to sufficient height.
- For surface wave propagation.

Horizontally polarized antennas are preferred:

- On most short-range skywave circuits (200–300 mi) due to lower noise and higher radiation angles.
- Where man-made noise propagated by surface waves may be a problem (near cities).
- Where the ground conductivity is sufficiently low, therefore less favorable reflection characteristics for vertical polarization.
- For high gain over long ranges (over 4000 km) for the rhombic antenna, where the horizontal antenna can be raised to the height required for the desired radiation angle (LP application) and that the horizontal antenna has a higher gain over ground than the corresponding vertical antenna.

For path distances greater than 2500 km only the vertically polarized LP or rhombic antennas can provide the necessary low angles of radiation (i.e., below 12° elevation angle of main lobe).

4.9 HF FACILITY LAYOUT

As mentioned previously HF point-to-point transmission facilities may consist of one-site, two-site, or three-site configurations. At small HF installations, where one or perhaps two transmitters operate simultaneously on a full-duplex basis, the companion receivers may be colocated with the transmitters if care is taken to reduce transmitter spurious outputs, and at least a 10% frequency separation is maintained between all transmitter and receiver frequencies that could simultaneously be in use.

Larger HF point-to-point facilities require as a minimum the separation of the transmitter location from the receiver location. The separation must be such that transmitter spurious outputs do not interfere with receiver operations. Separations are usually on the order of 3–45 km.

Often a three-site configuration is used. Both transmitter and receiver sites require many acres of land for antennas. For instance transmitting rhombic antennas should be separated one from another by 300 m in the direction of transmission and by 80 m at the sides. This is done to reduce intermodulation interference. Two or three rhombics are used for each transmit link. Twice that number may be required at receivers where space diversity is used. The third site, which we shall call the communication relay center (CRC), is usually located near or with a telephone/telegraph central office (i.e., in a city). The sites are interconnected by a multichannel telephone link, either multipair cable or LOS radiolink.

Both radio sites require good grounds. Further, the receiver site should be located at areas with a low level of RFI* in the HF band. Power lines have been found to be particularly troublesome. It is also advisable that the receiver site be situated as far as possible from highways and that all vehicles entering the site be inspected periodically to ensure sufficient ignition noise suppression. All metal, such as doors, benches tables,etc., should be well grounded at the transmitter and receiver sites.

4.10 GREAT-CIRCLE BEARING AND DISTANCE

For siting antennas on HF installations and for system calculations, great-circle bearing and distance are used. This involves relatively simple concepts of solid geometry and trigonometry. The method is described in several publications. One of the most available is *Reference Data for Radio Engineers* (Ref. 1).

---

* Radio frequency interference.

Another means of determining great-circle bearings and distances is by use of navigation tables and charts available from the U.S. Navy's Hydrographic Office. Some of these are listed below:

Navigation Tables for Navigators and Aviators, HO 206
Dead Reckoning Altitude and Azimuth Table, HO 211
Great-circle charts: *HO Chart 1280, North Atlantic Ocean*
                                HO Chart 1281, South Atlantic Ocean
                                HO Chart 1282, North Pacific Ocean
                                HO Chart 1283, South Pacific Ocean
                                HO Chart 1284, Indian Ocean

4.11 INDEPENDENT SIDEBAND (ISB) TRANSMISSION

Modern point-to-point HF with voice or composite voice/data/facsimile transmission requirements normally uses some form of ISB transmission. ISB utilizes SSB techniques in that the carrier is suppressed but affords transmission of separate information in each sideband. Up to four nominal 3-kHz voice channels can be transmitted simultaneously on an ISB system. The limit of four voice channels in a 12-kHz total bandpass is set by international agreement.

For independent sideband transmission, the upper sideband by convention is referred to as the *A* side and the lower sideband, the *B* side. If a particular circuit transmits all four voice channels, the on-air trans-

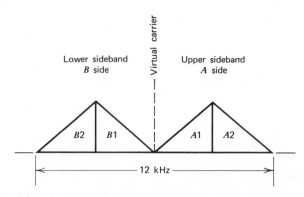

Figure 4.12 Voice channel assignment, standard four-channel ISB emission.

mitted spectrum is as shown in Figure 4.12. Following is a simplified block diagram of an ISB exciter:

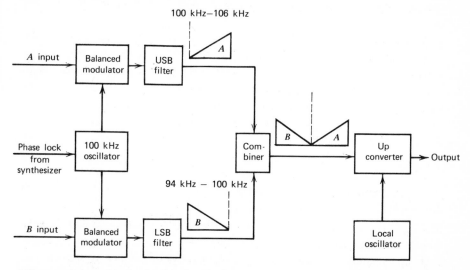

The receiving side is similar for the conventional SSB receiver. It provides two independent audio outputs, the information transmitted on the upper and lower sidebands.

The reader will take note that the *A* input and *B* input to the transmitter each contain two voice channels in the 6-kHz bandpass. One method of providing the two-channel multiplexed inputs is to use a voice multiplexer as a separate device ahead of the balanced modulator. Such a multiplexer passes the inboard channel (*A*1 or *B*1) directly and mixes the outboard channel (*A*2 or *B*2) with a 6425-Hz carrier; the lower sideband is selected and then combined with the direct channel. At the distant end a demultiplex operation is carried out to separate the translated channel and provide two normal 3-kHz voice channel outputs.

Another approach to HF voice channel multiplexing is to carry out the multiplexing at the transmitter IF using the proper subcarrier derived from a common synthesizer. The subcarrier for the outboard channel *B*2 is the virtual center frequency or carrier minus 6425 Hz, and for the *A*2 channel, center frequency plus 6425 Hz. Care must be taken that *B*2 be an erect channel and *A*2 an inverted channel. At the receiver demultiplexing is carried out in the IF as well. Carrier insertion is the IF center frequency plus 6425 Hz for the *A*2 channel, and the IF center frequency minus 6425 Hz for the *B*2 channel. All reinsertion frequencies derive from a common receiver synthesizer.

Any compatible type of information may be placed on the 3-kHz voice channel. Commonly transmitted are voice, telegraph/data, and facsimile. Special problems for the transmission of data/telegraph in HF are discussed in Chapter 8.

The severe fading in HF coupled with the continuous variation of the voice level itself have limited the quality of telephone service that can be provided. Signal levels may vary 60 or 70 dB in very short periods. With full carrier suppression automatic gain control on receivers does not function well. If one of the four available channels has either FM facsimile or telegraph/data tones, which have a fairly constant amplitude, AGC can be derived from such a channel, and level control can be fairly effective.

Older systems and some systems still in existence use a device called a VOGAD (voice operated automatic gain device). A VOGAD is supplied for each demodulated voice channel and is usually mounted near the receiver. By detecting average level of the voice syllables, some effective gain control is provided.

The outstanding weakness of the VOGAD (and similar equipment) is that it has no continuous level signal with which to sense continuously the variation of audio level and afford the necessary gain control. One approach is to split the voice channel providing a small bandpass at channel center which carries a sinusoidal tone with which AGC can act. This equipment is usually combined with the two-wire/four-wire HF telephone terminal. Such a terminal is normally configured as follows:

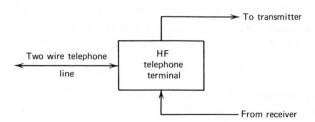

The gain control tone described above also provides a control on the voice terminal to control the transmit-to-receive function. This is done by frequency shifting the tone. A shift upward in frequency will lock the distant end in a receive condition and a shift downward in frequency will allow the distant end to transmit. This eliminates the problem of voice terminal lockup common on older systems. Lockup in the transmit or receive condition is usually caused by noise in the system.

The latest refinement of the tone control technique for gain has the trade name Lincompex. It utilizes a frequency modulated control tone at the high end of the HF voice channel that controls not only level but a compression and expansion function as well. Voice channel compression and expansion are discussed in Chapter 3 under compandors.

4.12 TRANSMITTER LOADING FOR ISB OPERATION

ISB transmitters as multichannel devices are sensitive to loading. Consider a 10,000-W ISB transmitter. While transmitting four voice channels simultaneously, we would ideally allow sufficient drive (excitation) such that each voice channel on the air provides 10,000/4 or 2500 W of power. We speak here of peak power or, more properly, peak envelope power (PEP). This would be a simple matter if we transmit only a simple sinusoidal signal in each voice channel. Outside of FSK, the transmitted signal applied to each voice channel is more complex. The drive on each voice channel input must be backed off, so on the average, each channel contributes less than 2500 W of on-air power. This is due to a peaking factor. If voice is being transmitted on all four ISB voice channels with an average PEP of 2500 W output from each, at some time several or all channels may reach a voice peak simultaneously and the 10,000 W would be well exceeded, probably causing the transmitter to blow a circuit breaker, arc over tubes, etc. Thus backoff of input signals must be carried out to ensure that the equipment stays within the peaking limits reasonably. It also must be remembered that the ratio of peak power to average power of a pure sine wave is 2:1; that is, the average power of a pure sine wave is one-half the peak power. A more detailed discussion of loading and peak factors is given in Chapter 3.

4.13 DIVERSITY TECHNIQUES ON HF

Two types of diversity reception are in fairly common use in HF systems today:

1. Space diversity.
2. In-band frequency diversity.

In practice diversity improvement is realized to the greatest extent only on digital systems. For composite voice/telegraph systems, diversity combining is used on the digital channel only.

For space diversity two antennas are used on each link. For full space diversity improvement the two antennas must be separated by at least six wavelengths (6λ). Combining is usually of the selective type and is normally carried out in the voice frequency carrier telegraph equipment located at the CRC. A typical space diversity system has the following configuration:

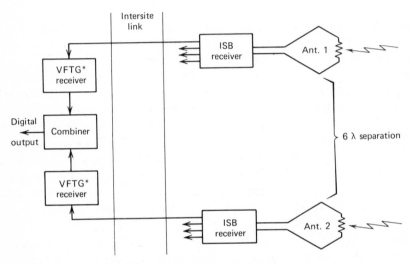

* Voice frequency telegraph (a digital demodulator); see Chapter 8.

Such a system may provide an improvement of bit error rate on digital transmission (usually carried in the A1 channel) from one to two orders of magnitude.

The reason for the improved error rate is that seldom is a fade correlated on antennas separated by at least six wavelengths. In other words, a signal received on one antenna may be in a deep fade, whereas on the second antenna the signal level will be high provided the fading follows a Rayleigh distribution.

In-band frequency diversity is somewhat less attractive as a diversity option and is effective on frequency selective fading only. It has been found that if two telegraph (data) channels derived from a single digital source (i.e., transmitting identical information) are separated by at least 600 Hz, an equivalent of a full order of diversity improvement may be

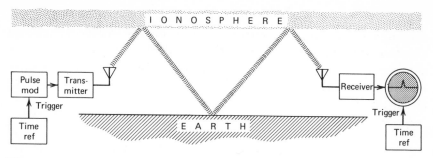

Figure 4.13 Synchronized oblique sounding.

derived on HF systems regarding frequency selective fading. The nominal maximum number of VFTG channels on HF is 16.* The subcarrier arrangement is such that the system may be reconfigured to eight channels of information permitting the pairing of channels and a 600-Hz separation between each pair member. A switch at the receiving terminal pairs already existing groups of pairs such that the system goes into a form of quadruple diversity when in-band frequency diversity is added to space diversity. (See Chapter 8 for more details on data/telegraph transmission.)

4.14   IONOSPHERIC SOUNDERS

Selection of the OWF (FOT) of a long-haul, point-to-point HF link is normally a "by guess and by God" procedure even with the use of CRPL forecasts, particularly during periods of high sunspot activity. Synchronized oblique ionospheric sounding removes much of the guesswork. Such a sounding system is shown schematically in Figure 4.13.

Oblique sounding as an adjunct to communications is carried out by a separate, pulsed radio system that operates in parallel with an HF radiolink or links. The system consists of a sounder transmitter at one end of a link and a receiver and display at the other end. The transmitter and receiver are clock-controlled with an accuracy of a few milliseconds per week. The sounder transmitter uses a separate antenna whereas the receiver may be multicoupled or share the same antenna with the communications service.

---

*The reader should note that some present-day systems using 120-Hz tone frequency spacing handle up to 22 telegraph channels. 120-Hz spacing is discussed in Chapter 8. Such a system is not suitable for tone diversity.

Sounding is carried out by the transmission of RF pulses on discrete frequencies. The transmission of pulses is a form of frequency scanning. One such equipment now in operation initiates a scan sequence with synchronized clocks at both ends of the link. The transmitter and receiver step in synchronization across the HF band at a rate of 10 frequency channels per second. The pulse train from the sounder transmitter consists of two RF pulses on each channel from the lowest frequency (4 MHz) to the highest (32 MHz) which are radiated by a broadband antenna to the distant end illuminating the ionosphere along the path. The distant end receiver displays the received pulses on an A-scan storage tube on which the horizontal axis is the frequency scale, marked off linearly from 4 to 32 MHz. The vertical scale is marked off in relative time delay in milliseconds.

Under usual circumstances not all the transmitted pulses are detected by the sounder receiver and displayed. Those not received include frequencies below the LUF and above the MUF. The display at the receiver is called an ionogram. The relative time delay is indicative of the number of hops a signal takes to arrive at the distant end. Multipath is evident in the ionogram by signal smearing or stretching. A typical ionogram is shown in Figure 4.14.

It would appear that the equivalent frequency scanning used by ionospheric sounders would cause significant interference to other HF services. Synchronized oblique sounders employ pulses on order of 1 ms in length with a PRF (pulse repetition frequency) of 20 pps. Channel spacing is on the order of 20 times (at midrange) the bandwidth of a typical communication system. It is impossible to hear the sounder on conventional HF receivers unless colocated with the sounder transmitter. The pulse width is considerably shorter than half of the shortest pulse width normally used on HF data circuits (see Chapter 8). Thus chances of interference are minimal.

It has been stated previously in this chapter that one of the most important considerations in HF pertains to frequency selection. Long-term propagation predictions are indispensable for circuit engineering. This involves decisions regarding antenna elevation and gain, diversity, takeoff angle, etc. The day-to-day and hour-by-hour selection of the MUF is the immediate need of the operator. There is considerable variation between actual up-to-the-minute, ionospheric conditions and those predicted 3 months previously. The oblique ionospheric sounder is a tool that gives the operator real time information on ionospheric conditions and may significantly reduce propagation outage on HF.

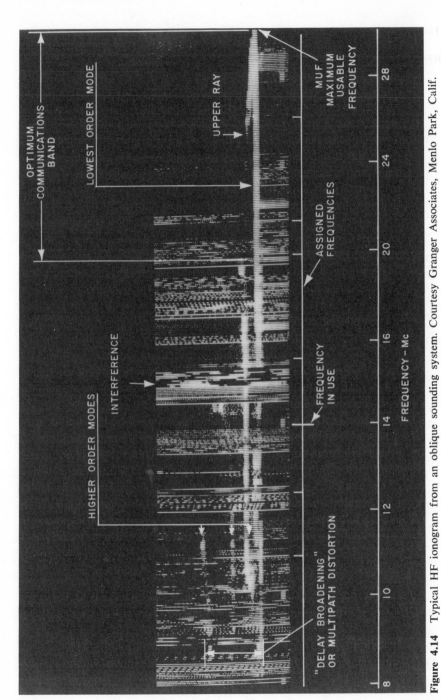

Figure 4.14 Typical HF ionogram from an oblique sounding system. Courtesy Granger Associates, Menlo Park, Calif.

175

REFERENCES AND BIBLIOGRAPHY

1. *Reference Data for Radio Engineers*, 5th ed., Howard W. Sams & Co., Indianapolis.
2. *U.S. Army Technical Manual, TM11.486-6*, "Electrical Communication Systems—Radio," 27 June 1958.
3. *Fundamentals of Single Sideband*, 3rd ed., Collins Radio Company, Cedar Rapids, Iowa.
4. Alfred F. Barghausen et al., *Predicting Long-Term Parameters of HF Sky-wave Telecommunication Systems*, Environmental Science Services Administration, ERL-110-ITS, Boulder, Colo., May 1969.
5. *Modern HF Antennas—Selection and Application*, 3rd ed., Granger Associates, Menlo Park, Calif.
6. *The Use of Ionosphere Sounders to Improve H-F Communications*, Technical Bulletin No. 3, Granger Associates, Menlo Park, Calif., March 1966.
7. *The Application Engineering Manual*, Defense Communications Agency, Washington, D.C., 1965.
8. *Uses of Time and Frequency in H-F Communications*, ITT Communication Systems, Inc., 1962.
9. *MF/HF Communication Antennas*, Addendum No. 1 to *DCA Circ. 330-175-1*, Defense Communications Agency, Washington, D.C., 1966.
10. *Handbook of Engineering, Siting and Installation of HF Rhombic Antennas*, T.O. 31R-1-7, U.S. Department of Defense (USAF), Washington, D.C.
11. *Siting of H-F Facilities*, T.O. 31-1-13, U.S. Department of Defense (USAF), Washington, D.C.
12. *Preliminary Analysis of Adaptive Communications*, Defense Communications Agency, Washington, D.C., 1963.
13. W. Henneberry, *Single Sideband Handbook*, Technical Materiel Corp., Mamaroneck, N.Y., 1963.
14. *SSB Communications*, NavShips 93225, U.S. Government Printing Office, Washington, D.C.
15. *Ionospheric Radio Propagation*, National Bureau of Standards Monograph 80, U.S. Government Printing Office, Washington, D.C.
16. *Handbook for CRPL Predictions based on Numerical Methods of Mapping*, NBS Standard Handbook No. 90, Boulder, Colo., 1962.
17. W. Sichak and R. Adams, "The Radio Path in Communication Systems Planning," North Jersey Section of the IEEE (session 8), Seminar on Overall Communication System Planning, 1964.
18. CCIR, Oslo, 1966, Rec. 371, Rep. 322, Rep. 246-1, Rep 344, Rec. 341, International Telecommunications Union, Geneva.

5 | RADIOLINK SYSTEMS (LINE-OF-SIGHT MICROWAVE)

5.1 INTRODUCTION

Let us define radiolink systems as those that fulfill the following requirements:

1. Signals follow a straight line or "line-of-sight" (LOS) path.
2. Signal propagation is affected by "free-space attenuation" and precipitation.
3. Use of frequencies greater than 150 MHz, thereby permitting transmission of more information per radio frequency (RF) carrier by use of a wider information baseband.
4. Use of angle modulation (i.e., FM or PM) (or spread spectrum and time sharing techniques).

Tables 5-1A and 5-1B show typical assignments for the frequency region through 13 GHz. Radio transmission above 13 GHz is covered in Chapter 10.

A valuable characteristic of line-of-sight transmission is that we can predict the level of a signal arriving at a distant receiver with known accuracy.

5.2 LINK ENGINEERING

Engineering a radiolink system involves the following steps:

1. Selection of sites (radio equipment plus tower locations) that are in "line-of-sight" of each other.

Table 5.1a   Letter Designations for Microwave Bands

Subband	Frequency (GHz)	Wavelength (cm)	Subband	Frequency (GHz)	Wavelength (cm)
	P Band			*X* Band (*continued*)	
	0.225	133.3	*d*	6.25	4.80
	0.390	76.9	*b*	6.90	4.35
			r	7.00	4.29
	L Band		*c*	8.50	3.53
			l	9.00	3.33
	0.390	76.9	*s*	9.60	3.13
p	0.465	64.5	*x*	10.00	3.00
c	0.510	58.8	*f*	10.25	2.93
l	0.725	41.4	*k*	10.90	2.75
y	0.780	38.4			
t	0.900	33.3		*K* Band	
s	0.950	31.6			
x	1.150	26.1		10.90	2.75
k	1.350	22.2	*p*	12.25	2.45
f	1.450	20.7	*s*	13.25	2.26
z	1.550	19.3	*e*	14.25	2.10
			c	15.35	1.95
	S Band		*u* [b]	17.25	1.74
			t	20.50	1.46
	1.55	19.3	*q* [b]	24.50	1.22
e	1.65	18.3	*r*	26.50	1.13
f	1.85	16.2	*m*	28.50	1.05
t	2.00	15.0	*n*	30.70	0.977
c	2.40	12.5	*l*	33.00	0.909
q	2.60	11.5	*a*	36.00	0.834
y	2.70	11.1			
g	2.90	10.3		*Q* Band	
s	3.10	9.67			
a	3.40	8.32		36.0	0.834
w	3.70	8.10	*a*	38.0	0.790
h	3.90	7.69	*b*	40.0	0.750
z [a]	4.20	7.14	*c*	42.0	0.715
d	5.20	5.77	*d*	44.0	0.682
			e	46.0	0.652
	X Band				
				V Band	
	5.20	5.77			
a	5.50	5.45		46.0	0.652
q	5.75	5.22	*a*	48.0	0.625
y [a]	6.20	4.84	*b*	50.0	0.600

Table 5.1a (Continued)

Subband	Frequency (GHz)	Wavelength (cm)	Subband	Frequency (GHz)	Wavelength (cm)
	V Band (*continued*)			*W* Band	
c	52.0	0.577		56.0	0.536
d	54.0	0.556		100.0	0.300
e	56.0	0.536			

Source: Ref. 1.
[a] C band includes S_z through X_y (3.90–6.20 GHz).
[b] K_1 band includes K_u through K_q (15.35–24.50 GHz).

Table 5.1b Microwave Radio Link Frequency Assignments for Fixed Service (i.e., Point-to-Point Communications and Some Other Nonmobile Applications)

450–470 MHz	5925–6425 (7250) MHz
890–960 MHz	7300–8400 MHz
1710–2290 MHz	10550–12700 MHz
2550–2690 MHz	14400–15250 MHz
3700–4200 MHz	

SPECIFIC ASSIGNMENTS FOR THE UNITED STATES

Band	GHz	Band	GHz
Military	1.710–1.850	Common carrier—space	5.925–6.425
Operational fixed	1.850–1.990	Operational fixed	6.575–6.875
STL[a]	1.990–2.110	STL[a]	6.875–7.125
Common carrier	2.110–2.130	Military	7.125–7.750
Operational fixed	2.130–2.150	Military	7.750–8.400
Common carrier	2.160–1.280	Common carrier	10.7–11.7
Operational fixed	2.180–2.200	Operational fixed	12.2–12.7
Operational fixed (TV only)	2.500–2.690	CATV-STL (CARS)[a, b]	12.7–12.95
Common carrier—space	3.700–4.200	STL[a]	12.95–13.2
Military	4.400–5.000	Military	14.4–15.25

[a] STL = Studio transmitter link.
[b] CATV = Community antenna television.

2. Selection of an operational frequency band from those set forth in Tables 5.1A and 5.1B, considering radio frequency interference environment and legal restraints.

3. Development of path profiles to determine radio tower heights. If tower heights exceed a certain economic limit, then Step 1 must be repeated, bringing the sites closer together or reconfiguring the path, usually along another route.

Making a profile, taking into consideration that microwave energy is

- Attenuated or absorbed by solid objects.
- Reflected from flat conductive surfaces such as water and sides of metal buildings.
- Diffracted around solid objects.
- Refracted or bent by the atmosphere. Often the bending is such that the beam may be extended beyond the optical horizon.

4. Path calculations, after setting a propagation reliability expressed as a percentage of time the received signal will be above a certain threshold level. Often this level is the FM improvement threshold of the FM receiver. To this level a margin is set for signal fading under all anticipated climatic conditions.

5. Making a path survey to ensure correctness of Step 1–4. It also provides certain additional planning information vital to the installation project or bid.

6. Equipment configuration to achieve the fade margins set in Step 4 most economically.

7. Establishment of a frequency plan and necessary operational parameters.

8. Installation.

9. Beam alignment, equipment line-up, checkout, and acceptance by a customer.

Reference will be made, where applicable, to these steps so that the reader will be exposed to practical radiolink problems.

Line-of-sight (LOS) microwave is often used synonymously with radiolink. Radiolink is preferred in this text because

- Microwave is harder to define, even LOS microwave.
- The term "radiolink" is more universal, particularly outside the United States, for the material that this chapter covers.

5.3 PROPAGATION

5.3.1 Free-Space Loss

Consider a signal traveling between a transmitter at A and a receiver at B:

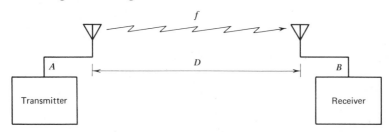

The distance between antennas is D and the frequency of transmission is f. Let D be in kilometers and f in megahertz; the the free-space loss in decibels may be calculated with the following formula:

$$L_{dB} = 32.44 + 20 \log D + 20 \log f \qquad (5.1a)$$

If D is in statute miles then

$$L_{dB} = 36.6 + 20 \log D + 20 \log f \qquad (5.1b)$$

(all logarithms are to the base 10).

Suppose the distance separating A and B were 40 km ($D = 40$). What would the free-space path loss be at 6 GHz ($f = 6000$)?

$$
\begin{aligned}
L &= 32.44 + 20 \ \log 40 + 20 \ \log 6 \times 10^3 \\
&= 32.44 + 20 \times 1.6021 + 20 \times 3.7782 \\
&= 32.44 + 32.042 + 75.564 \\
&= 140.046 \ \text{dB}
\end{aligned}
$$

Let us look at it another way. Consider a signal leaving an isotropic antenna.* At one wavelength (1λ) away from the antenna, the free-space attenuation is 22 dB. At two wavelengths (2λ) it is 28 dB; at four wavelengths it is 34 dB. Every time we double the distance, the free-space attenuation (or loss) is 6 dB greater. Likewise, if we halve the distance, the attenuation (or loss) decreases by 6 dB.

Figure 5.1 relates free-space path loss to distance at seven discrete frequencies between 450 and 14,000 MHz.

---

* An isotropic antenna is an ideal antenna with a reference gain of 1 (0 dB). It radiates equally in all directions (i.e., perfectly omnidirectional).

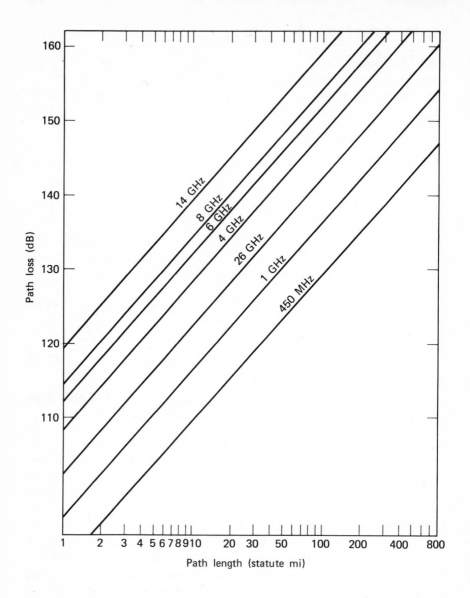

Figure 5.1 Path loss in decibels related to path length in statute miles for seven discrete radiolink frequencies.

5.3.2 Bending of Radio Waves above 100 MHz from Straight-Line Propagation

Radio waves traveling through the atmosphere do not follow true straight lines. They are refracted or bent. They may also be diffracted.

The velocity of an electromagnetic wave is a function of the density of the media through which it travels. This is treated by Snell's law, which provides a valuable relationship for an electromagnetic wave passing from one medium to another (i.e., from an air mass with one density to an air mass with another density). It states that the ratio of the sine of the angle of incidence to the sine of the angle of refraction is equal to the ratio of the respective velocities in the media. This is equal to a constant which is the refractive index of the second medium relative to the first medium.

The absolute refractive index of a substance is its index with respect to a vacuum and is practically the same value as its index with respect to air. It is the change in the refractive index that determines the path of an electromagnetic wave through the atmosphere, or how much the wave is bent from a straight line.

If radiowaves above 100 MHz traveled a straight line, the engineering of line-of-sight microwave (radiolink) systems would be much easier. We could then accurately predict the height of the towers required at repeater and terminal stations and exactly where the radiating device on the tower should be located (Steps 2–3, Section 5.2). Essentially what we are dealing with here, then, is a method to determine the height of a microwave radiator (i.e., an antenna or other radiating device) to permit reliable radiolink communication from one location to another.

To determine tower height, we must establish the position and height of obstacles in the path between stations with which we want to communicate by radiolink systems. To each obstacle height, we will add earth bulge. This is the number of feet or meters an obstacle is raised higher in elevation (into the path) owing to earth curvature or earth bulge. The amount of earth bulge in feet at any point in a path may be determined by the formula

$$h = 0.667 \, d_1 d_2 \tag{5.2}$$

where d_1 is the distance from the near end of the link to the point (obstacle location), and d_2 is the distance from the far end of the link to the obstacle location.

Let us make the equation more useful by making it directly applicable to the problem of ray bending. As the equation is presented above, the ray is unbent or a straight line.

Atmospheric refraction may cause the ray beam to be bent toward the earth or away from the earth. If it is bent toward the earth, it as as if we shrank earth bulge or lowered it from its true location. If the beam is bent away from the earth, it is as if we expanded earth bulge or raised it up toward the beam above its true value. This lowering or raising is handled mathematically by adding a factor K to the earth bulge equation. It now becomes

$$d_{\text{ft}} = \frac{0.667\, d_1 d_2}{K} \qquad (d_1 \text{ and } d_2 \text{ in mi}) \qquad (5.3a)$$

$$d_{\text{m}} = \frac{0.078\, d_1 d_2}{K} \qquad (d_1 \text{ and } d_2 \text{ in m}) \qquad (5.3b)$$

and

$$K = \frac{\text{effective earth radius}}{\text{true earth radius}} \qquad (5.4)$$

If the K factor is greater than 1, the ray beam is bent toward the earth, which essentially allows us to shorten radiolink towers. If K is less than 1, the earth bulge effectively is increased, and the path is shortened or tower

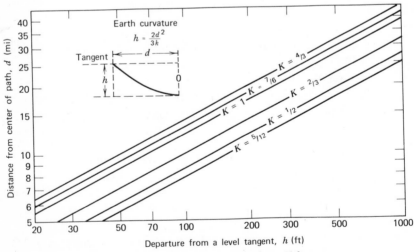

Figure 5.2 Earth curvature or earth bulge for various K factors.

height must be increased. Figure 5.2 gives earth curvature for various values of K (ft).

Many texts on radiolinks refer to normal refraction, which is equivalent to a K factor of $\frac{4}{3}$ or 1.33. It follows a rule of thumb that applies to refraction in that a propagated wave front (or beam) bends toward the region of higher density, that is, toward the region having the higher index of refraction. Older texts insisted that the K factor should nearly always be $\frac{4}{3}$. Care should be taken when engineering radiolinks that the $\frac{4}{3}$ theory (standard refraction or normal refraction) not be accepted "carte blanche" on many of the paths likely to be encountered. However, $K = \frac{4}{3}$ may be used for gross planning of radiolink systems. Figure 5.3 may be used to estimate tower heights with $K = \frac{4}{3}$ for smooth earth paths (no obstacles besides midpath earth bulge).

Another factor must be added to obstacle height to obtain an effective obstacle height. This is the Fresnel clearance. It derives from electromagnetic wave theory that a wave front (which our ray beam is) has expanding properties as it travels through space. These expanding properties result in reflections and phase transitions as the wave passes over an obstacle. The result is an increase or decrease in received signal level. The amount of additional clearance that must be allowed to avoid problems with the Fresnel phenomenon (diffraction) is expressed in Fresnel zones. Optimum clearance of an obstacle is accepted as 0.6 of the first Fresnel zone radius. The first Fresnel zone radius in feet may be calculated with the following formula:

$$R = \frac{13.58\sqrt{\lambda d_1 d_2}}{D} \tag{5.5a}$$

where λ = wavelength of signal (ft)

d_1 = distance from transmitter to path obstacle (statute mi)

d_2 = distance from path obstacle to receiver (statute mi)

$D = d_1 + d_2$ (total path length in statute mi)

To determine the first Fresnel zone radius when using metric units:

$$R_m = \frac{31.6\sqrt{\lambda d_1 d_2}}{D} \tag{5.5b}$$

where λ = wavelength of signal (m)

d_1, d_2, and D the same as in formula 5.5a but in kilometers

R_m in meters

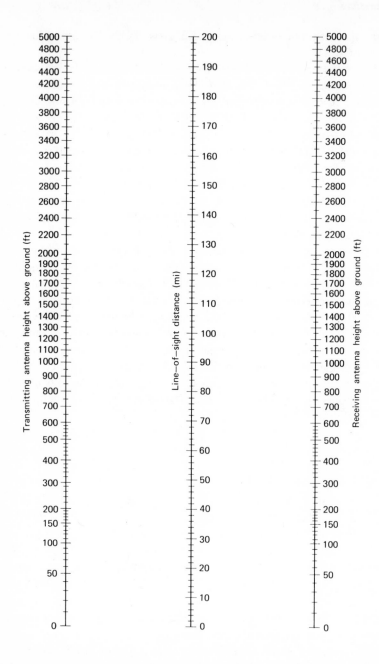

Figure 5.3 Line-of-sight distance for smooth spherical earth with $K = \frac{4}{3}$; distance in statute miles, tower heights in feet.

186

5.3.3 Path Profiling—Practical Application (Step 3)

After tentative repeater sites have been selected, path profiles are plotted on rectangular graph paper. Obstacle information is taken from topographical (topo) maps. The best topo maps are 1:7500, However, 1:15000 can serve in a pinch. Both types of maps are readily available for the United States (lower 48) and most of Canada. Many other locations can provide topo maps no better than 1:25000. Profiles taken from these maps are often inaccurate. The 1:7500 maps must often be pasted or taped together so that two sites to be interconnected by radiolink are on the same map sheet.

Set up the rectangular graph paper to express obstacle height on the Y scale in feet or meters versus obstacle location relative to the two sites that have tentatively been selected. One convenient scale to use for X is 2 mi to the inch or 2 km to the centimeter. More accuracy may be achieved by pasting graph paper together and using 1:1 (i.e., 1 in. equivalent to 1 mi or 1 cm equivalent to 1 km).

For most applications 1 in. $\simeq$ 100 ft (1 cm $\simeq$ 10 m) is satisfactory for the Y scale. When there is more than 1000-ft (300-m) elevation change in a link, 200 or even 500 ft to the inch may be required for obstacle height. All heights are plotted above mean sea level (MSL).

Draw a straight line on the map connecting the two adjacent radiolink sites. Carefully trace with your eye or thumb down the line from one site to the other marking all obstacles or obstructions and possible points of reflection such as bodies of water, marshes, etc., assigning consecutive letters to the obstacles.

Plot the horizontal location of each point on the graph paper. Mark path midpoint, which is the point of maximum earth bulge and should be marked as an obstacle. Determine the K factor by one of the following methods:

1. Refer to a sea level refractivity profile chart (see Figure 5.4). Select the appropriate N_0 for the area of interest. Apply the N_0 selected to Figure 5.5, with the midpath elevation and refractivity N_0. Read off the corresponding K factor.

2. Lacking a refractivity contour chart for the area, such as shown in Figure 5.4, plot using three K factors, 1.33, 1.0, and 0.5. A later field survey will help to decide which factor is valid. For instance in coastal regions, over-water paths, and damp regions, assume the lowest value. In most dry regions (nondesert) the so-called normal value (1.33) may be assumed. Table 5.2 will help as a guide to determining K factor.

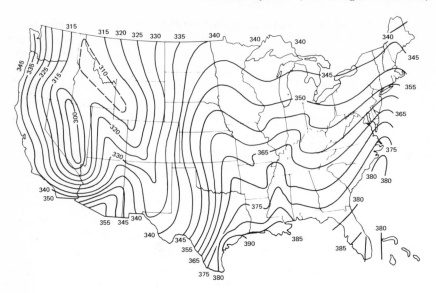

Figure 5.4 Sea level refractivity (N_0) index for the continental United States—maximum for worst month (August).

Table 5.2   K Factor Guide[a]

	Propagation Conditions				
	Perfect	Ideal	Average	Difficult	Bad
Weather	Standard atmosphere	No surface layers or fog	Substandard, light fog	Surface layers, ground fog	Fog moisture over water
Typical	Temperate zone, no fog, no ducting, good atmospheric mix day and night	Dry, mountainous, no fog	Flat, temperate, some fog	Coastal	Coastal, water, tropical
K factor	1.33	1–1.33	0.66–1.0	0.66–0.5	0.5–0.4

[a] For 99.9–99.99 path reliability.

For each obstacle point compute d_1, the distance to one repeater site, and d_2, the distance to the other. Compute the equivalent earth curvature for each point with the equation (equation 5.3):

$$\text{E.C.} = \frac{0.667\,(d_1 d_2)}{K}$$

Compute Fresnel zone clearance using Figures 5.6 and 5.7. Figure 5.6 gives the Fresnel clearance for midpath for various bands of frequencies. For other obstacle points compute percentage of total path length. For instance, on a 30-mi path, midpoint is 15 mi (50%). However, a point 5 mi from one site is 25 mi from the other, or represents 5/30 and 25/30, or 1/6 and 5/6. Converted to percentage, $1/6 = 16.6\%$ and $5/6 = 83\%$. Apply this percentage on the X axis of Figure 5.7. In this case 16.6% or

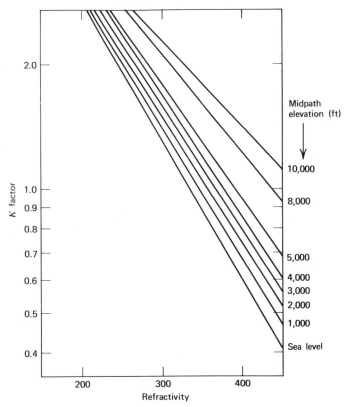

Figure 5.5 K factor scaled for midpath elevation above mean sea level.

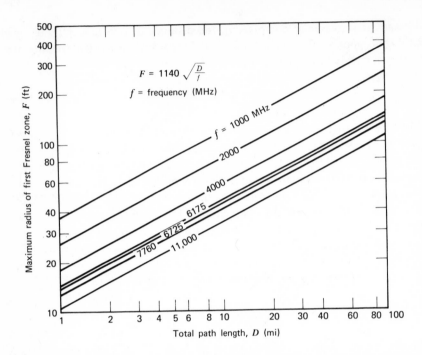

$$F = 1140 \sqrt{\frac{D}{f}}$$

f = frequency (MHz)

f = 1000 MHz

2000

4000

6175

6725

7760

11,000

Maximum radius of first Fresnel zone, F (ft)

Total path length, D (mi)

Figure 5.6 Midpath Fresnel clearance for first Fresnel zone. 0.6 of this value is used in the calculations or the value is calculated in conjunction with Figure 5.7. Courtesy GTE Lenkurt Incorporated, San Carlos, Calif.

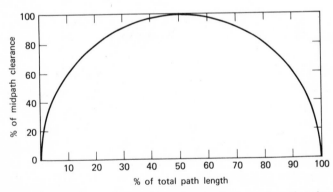

% of midpath clearance

% of total path length

Figure 5.7 Conversion from midpath Fresnel zone clearance for other-than midpath obstacle points.

83% is equivalent to 76% of midpath clearance. If midpath Fresnel clearance were 40 ft, only 30.4 ft of Fresnel clearance would be required at the 5-mile point.

Set up a table on the profile chart as follows:

Obstacle Point	d_1	d_2	E.C.	Fresnel
A				
B				
C				
D				
E				

On the graph paper plot the height above sea level of each obstacle point. To this height add the E.C. (earth curvature), the F (Fresnel zone clearance), and if tree growth exists, add 40 ft for trees and 10 ft for additional growth. If undergrowth exists assign 10 ft for vegetation. A path survey will confirm or deny these figures or will allow adjustment.

Minimum tower heights may now be determined by drawing a straight line from site to site through highest obstacle points. These often cluster around midpath. Figure 5.8 is a hypothetical profile exercise.

5.3.4   Reflection Point

From the profile, possible reflection points may be obtained. The objective is to adjust tower heights such that the reflection point is adjusted to fall on land area where the reflected energy will be broken up and scattered. Bodies of water and other smooth surfaces cause reflections which are undesirable. Figure 5.9 will assist in adjusting the reflection point. It uses a ratio of tower heights, h_1/h_2, and the shorter tower height is always h_1. The reflection area lies between a K factor of grazing ($K = 1$) and a K factor of infinity. The distance expressed is always from h_1, the shorter tower. By adjusting the ratio h_1/h_2 the reflection point can be moved. The objective is to be sure the reflection point does not fall on an area of smooth terrain or on water, but rather on land area where the reflected energy will be broken up or scattered (i.e., by wooded areas, etc.).

For a highly reflective path, space diversity operation may be desirable to minimize the effects of multipath reception (see Section 5.6).

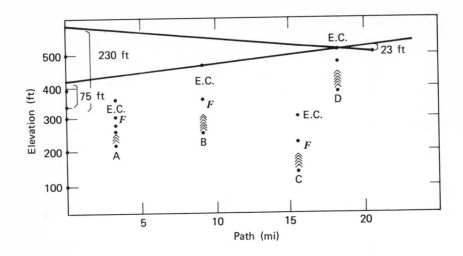

Figure 5.8 Practice path profile.

		d_1	d_2	F (Fresnel) (ft)	E.C. (ft)
Tree conditions: 40 + 10 ft growth	A	3.5	19.0	30	49
Frequency band: 6 GHz	B	10	12.5	41	91.7
Midpath Fresnel (0.6)	C	17	5.5	36	68.6
= 42 ft	D	20	2.5	25.2	36.6

x in miles, *y* in feet   Assume that $K = 0.9$.

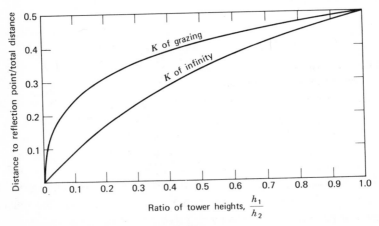

Figure 5.9 Calculation of reflection points.

5.4 PATH CALCULATIONS (Step 4)

5.4.1 General

The next step in path engineering is to carry out path calculations. Essentially, this entails the determination of equipment parameters and configurations to meet a minimum performance requirement. Such performance requirements are usually related to noise in an equivalent voice channel and/or related to signal-to-noise ratio. Either way, the requirement is given for a percentage of time. For a single path this may be stated as 99, 99.9, or 99.99% of the time said performance exceeds a certain minimum. This is often called propagation reliability. Table 5.3 states reliability percentages versus outage time, outage time being the time that the requirement will not be met.

Table 5.3 Reliability Versus Outage Time

Reliability (%)	Outage Time (%)	Outage Time Per Year	Per Month (avg.)	Per Day (avg.)
0	100	8760 h	720 h	24 h
50	50	4380 h	360 h	12 h
80	20	1752 h	144 h	4.8 h
90	10	876 h	72 h	2.4 h
95	5	438 h	36 h	1.2 h
98	2	175 h	14 h	29 min
99	1	88 h	7 h	14 min
99.9	0.1	8.8 h	43 min	1.4 min
99.99	0.01	53 min	4.3 min	8.6 s
99.999	0.001	5.3 min	26 s	.86 s

5.4.2 Basic Path Calculations

FM RECEIVER THRESHOLD—THE STARTING POINT

For path calculations the starting point is the FM receiver. The following question must be answered: what signal level coming into the receiver will give the desired performance?

For the basic approach the case must be simplified. Figure 5.10 is a curve comparing input carrier level to output signal-to-noise ratio of a

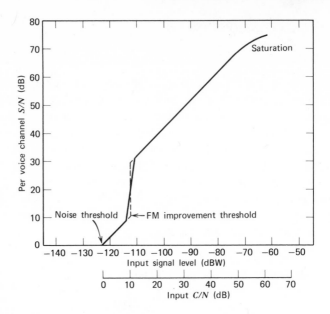

Figure 5.10 Input C/N in decibels versus output S/N (per voice channel) for a typical FM radiolink receiver.

typical radiolink FM receiver. Assume that the noise threshold for this receiver is at -122.5 dBW. Thus at this point the input carrier-to-noise ratio is zero. At -112.5 dBW input carrier level, the carrier-to-noise (C/N) ratio is 10 dB. Carrier-to-noise ratio in decibels is shown below the input carrier level in Figure 5.10.

The relationship between the input carrier-to-noise ratio of an FM receiver and the output signal-to-noise ratio is as follows. At noise threshold, for every 1-dB increase of carrier-to-noise ratio, the signal-to-noise ratio of the output increases approximately 1 dB. This occurs up to a carrier-to-noise ratio of 10 dB or a little greater (11 or 12 dB), when a "capture effect" takes place and the output signal-to-noise ratio suddenly jumps to 30 dB. This point of "capture" is called the FM improvement threshold and is shown in Figure 5.10. Beyond this point the signal-to-noise ratio at the receiver output improves again by 1 dB for every increase of 1 dB of input carrier-to-noise ratio, up to a point where compression starts to take effect (saturation). For instance if the input carrier-to-noise ratio were 15 dB, we might expect the output signal-to-noise ratio to be approximately 35 dB. This assumes that the FM system has been "adjusted" properly (e.g., there is sufficient deviation to effect FM improvement; see Section 5.5.3).

For many radiolink systems, an input carrier-to-noise ratio of 10 dB is used as a starting point for path calculations. The reason is obvious, for it is at this point where we start to get the improved signal-to-noise ratio. For future discussion in this chapter, we will use the same point of departure. We can calculate this point. It is the noise threshold level, calculated in Chapter 1, plus 10 dB. FM improvement threshold is

$$\text{Input level (dBW)} = 10 \log kTB_{\text{if}} + NF_{\text{dB}} + 10 \text{ dB} \qquad (5.6)$$

where T = noise temperature ($°K$)

B_{if} = noise bandwidth (bandwidth of the IF) (Hz)

K = Boltzman's constant (1.38×10^{-23} W-s/deg)

NF = receiver noise figure (dB)

Simplified for an uncooled receiver (i.e., T = 290°K or 62°F);

$$\text{Input level (dBW)} = -204 \text{ dBW} + NF + 10 \text{ dB} + 10 \log B_{\text{if}} \qquad (5.7)$$

Figure 5.11 gives values for threshold for several receiver noise figures. However, threshold may be computed rapidly by directly subtracting the receiver noise figure, the 10 dB, and then working with the bandwidth. If the bandwidth were only 10 Hz, we would subtract (algebraically add) 10 dB; if it were 20 Hz, we would subtract 13 dB; 100 Hz, 20 dB; 1000 Hz, 30 dB; 10 MHz, 70 dB; 20 MHz, 23 dB; etc.

Consider an uncooled receiver with a 14-dB noise figure and 4-MHz IF bandwidth. What would the FM improvement threshold be?

$$\text{Threshold}_{\text{dBW}} = -204 + 14 + 10 + 66$$
$$= -114 \text{ dBW}$$

This is the necessary input level to reach an FM improvement threshold. The required IF bandwidth may be estimated by Carson's rule, which states:

$$B_{\text{if}} = 2 \text{ (highest modulating baseband frequency}$$
$$+ \text{ peak FM deviation)}$$

Consider the following example: a video signal with a 4.2-MHz baseband modulates an FM transmitter with a peak deviation of 5 MHz. What is B_{if}?

$$B_{\text{if}} = 2 \text{ (4.2 MHz + 5 MHz)}$$
$$= 18.4 \text{ MHz}$$

FM improvement theshold may be calculated when given the peak deviation and the bandwidth of the information baseband as well as the receiver noise figure (or noise temperature; see Section 1.5.9).

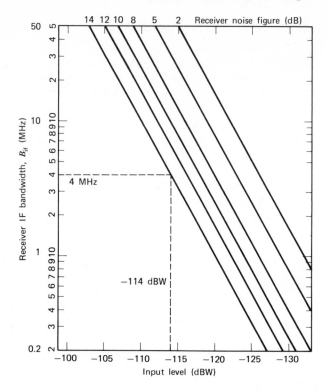

Figure 5.11 FM Improvement threshold for several receiver noise figures.

For the most basic path calculation, consider a receiver with an FM improvement theshold of -114 dBW, a free-space path loss of 140 dB, an isotropic antenna* at both ends, and lossless transmission lines. What would the transmitter output have to be to provide a -114 dBW input level to the receiver?

$$\text{Output}_{dBW} = -114 + 140 = +26 \text{ dBW}$$
$$+26 \text{ dBW} \simeq 400 \text{ W}$$

For this case both antennas have unity gain and the transmission lines are lossless. Now extend the example for 2.0-dB line losses and 20-dB antenna gain at each end.

$$\text{Output}_{dBW} = -114 \text{ dBW} + 2.0 - 20 \text{ dB} + 140 - 20 \text{ dB} + 2.0$$
$$= -10 \text{ dBW} = 0.1 \text{ W}$$

* Unrealistic, but useful for discussion.

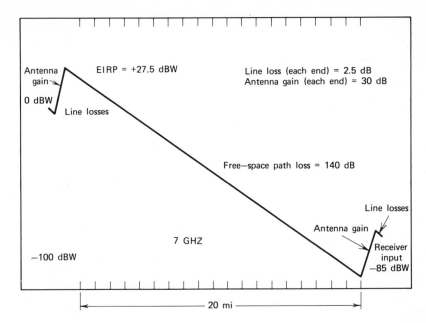

Figure 5.12 Radiolink gains and losses (simplified). Transmitter output $= 0$ dBW.

Path calculations, therefore, deal with adding gains and losses to arrive at a specified system performance.

As the discussion progresses we shall see many means to attain gain and the various ways losses occur. Figure 5.12 shows graphically the gains and losses as a radiolink signal progresses from transmitter to receiver. Section 5.8 deals with detailed path calculations.

5.4.3 The Mechanism of Fading—An Introductory Discussion

GENERAL

Up to this point we have considered that the received signal level remains constant. Truly on most paths, particularly on shorter ones, at lower frequencies, much of the time this holds true. When a receive level varies from the free-space calculated level for a given far-end transmitter output, the result is called fading. Fading due to propagation mechanisms involves refraction, reflection, diffraction, scattering, focusing attenuation, and other miscellaneous causes.† Such factors, when associated with fading,

† Above 10 GHz rainfall in also an important factor, see Section 5.11 and Chapter 10.

relate to certain conditions classified as meteorological phenomena and terrain geometry. The radiolink engineer must be alert to these factors when planning specific links and during site survey phases.

The following paragraphs cover the two general types of fading, multipath and power.

MULTIPATH FADING

This type of fading is due to interference between a direct wave and another wave, usually a reflected wave. The reflection may be from the ground or from atmospheric sheets or layers. Direct path interference may also occur. It may be caused by surface layers of strong refractive index gradients or horizontally distributed changes in the refractive index.

Multipath fading may display fades in excess of 30 dB for periods of seconds or minutes. Typically this form of fading will be observed during quiet, windless, and foggy nights, when temperature inversion near the ground occurs and there is not enough turbulence to mix the air. Thus stratified layers, either elevated or ground based, are formed. Two-path propagation may also be due to specular reflections from a body of water, salt beds, or flat desert between the transmitting and receiving antennas. Deep fading of this latter type usually occurs in daytime on over-water paths or other such paths with a high ground reflection. Vegetation or other "roughness" found on most radiolink paths breaks up the reflected components rendering them rather harmless. Multipath fading at its worst is independent of obstruction clearance, and its extreme condition approaches a Rayleigh distribution.

POWER FADING

Dougherty (Ref. 26) defines power fading as a

. . . partial isolation of the transmitting and receiving antennas because of

- Intrusion of the earth's surface or atmospheric layers into the propagation path (earth bulge or diffraction fading).
- Antenna decoupling due to variation of the refractive index gradient.
- Partial reflection from elevated layers interpositioned between terminal antenna elevations.
- "Ducting" formations containing only one of the terminal antennas.
- Precipitation along the propagation path . . . (see Section 5.11).

Power fading is characterized by marked decreases in free-space signal level for extended time periods. Diffraction may persist for several hours with fade depths of 20–30 dB.

5.5 FM RADIOLINK SYSTEMS

5.5.1 General

Figures 5.13*a* and 5.13*b* are simplified block diagrams of a radiolink transmitter and a radiolink receiver. The type of modulation used is FM. The input information to the transmitter (baseband) and output from the receiver is considered to be groupings of nominal 4-kHz telephone channels in an SSBSC frequency division multiplex configuration. This type of multiplexing is discussed in Chapter 3. Orderwire plus alarm information is inserted from 300 Hz to 12 kHz for many installations accepting an FDM multitelephone channel input. Thus the composite baseband will be made up of the following:

300 Hz–12 kHz	Orderwire/alarms
12 kHz–60 kHz	A 12-channel group of FDM multichannel information
or	
300 Hz–12 kHz	Orderwire/alarms
60 kHz–*	Supergroups/mastergroups of multichannel information
300 Hz–12 kHz	Orderwire/alarms
312 kHz–*	Supergroups/master groups of multichannel information

Other systems use a 12-kHz spectrum with an SSB signal at 3.25 MHz for alarms and orderwire. Video plus program channel subcarriers is another information input which will be discussed later. The orderwire/alarms input for video is necessarily inserted above the video in the baseband.

5.5.2 Preemphasis-Deemphasis

The output characteristics of an FM receiver without system preemphasis-deemphasis, with an input FM signal of a modulation of uniform amplitude, are such that it has linearly increasing amplitude with increasing baseband frequency. This is shown in Figure 5.14. Note the ramplike or triangular noise of the higher frequency baseband components. The result is decreasing signal-to-noise ratio with increasing baseband frequency. The desired

---

* See Chapter 3 for specific baseband makeup.

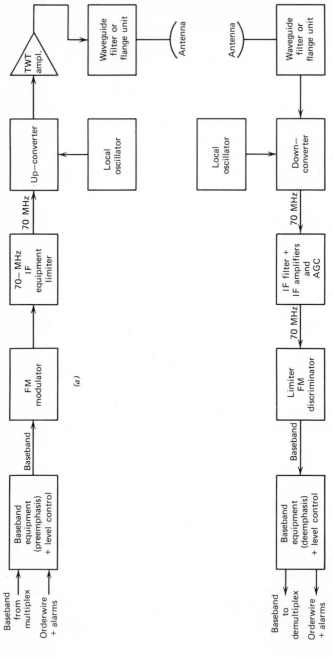

Figure 5.13 (a) Typical radiolink terminal transmitter; (b) typical radiolink terminal receiver.

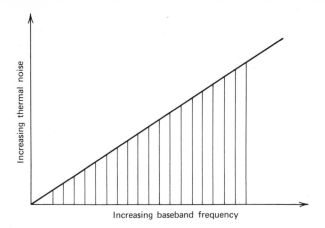

Figure 5.14 Sketch of increasing thermal noise from the output of an FM receiver in a system without preemphasis/deemphasis.

receiver output is a constant signal-to-noise ratio across the baseband. Preemphasis at the transmitter and deemphasis at the receiver achieve this end.

Preemphasis is accomplished by increasing the peak deviation during the FM modulation process for higher baseband frequencies. This increase of peak frequency deviation is done in accordance with a curve designed to effect compensation for the ramplike noise at the FM receiver output. CCIR has recommended standardization on the curve shown in Figure 5.15 for multichannel telephony (CCIR Rec. 275-2, New Delhi, 1970), and for video transmission a preemphasis curve may be found in CCIR Rec. 405-1 (Figure 5.16).

The preemphasis characteristic is achieved by applying the modulating baseband to a passive network that "forms" the input signal. At the receive end after demodulation, the baseband signal is applied to a de-emphasis network which restores the baseband to its original amplitude configuration. This is shown diagrammatically in Figure 5.17.

5.5.3 The FM Transmitter

The combined baseband (orderwire and FDM multiplex line frequency) frequency modulates a wave. The modulated wave is at IF or converted to IF, and thence up-converted to the output frequency. The up-converted

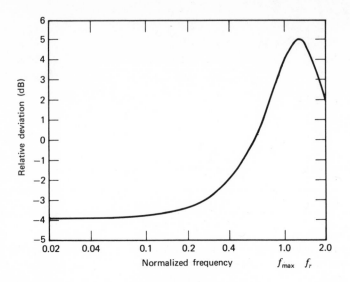

Figure 5.15 Preemphasis characteristic for multichannel telephony (CCIR Rec. 275). Courtesy International Telecommunications Union—CCIR.

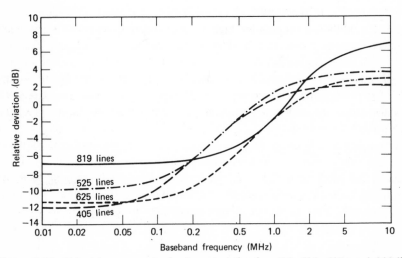

Figure 5.16 Television preemphasis characteristics for 405; 525; 625; and 819-line systems. 0 dB corresponds to a deviation of 8 MHz for a 1 volt peak-to-peak signal (CCIR Rec. 405). Courtesy International Telecommunications Union—CCIR.

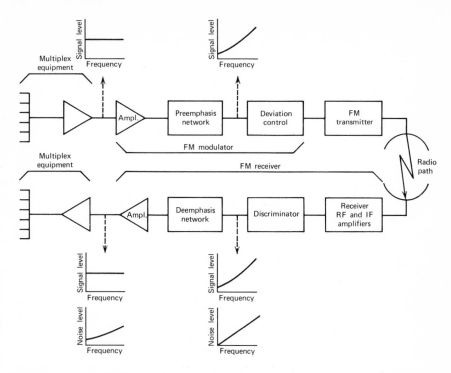

Figure 5.17 Simplified block diagram of an FM radiolink system showing the effects of preemphasis on the signal and thermal noise.

signal may be applied directly to the antenna system for radiation or amplified further. The additional amplification is usually carried out by a TWT for radiolink systems. Outside of the TWT, modern radiolink transmitters are all solid state. However, some equipment is still available in which the input signal directly modulates a klystron whose output is radiated.

Radiolink transmitter outputs have been fairly well standardized as follows:

0.1 W	−10 dBW	(usually for transmitters operating at higher frequencies)
1.0 W	0 dBW	
10 W	+10 dBW	(for transmitters operating at lower frequencies or those with a TWT final amplifier)

The percentage of modulation or modulation index is a most important parameter for radio transmitters. The percentage of modulation is used to describe the modulation in AM transmitters. Modulation index is the equivalent parameter for FM transmitters. The modulation index for frequency modulation is

$$M = \frac{F_d}{F_m} \tag{5.8}$$

where F_d = frequency deviation
$\quad F_m$ = maximum significant modulation frequency

Keep in mind that frequency deviation is a function of the input level to the transmitter—the modulation input.

One of the primary advantages of FM over other forms of modulation is the improvement of output signal-to-noise ratio for a given input as illustrated in Figure 5.18 (also see Figure 5.10). However, this advantage

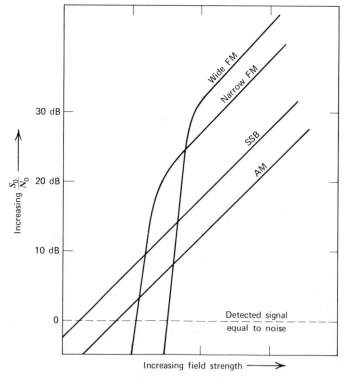

Figure 5.18 The advantage of FM over AM. Relative signal-to-noise ratios as a function of field strength. Courtesy Institute of Telecommunication Sciences, Office of Telecommunications, U.S. Department of Commerce.

is accomplished only if the signal level into the receiver is greater than FM improvement threshold (see Section 5.4.2).

If the modulation index is in excess of 0.6, FM systems are superior to AM, and the rate of improvement is proportional to the square of the modulation index (*NBS Tech. Note 103*).

An important parameter for an FM transmitter is its deviation sensitivity, which is usually expressed in volts per megahertz. Suppose a deviation sensitivity were given as 0.05 V rms/MHz; then if 0.05-V signal level appeared at the transmitter input, the output wave would deviate 1 MHz (or ± 1 MHz above and below the carrier).

Another important parameter given for FM wide band transmitters is the rms deviation per channel at test tone level. Table 5.4 relates channel capacity and rms deviation.

Table 5.4   CCIR Recommended Deviations

Voice Channel Capacity	Rms Deviation per Channel at Test-Tone Level (kHz)
120	50, 100, 200
300	200
600	200
960	200
1800	140 (provisional)

Source: CCIR Rec. 404-2.

For systems using SSBSC FDM basebands, peak deviation, D, may be calculated as follows:

$$D = 4.47 \, d \left(\log^{-1} \frac{-15 + 10 \log N}{20} \right) \qquad (5.9a)$$

(N is 240 or more)

or

$$D = 4.47 \, d \left(\log^{-1} \frac{-1 + 4 \log N}{20} \right) \qquad (5.9b)$$

(N is between 12 and 240)

where $D = $ peak deviation (kHz)
$d \ = $ per channel rms test-tone deviation (kHz)
$N = $ number of SSBSC voice channels in the system

(See Section 3.4.4 for more explanation of the loading of FDM systems.)

Example   A 300-channel radiolink system would have a 200 kHz/channel rms deviation according to Table 5.4. Then

$$D = (4.47)\,(200)\left(\log^{-1}\frac{9.77}{20}\right) = 2753 \text{ kHz}$$

The question arises regarding how much deviation will optimize FM transmitter operation. We know that as the input signal to the transmitter is increased, the deviation is increased. Again, as the deviation increases, the FM improvement threshold becomes more apparent. In effect we are trading off bandwidth for thermal noise improvement. However, with increasing input levels, the intermodulation noise of the system begins to increase. There is some point of optimum input to a wide band FM transmitter where the thermal noise improvement in the system has been optimized and where the intermodulation noise is not excessive. This concept is shown in Figure 5.19.

Frequency stability requirements of FM systems are far less severe than for equivalent SSB systems (see Chapter 4); $\pm\,0.00.1\%$/month is usually sufficient.

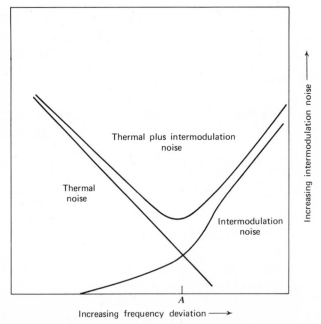

Figure 5.19   Optimum setting for deviation shown by the letter A.

5.5.4   The Antenna System

GENERAL

For conventional radiolinks the antenna subsystem offers more room for trade-off to meet minimum system requirements than any other subsystem. Basically the antenna system looking outward from a transmitter must have

- Transmission line (waveguide or coaxial line).
- Antenna: a reflecting surface or device.
- Antenna: a feed horn or other feeding device.

In addition the antenna system may have

- Circulators.
- Directional couplers.
- Phasers.
- Passive reflectors.
- Radomes.

ANTENNAS

Below 700 MHz antennas used for point-to-point radiolink systems are often a form of Yagi and are fed with coaxial transmission line. Above 700 MHz some form of parabolic reflector-feed arrangement is used. The 700 MHz is no hard and fast dividing line. Above 2000 MHz the transmission line is usually waveguide. The same antenna is used for transmission and reception (see the "diversity" subsection below).

An important antenna parameter is radiation efficiency. Assuming no losses, the power radiated from an antenna would be equal to the power delivered to the antenna. Such power is equal to the square of the rms current flowing on the antenna times a resistance, called the radiation resistance:

$$P = I_{rms}^2 R \tag{5.10}$$

where P = radiated power (W)
R = radiation resistance (Ω)
I = current (A)

In practice, all the power delivered to the antenna is not radiated in space. The radiation efficiency is defined as the ratio power radiated to the total power delivered to an antenna.

To derive a more realistic equation to express power, we divide the resistance, which we shall call the terminal resistance, into two component parts:

$$R = \text{radiation resistance}$$
$$R_1 = \text{equivalent terminal loss resistance}$$

so that

$$P = I_{rms}^2(R + R_1) \tag{5.11}$$

and

$$\text{Radiation efficiency } (\%) = \frac{R}{R + R_1} \times 100 \tag{5.12}$$

Antenna gain is a fundamental parameter in radiolink engineering. Gain is conventionally expressed in decibels and is an indication of the antenna's concentration of radiated power in a given direction. Antenna gain expressed anywhere in this work is gain over an isotropic. An isotropic is a theoretical antenna with a gain of 1 (0 dB). In other words, it is an antenna that radiates equally in all directions.

For parabolic reflector type antennas, gain is a function of the diameter of the parabola D and the frequency f. Theoretical gain is expressed by the formula

$$G_{dB} = 20 \log F_{MHz} + 20 \log_{10} D_{ft} - 52.6$$
$$\text{(for a 54\% surface efficiency)} \tag{5.13}$$

Figure 5.20 is a graph from which the gain of a parabolic reflector type of antenna may be derived for several discrete reflector diameters.

Note. In practice we assume surface efficiencies of usually around 55% for radiolink systems. Chapter 7 discusses antennas with improved efficiencies.

For uniformly illuminated parabolic reflectors, an approximate relation between half-power beamwidth and gain is expressed by the following:

$$Q \cong \frac{142}{\sqrt{G}} \text{ degrees} \tag{5.14}$$

where Q is the half-power beamwidth and

$$\sqrt{G} = \text{antilog}\left(\frac{G_{dB}}{20}\right)$$

In actual practice absolutely uniform illumination is not used so that side lobe levels may be reduced using a tapered form of illumination. Expect beamwidths to be 0.1–0.2° wider in practical applications.

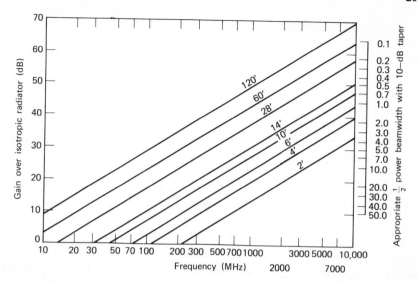

Figure 5.20 Parabolic antenna gains for discrete reflector diameters. Courtesy Institute of Telecommunication Sciences, Office of Telecommunications, U.S. Department of Commerce.

Beamwidths are narrow in radiolink systems. Table 5.5 illustrates their narrowness. From this it is evident that considerable accuracy is required in pointing the antenna at the distant end.

Table 5.5 Antenna Beamwidths

Gain (dB)	Half-Power Beamwidth
30	5°
35	3°
44	1°

POLARIZATION

In radiolink systems antennas use linear polarization and, depending on the feed, may be horizontally or vertically polarized, or both. That is, an antenna radiating and receiving several frequencies at once often will radiate adjacent frequencies with opposite polarizations, or the received polarization is opposite to the transmit. Isolation of 20 dB or better may be expected between polarizations on well designed installations allowing closer interworking at installations using multi-RF carrier operation.

5.5.5 The FM Receiver

In most applications the radiolink receiver shares the same antenna and waveguide as its companion transmitter to/from a common distant end. Figure 5.13*b* is a simplified block diagram of a typical radiolink receiver (ideal configuration). The receiver may or may not be connected to the common waveguide manifold by means of a circulator. It also may use a bandpass or preselector filter. Both the circulator and preselector filter reduce the effects of adjacent transmitter energy to a negligible amount in the receiver front end. Radiolink receivers operating at lower frequencies use coaxial transmission lines instead of waveguide.

From the manifold or via a circulator and/or preselector the incoming FM signal next looks into a mixer or down-converter. This unit heterodynes the received signal with the local oscillator signal to produce an intermediate frequency (IF). Most installations have standardized on a 70-MHz IF. However, CCIR discusses a 140-MHz IF for systems designed to carry voice channels in excess of 1800 in a standard CCITT FDM configuration.

From the mixer or down-converter output, the IF is fed through several amplification stages, often through a phase equalizer to correct delay distortion introduced by IF (and RF) filters. IF gains commonly are on the order of 80–90 dB.

The output of the IF is fed to a limiter-discriminator which is the FM detector or demodulator. The output of the discriminator is the composite baseband made up of the information baseband plus orderwire/alarms signals.

After demodulation the composite baseband is passed through a deemphasis network (see Figure 5.17) and is amplified. Thence the composite signal is split, the information baseband being directed to the demultiplex equipment and the orderwire (service channel) and alarms to the orderwire equipment and alarm display.

5.5.6 Diversity Reception

GENERAL

Various types of diversity reception are used widely on point-to-point HF systems, on transhorizon microwave (tropo), and to an increasing extent on radiolinks (LOS microwave).

Diversity is attractive for the following reasons:

- Tends to reduce depth of fades on combined output.
- Provides improved equipment reliability (if one diversity path is lost due to equipment failure, other path(s) remain in operation).
- Depending on the type of combiner in use, combined output S/N is improved over that of any single signal path.

Diversity reception is based on the fact that radio signals arriving at a point of reception over separate paths may have noncorrelated signal levels. More simply, at one instant of time a signal on one path may be in a condition of fade while the identical signal on another path may not.

First one must consider what are separate paths and how "separate" must they be (CCIR Rep. 376-1). The separation may be in

1. Frequency.
2. Space (including angle of arrival and polarization).
3. Time (a time delay of two identical signals on parallel paths).
4. Path (signals arrive on geographically separate paths).

The most common forms of diversity in radiolink systems are those of frequency and space. A frequency diversity system utilizes the phenomenon that the period of fading differs for carrier frequencies separated by 2–5%. Such a system employs two transmitters and two receivers, with each pair tuned to a different frequency (usually 2–3% separation since the frequency band allocations are limited). If the fading period at one frequency extends for a period of time, the same signal on the other frequency will be received at a higher level, with the resultant improvement in propagation reliability.

As far as equipment reliability is concerned, frequency diversity provides a separate path, complete and independent, and consequently, one whole order of reliability has been added. Besides the expense of the additional equipment, the use of additional frequencies without carrying additional traffic is a severe disadvantage to the employment of frequency diversity, especially when frequency assignments are even harder to get in highly developed areas where the demand for frequencies is greatest.

One of the main attractions of space diversity is that no additional frequency assignment is required. In a space diversity system, if two or more antennas are spaced many wavelengths apart (in the vertical plane), it has been observed that multipath fading will not occur simultaneously at both antennas. Sufficient output is almost always available from one of the antennas to provide useful signal to the receiver diversity system. The use of two antennas at different heights provides a means of compensating, to a certain degree, for changes in electrical path differences between direct and reflected rays by favoring the stronger signal in the diversity combiner.

Diversity combiners are discussed below. The antenna separation required for optimum operation of a space diversity system may be calculated using the following formula:

$$S = \frac{3\lambda R}{L} \tag{5.15}$$

where S = separation (m)
   R = effective earth radius (m)
   λ = wavelength (m)
   L = path length (m)

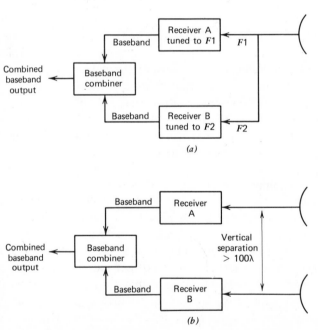

(a)

(b)

Figure 5.21 (a) Simplified block diagram of a frequency diversity configuration $F1 < F2 + 0.02F2$. (b) simplified block diagram of a space diversity configuration.

However, any spacing between 100 and 200λ is usually found to be satisfactory. The goal in space diversity is to make the separation of diversity antennas such that the reflected wave travels a half wavelength further than the normal path.

Figures 5.21a is a simplified block diagram of a radiolink frequency diversity system and 5.21b of a space diversity system. Polarization diversity is discussed in Section 6.9.6. Figure 5.22 shows approximate interference fading for nondiversity versus frequency diversity systems for various percentages of frequency separation.

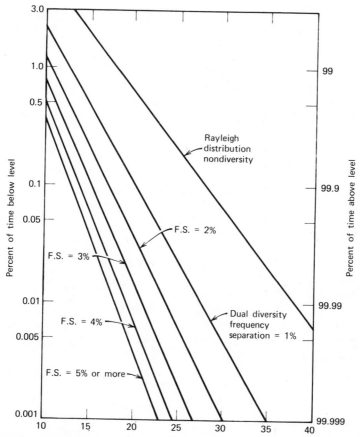

Figure 5.22 Approximate interference fading distribution for a nondiversity system with Rayleigh fading versus frequency diversity systems for various percentages of frequency separation (F.S.). (*Note:* Signal level referred to unfaded level for radiolinks; signal level referred to median for tropo; see Chapter 6.)

DIVERSITY COMBINERS

General

A diversity combiner combines signals from two or more diversity paths. Combining is traditionally broken down into two major categories:

- Predetection.
- Postdetection.

The classification is made according to where in the reception process the combining takes place. Predetection combining takes place in the IF. However, at least one system (see note*) performs combining at RF. With the second type, combining is carried out at baseband (i.e., after detection).

For predetection combining, phase control circuitry is required unless some form of path selection is used.

Figures 5.23a and 5.23b are simplified functional block diagrams of radiolink receiving systems using predetection and postdetection combiners.

Types of Combiners

Three types of combiners find more common application in radiolink diversity systems. These are

- Selection combiner.
- Equal gain combiner.
- Maximal ratio combiner (ratio squared).

The selection combiner uses but one receiver at a time. The output signal-to-noise ratio is equal to the input signal-to-noise ratio from the receiver selected for use at the time.

The equal gain combiner simply adds the diversity receiver outputs, and the output signal-to-noise ratio of the combiner is

$$\frac{S_o}{N_o} = \frac{S_1 + S_2}{2N} \tag{5.16}$$

where N = receiver noise.

The maximal ratio combiner uses a relative gain change between the

---

* *Note:* System manufactured by STC (UK).

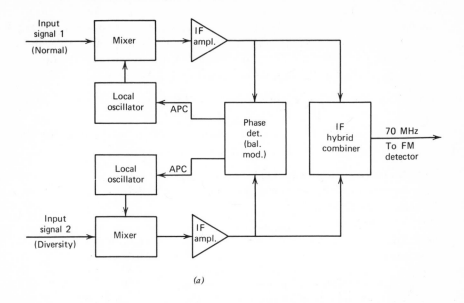

(a)

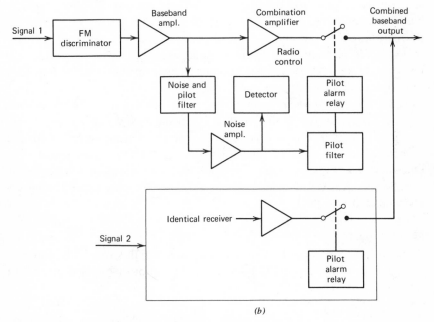

(b)

Figure 5.23 (a) Predetection combiner (APC = automatic phase control); (b) postdetection combiner (maximal ratio squared).

215

output signals in use. For example, let us assume that the stronger signal has unity output and the weaker signal has an output proportional to G (gain); it then can be shown that $G = S_2/S_1$ such that the signal gain is adjusted to be proportional to the ratio of the input signals. We then have

$$\left(\frac{S_o}{N_o}\right)^2 = \left(\frac{S_1}{N}\right)^2 + \left(\frac{S_2}{N}\right)^2 \qquad (5.17)$$

where $N =$ receiver noise. For the signal-to-noise ratio equation for the latter two combiners, we assume the following:

- All receivers have equal gain.
- Signals add linearly; noise adds on an rms basis.
- Noise is random.
- All receivers have equal noise outputs (N).
- The output (from the combiner) signal-to-noise ratio, S_o/N_o, is a constant.

Figure 5.24 shows graphically a comparison of the three types of combiners. In the figure we can see the gain that we can expect in output signal-to-noise ratio for various orders of diversity for the three types of combiners discussed, assuming a Rayleigh distribution. The order of diversity refers to the number of independent diversity paths. If we were to use space *or* frequency diversity alone, we would have two orders of diversity. If we used space *and* frequency diversity, we would then have four orders of diversity.

The reader should bear in mind that the efficiency of diversity depends on the correlation of fading of the independent diversity paths. If the correlation coefficient is zero (i.e., there is no relationship in fading for one path to another), we can expect maximum diversity enhancement. The efficiency of a diversity system drops by half with a correlation coefficient of 0.8, and nearly full efficiency can be expected with a correlation coefficient of 0.3.

PILOT TONES

A radio continuity pilot is provided between radio treminals which is independent of the multiplex pilot tones. The pilot or pilots provided are used for

    Gain regulating
    Monitoring
    Frequency comparison
    Measurement of level stability
    Control of diversity combiners

The latter application involves the simple sensing of continuity by a diversity combiner. The presence of the continuity tells the combiner that the diversity path is operative. The problem is that the most commonly used postdetection combiners, the maximal ratio type, selection type, and others, use noise as the means to determine the path contribution to the combined output. The path with the least noise, as in the case of the maximal ratio combiner, provides the greatest path contribution.

If, for some reason, a path were to fail, it would be comparatively noiseless and would provide 100% contribution. Thus we would have a no-signal output. In this case the continuity pilot tells the combiner that the path is a valid one.

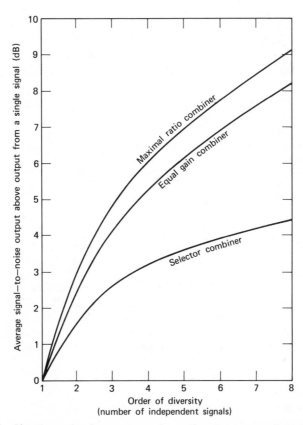

Figure 5.24 Signal-to-noise improvement in a diversity system for various orders of diversity. Courtesy Institute of Telecommunication Sciences, Office of Telecommunications, U.S. Department of Commerce.

Pilots are inserted previous to modulation and are stopped (eliminated) at baseband output and are reinserted anew if another radiolink is added. Table 5.6 taken from CCIR Rec. 401-2, provides a list of recommended radio continuity pilots.

Table 5.6   Radio Continuity Pilot Tones

System Capacity (Channels)	Limits of Band Occupied by Telephone Channels (kHz)	Frequency Limits of Baseband (kHz)[a]	Continuity Pilot Frequency (kHz)	Deviation (rms) Produced by the Pilot (kHz)[b,e]
24	12–108	12–108	116 or 119	20
60	12–252	12–252	304 or 331	24, 50, 100[c]
	60–300	60–300		
120	12–552	12–552	607[d]	25, 50, 100[c]
	60–552	60–552		
300	60–1300	60–1364	1499, 3200[f] or 8500[f]	100 or 140
600	60–2540	60–2792	3200 or 8500	140
	64–2660			
960⎫ 900⎭	60–4028⎫ 316–4188⎭	60–4287	4715 or 8500	140
1260⎫ 1200⎭	60–5636⎫ 60–5564⎬ 316–5564⎭	60–5680	⎰6199 ⎱8500	100 or 140 140
1800	312–8204	300–8248	9203	100
	316–8204			
2700	312–12,388	308–12,435	13,627	100
	316–12,388			
Television			⎰8500 ⎱9023[g]	140 100

Source: CCIR Rec. 401.

[a] Including pilot or other frequencies which might be transmitted to line.

[b] Other values may be used by agreement between the administrations concerned.

[c] Alternative values dependent on whether the deviation of the signal is 50, 100, or 200 kHz (Rec. 404-2).

[d] Alternatively 304 kHz may be used by agreement between the administrations concerned.

[e] This deviation does not depend on whether or not a preemphasis network is used in the baseband.

[f] For compatibility in the case of alternate use with 600-channel telephony systems and television systems.

[g] The frequency 9023 kHz is used for compatibility purposes between 1800-channel telephony systems and television.systems, or when the establishment of multiple sound channels so indicates.

5.5.7   Transmission Lines and Related Devices

WAVEGUIDE AND COAXIAL CABLE

Waveguide may be used on installations operating above 2 GHz to carry the signal from the radio equipment to the antenna (and vice versa). For those systems operating above 4 GHz it is mandatory from a transmission efficiency point of view. Coaxial line is used on those systems operating below 2 GHz.

From a systems engineering aspect, the concern is loss with regard to a transmission line. Figures 5.25 and 5.26 identify several types of commonly used waveguide and coaxial cable, with loss versus frequency.

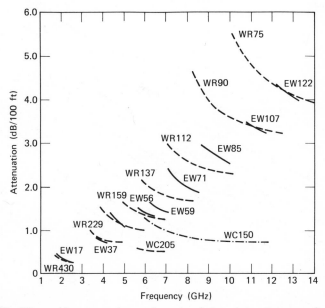

Figure 5.25   Waveguide attenuation and power rating data. Courtesy the Andrew Corp.

Average Power Ratings

Heliax Elliptical Waveguides			Rigid Rectangular Waveguides			
Andrew Type No.	Frequency (GHz)	Power (kW)	Type Numbers EIA	IEC	Frequency (GHz)	Power (kW)
EW17	2.0	27.0	WR430	R22	2.0	22.4
EW37	4.0	7.3	WR229	R40	4.0	5.4
EW44	4.7	3.9	WR187	R48	4.7	3.1
EW56	6.0	3.0	WR159	R58	6.0	2.5
EW59	6.5	2.5	WR137	R70	6.5	1.5
EW71	7.7	1.7	WR112	R84	7.7	0.9
EW85	9.2	1.2	WR112	R84	9.2	1.0
EW107	11.2	0.9	WR90	R100	11.2	0.6
EW122	12.7	0.7	WR75	R120	12.7	0.4

Note: Ratings based on temperature rise of 25°C (45°F) with an ambient temperature of 55°C (130°F) and VSWR of 1.0. All waveguides shown are fabricated of high conductivity copper.

Rigid Circular Waveguides			
Type Numbers EIA	IEC	Frequency (GHz)	Power (kW)
WC205	C40	6.5	8.4
WC150	C56	9.2	4.2

Waveguide installations are always maintained under dry air or nitrogen pressure to prevent moisture condensation within the guide. Any constant positive pressure up to 10 lb/in.^{2} (0.7 kg/cm^2) is adequate to prevent "breathing" during temperature cycles.

Waveguide may be rectangular, elliptical, or circular. Nearly all older installations used rectangular waveguide exclusively. Today its use is limited to routing in tight places where space is limited. However, bends

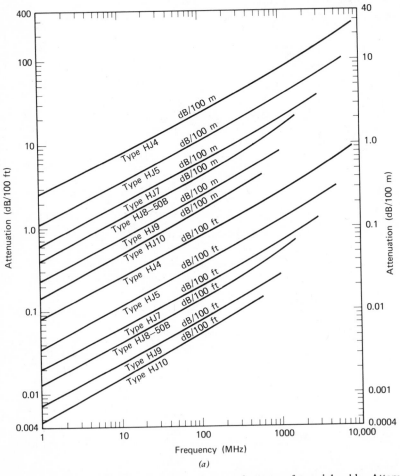

(a)

Figure 5.26 Attenuation per unit length for certain types of coaxial cable. Attenuation curves based on VSWR 1.0:1, for 75-Ω cables, multiply by 0.95. 50 Ω; copper conductors; (a) air dielectric, (b) foam dielectric. Courtesy the Andrew Corporation.

are troublesome and joints add 0.06-dB loss each to the system. Optimum electrical performance is achieved by using the minimum number of components. Therefore it has become the practice that wherever a single length is required, elliptical waveguide is used from the antenna to the radio equipment without the addition of miscellaneous flex-twist or rigid sections which are used for rectangular guide.

Circular waveguide offers generally lower loss plus dual polarized capability such that only one waveguide run up the tower is necessary for

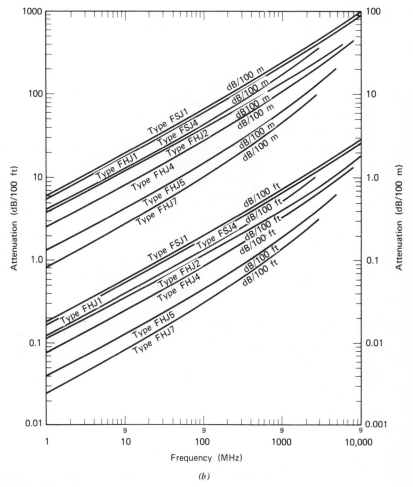

Figure 5.26 (*Continued*)

dual polarized installations. Circular waveguide is used when the run is so long that excessive loss is introduced.

TRANSMISSION LINE DEVICES

A ferrite load isolator is a waveguide component which provides isolation between a signal source and its load, reducing ill effects of VSWR and often improving stability as a result. Most commonly load isolators are used with transmitting sources absorbing much of the reflected energy from high VSWR. Owing to the ferrite material with its associated permanent magnetic field, ferrite load isolators have a unidirectional property. Energy traveling toward the antenna is relatively unattenuated, whereas energy traveling back from the antenna undergoes fairly severe attenuation. The forward and reverse attenuations are on the order of 1 and 40 dB, respectively.

A waveguide circulator is used to couple two or three microwave radio equipments to a single antenna. The circulator shown in Figure 5.27 is a four-port device. It consists essentially of three basic waveguide sections combined into a single assembly. The center section is a ferrite nonreciprocal phase shifter. An external permanent magnet causes the ferrite material to exhibit phase shifting characteristics. Normally an antenna transmission line is connected to one arm and either three radio equipments are connected to the other three arms, or two equipments

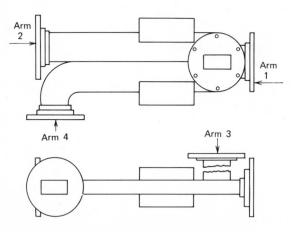

Figure 5.27 A waveguide circulator (four ports).

and a shorting plate. Attenuation in a clockwise direction from arm to arm is low, on the order of 0.5 dB, whereas in the counterclockwise direction from arm to arm it is high, on the order of 20 dB.

A power splitter is a simple waveguide device that divides the power coming from or going to the antenna. A 3-dB power split divides power in half; such a device could be used, for instance, to radiate the power from a transmitter in two directions. A 20-dB power split, or 30 dB, has an output that serves to sample the power in a transmission line. Often such a device has directional properties (therefore called a directional coupler) and is used for VSWR measurements allowing measurement of forward and reverse power.

Magic tees or hybrid tees are waveguide devices used to connect several equipments to a common waveguide run.

PERISCOPIC ANTENNA SYSTEMS

Instead of a long waveguide run up a radiolink tower, a periscopic antenna system uses a short run to an antenna mounted near the tower, usually on the roof of the repeater building. A plane or semicurved reflector is mounted on the tower. The antenna is aimed upwards at the reflector, and the focused energy radiated by the antenna is reflected to the distant end.

A reflector will intercept all the main beam of radiated energy of a parabolic antenna if

- The reflector is situated in the near field.
- The cross-sectional area of the reflector is equal or slightly greater than the projected area of the antenna.

The further the reflector is moved from the near-field boundary, the larger the cross-section of the beamwidth and the larger the size of the reflector required to properly reflect all the beam energy. Figure 5.28 gives some near-field boundaries.

Reflectors so used may give a gain or a loss in system (path) calculations, depending on the operating frequency, separation distance from the antenna to the reflector, and the size of the reflector. Figure 5.29 gives gain/loss in decibels for various reflector/antenna/frequency configurations. Often the curving is done after orientation on the distant end to improve energy focus.

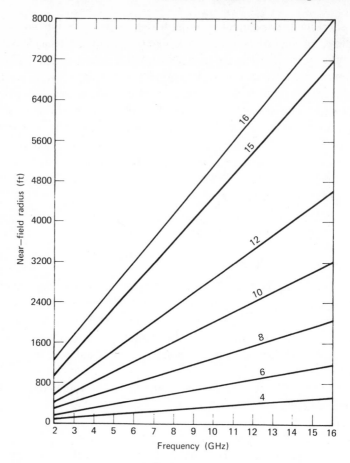

Figure 5.28 Near-field radius for parabolic antennas (curves are for parabola diameters in feet). $R = 2\ D^2/\lambda$, where $D =$ dish diameter (ft) and $\lambda =$ wavelength (ft); or $R = 0.002\ D^2 F_{MHz}$. Courtesy Collins Radio Group, Rockwell International Corp.

5.6 LOADING OF A RADIOLINK SYSTEM

5.6.1 General

Noise on a radiolink system can essentially be broken down into inter-modulation noise and residual noise (basically thermal noise). The important fact is that the intermodulation noise increases with load (i.e., increased traffic), and when a certain "break point" of load handling capability is exceeded, intermodulation noise becomes excessively high. On the other hand, residual noise is not affected by the amount of traffic or traffic load of the system. FM radio overcomes much of the residual noise that appears in a radio system by distributing it over a wide radio bandwidth. Such an exchange of bandwidth for lower noise is also proportional to the signal level appearing at the system receiver. At periods of low signal (i.e., during a radio fade) thermal or residual noise becomes important.

5.6.2   Noise Power Ratio (NPR)

Noise power ratio has come into wide use in radiolink systems to describe intermodulation noise performance. NPR gives an excellent indication

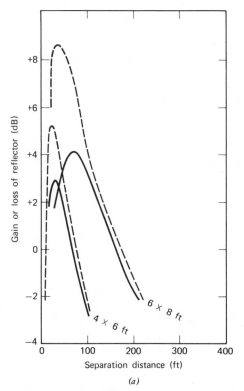

(a)

Figure 5.29   Near-field radius for parabolic antennas (curves are for parabola flat reflector; dashed line, curved reflector. (*a*) With 2-ft parabola; (*b*) with 4-ft parabola; (*c*) with 6-ft parabola; (*d*) with 8-ft parabola.

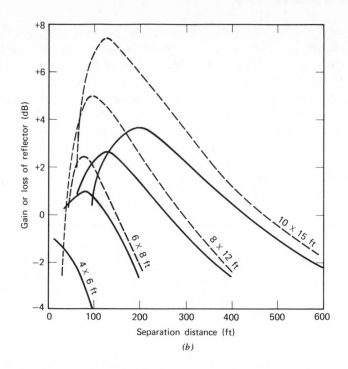

(b)

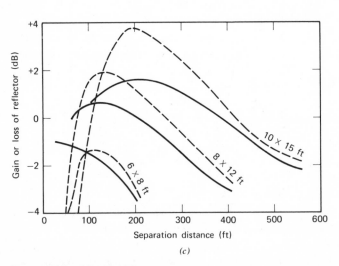

(c)

Figure 5.29 (*Continued*)

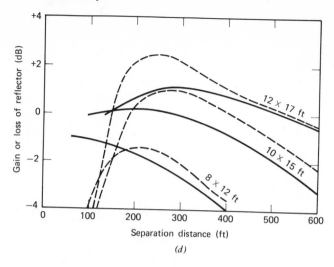

Figure 5.29 (*Continued*)

of performance of intermodulation noise when measured under standard fixed conditions. NPR is measured in decibels, and we can say of a radio-link transceiver combination that it has a specified NPR under certain load conditions.

Noise power ratio is a meaningful measurement for radiolink systems carrying multichannel FDM telephone baseband information. NPR can be defined as the ratio, expressed in decibels, of the noise in a test channel with all channels loaded with white noise to noise in the test channel with all channels except the test channel fully noise loaded.

5.6.3 Basic NPR Measurement

To measure NPR on a baseband-to-baseband basis, a radiolink transmitter is connected back-to-back with a receiver using proper waveguide attenuators to simulate real path conditions. A white noise generator is connected to the transmitter baseband input. The white noise generator produces a noise spectrum that approximates a spectrum produced in a multichannel (FDM) multiplex system. The output noise level from the generator is adjusted to a desired *composite noise baseband power*. A notched filter is then switched in to clear a narrow slot in the spectrum of the noise signal, and a noise analyzer is connected at the output of the

Table 5.7 Recommended NPR Measurement Frequencies

System Capacity (Channels)	Limits of Band Occupied by Telephone Channels (kHz)	Effective Cutoff Frequencies of Band-limiting Filters (kHz) — High Pass	Low Pass	Frequencies of Available Measuring Channels (kHz)						
60	60–300	60 ± 1	300 ± 2	70	270					
120	60–552	60 ± 1	552 ± 4	70	270	534				
300	60–1300 / 64–1296	60 ± 1	1296 ± 8	70	270	534	1248			
600	60–2540 / 64–2660	60 ± 1	2600 ± 20	70	270	534	1248	2438		
960	60–4028 / 64–4024	60 ± 1	4100 ± 30	70	270	534	1248	2438	3886	
900	316–4288	316 ± 5	4100 ± 30			534	1248	2438	3886	
1260	60–5636	60 ± 1	5600 ± 50	70	270	534	1248	2438	3886	5340
1200	60–5564 / 316–5564	316 ± 5	5600 ± 50			534	1248	2438	3886	5340
1800	312–8120 / 312–8204 / 316–8204	316 ± 5	8160 ± 75			534 / 7600	1248	2438	3886	5340
2700	312–12 336 / 316–12 388 / 312–12 388	316 + 5	12 360 ± 100			534 / 7600	1248	2438 / 11 700	3886	5340

Source: CCIR Rec. 399-1, New Delhi, 1970.

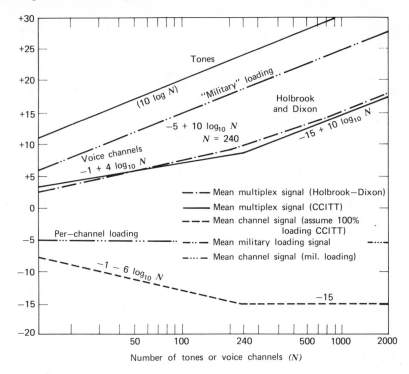

Figure 5.30 Curves for the loading of multichannel communication systems. Courtesy Marconi Instruments.

system. The analyzer is used to measure the ratio of the noise in the illuminated (noise loaded) section of the baseband to the noise power in the cleared slot. The slot noise level is equivalent to the total noise (residual plus intermodulation) that is present in the slot bandwidth. Slot bandwidths are the width of a standard voice channel and are taken at the upper, middle, and lower portions of the baseband. Table 5.7, taken from CCIR Rec. 399-1, gives standards for slot locations for various FDM channel configurations.

The composite noise power is taken from one of the following formulas for N telephone channels in an FDM (SSBSC) configuration (see Chapter 3):

$$P(\text{dBm0}) = -1 + 4 \log_{10} N \qquad (5.18a)$$
$$\text{when } N < 240 \text{ channels}$$

$$P(\text{dBm0}) = -15 + 10 \log_{10} N \qquad (5.18b)$$
$$\text{when } N > 240 \text{ channels}$$

$$P(\text{dBm0}) = -5 + 10 \log_{10} N \qquad (5.18c)$$
$$\text{for certain U.S. military systems}$$
$$\text{with heavy data usage}$$

Figure 5.30 shows curves for multichannel loading for all three applications above plus that for Holbrook-Dixon (see Chapter 3, particularly Section 3.4.4, for more information on loading FDM systems).

$$\text{NPR} = \text{composite noise power (dB)}$$
$$- \text{ noise power in the slot (dB)} \qquad (5.19)$$

A good guide for NPR for high capacity radiolink systems with a 300-channel capacity (1 hop) should be 55 dB; for 1200 voice channels, in excess of 50 dB.

5.6.4 Derived Signal-to-Noise Ratio

Given the NPR of a system, we can then compute per voice channel test tone to noise ratio.

$$\frac{S}{N} = \text{NPR} + \text{BWR} - \text{NLR} \qquad (5.20)$$

BWR (bandwidth ratio)
$$= 10 \log \frac{(\text{occupied baseband bandwidth})}{(\text{voice channel bandwidth})} \qquad (5.21)$$

NLR (noise load ratio) $= P$,
Taken from the load equation, 5.18

The signal-to-noise ratio as given is unweighted. For F1A or psophometric weighting add 3 dB (when reference is 1000 Hz). For an 800-Hz reference, add 2.5 dB. Figure 5.31 is an example of this application for a 240-channel system.

5.6.5 Conversion of Signal-to-Noise Ratio to Channel Noise

Using the S/N calculated above,

$$\text{Noise power in dBa0} = 82 - \frac{S}{N} \qquad (5.22a)$$

$$\text{Noise power in dBrnC} = 88.5 - \frac{S}{N} \qquad (5.22b)$$

$$\text{Noise power in pW} = \frac{10^9}{\text{antilog } S/N} \qquad (5.22c)$$

$$\text{Noise power in pWp} = \frac{10^9 \times 0.56}{\text{antilog } S/N} \qquad (5.22d)$$

5.7   OTHER TESTING TECHNIQUES

5.7.1   Out-of-band Testing

NPR tests require taking a system out of traffic. Out-of-band testing permits a continuous monitor of intermodulation and residual noise without interrupting normal traffic. In this arrangement normal traffic constitutes the loading of the system and a noise receiver continuously monitors the noise in a channel which is approximately 10% outside the normal base-

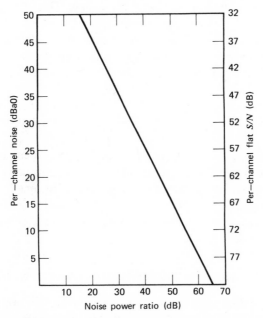

Figure 5.31   Conversion curve NPR to S/N for a 240-channel system. Courtesy Marconi Instruments.

band. Bandstop filters at the out-of-band frequency slots are placed permanently in the circuit at the transmitter. The noise receiver indication then is due only to system intermodulation and residual noise.

Often a three-channel pen recorder is used, where the first pen will show the variations of a system noise with time The second pen is used to measure traffic level and the third pen measures the AGC (automatic gain control) voltage of the radiolink receiver in question.

5.8 DETAILED PATH CALCULATIONS

Figure 5.32 is a path calculation sheet. It illustrates one way of performing detailed calculations. Section 5.5.4 discusses simple calculations. Before proceeding, a review of that paragraph is advised.

The following is an explanation of each entry or the step in the path calculation process.

Enter the names of the two sites which the radiolink interconnects. Enter equipment type, operating frequency, and receiver noise figure.

Enter the type of traffic, such as FDM-SSBSC or video plus program channels. This will determine the type of preemphasis (see Section 3.5.2). Enter highest modulating baseband frequency. This will determine preemphasis network values and preemphasis improvement (step U and Figure 5.15 and 5.16). The noise "equalizing" of preemphasis adds 1–3 dB to the channel S/N ratio. This improvement may be found using Figure 5.33. Enter the peak frequency deviation (see Section 5.5.3) and B_{if} (see 3.4.2).

Enter whether diversity is used, or if none, so state or enter if hot standby is used (see Sections 5.5.6 and 5.12). Enter radio pilot frequency (Section 5.6).

For the record, and to properly associate this calculation sheet with its companion profile for the same link, enter identification of the related path profile. From the profile we can get the path length, which will be entered below. Enter true bearing station A to station B and station B to station A. This information will be obtained from the maps used while preparing the profile.

5.9 DETERMINATION OF FADE MARGIN (Part of Step D)

Fading is a random increase in path loss during abnormal propagation conditions. During such conditions path loss may increase 10, 20, 30 or more dB for short periods. The objective of this subsection is to assist in

Figure 5.32 Radiolink path calculations sheet.

Site.................. To site...............
Equipment type........ Frequency (MHz)............ Rec. noise fig.....dB
Type of traffic.......... Highest baseband freq..............
Frequency dev.......... B_{if}........................
Diversity or hot standby.................... Radio pilot............
Ref. path profile.................... True bearing A-B............ °
Path length (st. mi or km)......... True bearing B-A.......... °

Step		Add or Subtract	Unit	Comments
A	Path loss (free space)	Minus		
B	Connector losses (sum)	Minus		Sum of transmit and receive
C	Circulator losses (sum)	Minus		Sum of transmit and receive
D	Power split losses (IF any)	Minus		
E	Directional coupler losses (sum)	Minus		Sum of transmit and receive
F	Transmission line losses (sum)	Minus		Sum of transmit and receive
G	(Other losses)	Minus		
H	(Other losses)	Minus		
I	Sum of losses			
J	Transmit power (dBW)	Plus		Power output to flange
K	Transmit ant. gain	Plus		
L	Transmit reflect (gain or loss)	Plus or Minus		(Gains are +, losses are −)
M	Receive ant. gain	Plus		
N	Receive reflect (gain or loss)	Plus or Minus		
O	Sum of gains			
P	Add step I + step O		dBW	Input level to receiver
Q	Receiver noise threshold		dBW	
R	FM improv. threshold (pWp)		dBW	$R = 10$ dB above Q
S	Fade margin to FM improv.		dB	Step P − Step R
T	Unfaded signal-to-noise ratio		dB	
U	Preemphasis improvement		dB	
V	Diversity improvement		dB	
W	Channel S/N (or video S/N)		dB	
X	Calculated path reliability		%	
Y	Channel noise contrib. (radio)		pWp	(see CCIR Rec. 395-1)

Step A. Enter path loss in decibels as calculated in Section 5.3.1.

Step B. Connector losses. One rule of thumb is to add 0.06 dB per joint. If we had 10 joints in a run, we would enter 0.6 dB in this box. Sum for both transmit and receive.

Step C. Circulator losses (see Section 5.5.7). Insert insertion loss. 0.5 dB is often a good figure. This figure inserted is the sum of the near-end transmit and far-end receive circulator insertion losses.

Step D. Power split loss. Insert insertion loss in desired direction plus split attenuation. Sum transmit and receive.

Step E. Directional coupler loss—see Section 5.5.7. Insert the insertion loss of directional couplers. Sum of transmit and receive installations.

Step F. Transmission line losses. Again enter the sum of the near-end transmit and far-end receive transmission line losses. See Section 5.5.7 and Tables 5-23 and 5-24. A good conversion figure to remember is

$$1 \text{ dB}/100 \text{ ft} = 3.28 \text{ dB}/100 \text{ m}$$

233

Steps G, H.	Enter here other losses, if any. For instance, you may be called on to use a transition from rectangular to circular waveguide or coax to waveguide. Enter those losses here for the transitions.
Step I.	Box I = sum of boxes $A–H$.
Step J.	Transmit power. Refer to Section 5.5.3. Keep in mind that power outputs are usually to the flange of the output device. In other words, care should be taken in reading equipment manufacturers' specifications.
Step K.	Insert transmit antenna gain. Be sure the gain is given in dBi or decibels over an isotropic and not over a dipole. See Section 5.5.4 and Figure 5.20.
Step L.	If a reflector is used (i.e., a periscopic system) insert the gain as a + entry, or the loss as a − entry. See Section 5.5.7.
Steps M, N.	The same as Steps K and L, respectively, but for the receive side.
Step O.	Add the gains. If the reflector(s) give losses, add algebraically.
Step P.	The result of this addition (always add algebraically) will give the input to the receiver.
Step Q.	Calculate the receiver noise threshold. An explanation of this calculation is given in Section 5.4.2.
Step R.	Add −10 dB aglebraically to the level in Step Q and insert this figure. See Section 5.4.2.
Step S.	Add algebraically Step P − Step R. In other words, this is the number of decibels that the unfaded input level to the receiver, Step P, is above Step R, the FM improvement threshold, where the psophometrically weighted channel noise is in pWp. See Figure 5.32 and Sections 5.6.4 and 5.6.5. The fade margin should be adjusted in accordance with paragraphs Sections 5.4.3 and 5.9. A more direct approach is to adjust the figure as a first step. To economically achieve the required fade margin, we adjust steps K, L, M, and N. Assuming the figure is too low, antenna sizes can be increased; losses can be reduced using a lower loss waveguide. Other steps that may be taken are to increase transmitter output by using a different transmitter, by using a receiver with an improved noise figure.
Step T.	Enter the unfaded signal-to-noise ratio. If we can assume a 1-for-1 increase of carrier-to-noise ratio, decibels, for signal-to-noise ratio, once the signal input is above FM improvement threshold, then S/N is approximately equal to C/N + 20 dB. A more accurate result can be found using information in Section 5.6.
Step U.	Preemphasis improvement. This figure in decibels can be taken from Figure 5-33 and is discussed in Section 5.5.2.
Step V.	Diversity improvement. This figure in decibels may be taken from information provided in Section 5.5.6 and Figure 5.24.
Step W.	Channel or video signal-to-noise ratio (unfaded). This derives from steps $T–V$ and is discussed in Section 5.6.
Step X.	Enter path reliability. This is usually taken to be the percentage of time that a path (hop) will be above FM improvement threshold. Refer to Section 5.10. Step X may have to be adjusted upwards when total line reliability is considered. If this is the case, Step S, the fade margin, will have to be increased accordingly.
Step Y.	Enter the channel noise contribution (radio). See Section 5.6. This figure is added to the multiplex (if any is on the hop) noise for total hop noise contribution. Total link, system noise is covered in Section 5.18.

Figure 5.32

234

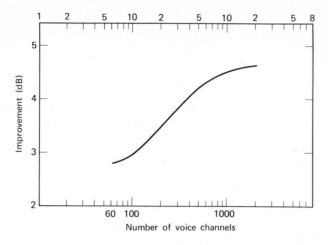

Figure 5.33 Preemphasis improvement in decibels for a given number of FDM voice channels.

setting a margin or overbuild in system design to minimize the effects of fading.

The factors involved in fading phenomena are many and complex. They have been discussed in some detail in Section 5.4.3. In general, we can state that the depth of fades and the number of fades per unit time vary directly with path length and operating frequency. The depth of a fade can be nearly any amount, but the deeper the fade, the less frequently it occurs and the shorter its duration when it does occur.

Figures 5.34a–c relate fade margin with distance or length of a hop for three frequency bands for propagation reliabilities of 99.0, 99.9, and 99.99%. Table 5.3 provides a cross reference of reliability expressed as a percentage and outage time. The reliability selected is a design decision. CCIR Rec. 395-1, dealing with noise over real (radio) links and total overall system reliability, discussed in Section 5.10, will have impact on that choice.

Let us try some examples. A radiolink is operating in the 2-GHz band with a 99.9% hop reliability. The hop is 28 mi long. What is the fade margin to enter in Step S, Figure 5.32? Refer to Figure 5.34a, the 99.9% curve. From this we derive a fade margin of 27 dB.

A radiolink is operating in the 11.2 GHz band. It is desired to have a 99.99% propagation reliability on the hop, which is 20 mi long. Refer to Figure 5.34c, upper curve. From this curve we see that a 38-dB fade margin is required (neglecting rainfall attenuation considerations).

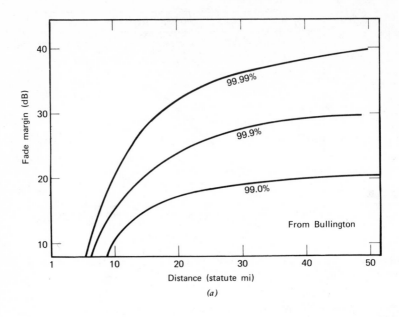

(a)

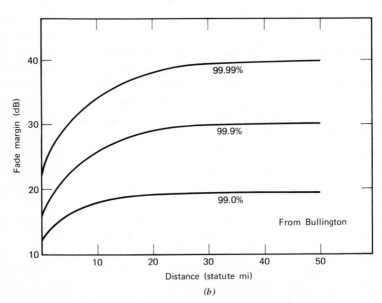

(b)

Figure 5.34 Fade margin related to distance or length of hop for three frequency bands: (*a*) 2 GHz, (*b*) 6 GHz, (*c*) 11.2 GHz (Ref. 20).

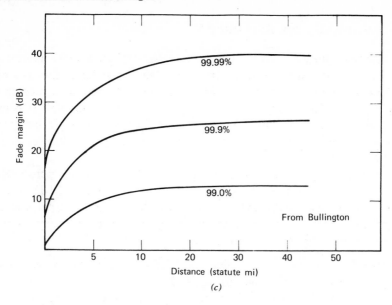

Figure 5.34 (*continued*)

Considerable effort has been made to refine fade margin predictions without resorting to live path tests. Pearson of Standard Telephone Laboratories (UK) has refined the fade margin prediction procedure by relaxing or increasing the margins of fade as a function of intervening terrain roughness. He treats basically the terrain reflectivity and, to an extent, resulting atmospheric mixing. From this information, the degree of multipath fading can be predicted. For example, wooded areas and the like tend to break up signal reflections, reducing the chances of multipath fading. A description of the "Pearson method" may be found in Ref. 30.*

Diversity also reduces fade margin requirements as established in Figures 5.34a–c. Diversity tends to reduce the depth of fades. Figure 5.22 shows how fade margins may be reduced with diversity. For instance, assume a Rayleigh distribution (random fading) with a 35-dB fade margin with no diversity; then using frequency diversity with a frequency separation of 2%, only a 23.5-dB fade margin would be required to maintain the same path reliability (percent of time level is exceeded).

* For more information on fading, the reader should consult *ESSA TECH Rep. ERL 148-ITS 97* (Institute of Telecommunication Sciences, Boulder, Colo.).

5.10 PATH AND LINK RELIABILITY

A planning engineer is interested in system reliability. He is more interested in setting a limit, expressed in a percentage, that a telephone conversation will be degraded by noise or will drop out entirely. He may set a reliability of 99.9% on a link between points X and Y. Suppose this link is composed of 10 radiolink hops. What then will be the per-hop reliability? We consider here only propagation outages, not equipment outages (i.e., failures).

The method described here to assign a per-hop reliability percentage assumes that fades on separate hops will not be correlated. Thus the outage time of each individual hop will add directly to the total system outage time. This is the worst condition because there usually is some correlation of fades on different paths.

Example 1. With a system reliability of 99.9% we will have 0.1% outage/year. This corresponds to 0.1% for 10 radiolinks in tandem or 0.01% per hop. Thus the per-hop reliability required is 99.99% to maintain a system reliability of 99.9%

Example 2. Assume a system reliability of 99.99%. The system consists of seven hops in tandem. 99.99% is equivalent to 0.01% outage, which corresponds to a per-hop outage of $0.01/7 = 0.00143\%$, which is equivalent to 99.9986% reliability per hop.

5.11 RAINFALL AND OTHER PRECIPITATION ATTENUATION

For radiolink systems rainfall and other precipitation attenuation are not significant below 10 GHz. The cause of attenuation of microwave energy passing through rain is attributed to both absorption and scattering. Absorption attenuation is predominant in clouds and fogs. Scattering attenuation is predominant in rain. Worst-case cloud and fog attenuation is 1/10 that of rainfall attenuation below 15 GHz. Fog, cloud, and water vapor attenuation are covered in Chapter 10.

Rainfall attenuation is a function of rainfall density, usually given in millimeters per hour, and frequency. Figure 5.35 is a curve giving atttenuation in decibels per mile for typical rainfall densities as a function of frequency.

We are dealing here not with the total quantity of rain but with periods of heavy intensity. One short cloudburst can cause more outage than weeks of light drizzle.

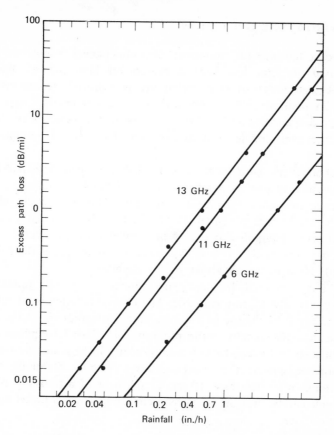

Figure 5.35 Excess path loss due to rainfall (after Ryde and Ryde).

For the continental United States extensive statistical data are available on rainfall.* The country has been divided into zones, and incidence of rainfall and cloudburst activity are given statistically. This information is mandatory for good planning of radiolinks in the 11–13- and 15-GHz bands. In certain areas we can expect an increase of free-space attenuation of 3 dB/mi for that part of a path affected by the downfall or cloudburst. For example, in the Florida panhandle, by one set of standards, radiolink sections (hops) have been limited to 8 mi (12.8 km) due to rain attenuation. In this particular case the system was designed for CATV operation in the 12.9-GHz band.

* Available from U.S. Weather Bureau and U.S. Government Printing Office.

5.12 HOT-STANDBY OPERATIQN

Radiolinks often provide transport of multichannel telephone service and/or point-to-point broadcast television on high priority backbone routes. A high order of route reliability is essential. Route reliability depends on path reliability (propagation) and equipment/system reliability. The first item, path reliability, has been discussed above; namely, a systematic overbuild is used to overcome fades or to minimize their effects.

The second item, equipment/system reliability, deals with making the equipment and system as a whole more reliable by minimizing system downtime to maximize equipment availability. One way to do this most effectively is to provide a parallel system. We did this in frequency diversity. There all active equipment operated in parallel with two distinct systems carrying the same traffic. This is expensive, but necessary if the reliability is desired. Here we speak about route reliability.

Often the additional frequency assignments to permit operation in frequency diversity are not available. The equivalent equipment reliability may be achieved by the use of a hot-standby configuration. As the expression indicates, hot standby is the provision of a parallel equipment configuration such that it can be switched in nearly instantaneously on failure of operating equipment. The switchover takes place usually on the order of microseconds, often less than 10 μs. Changeover of a transmitter and/or receiver line can be brought about by a change in (over a preset amount), for a transmitter,

- Frequency.
- RF power.
- Demodulated baseband (radio) pilot level.

and for a receiver,

- AGC voltage.
- Squelch.
- Incoming pilot level.

Hot-standby protection systems provide sensing and logic circuitry for the control of waveguide switches (or coax) on transmitters and output signals on receivers. The use of a combiner is common on the receiver side with both receivers on line at once.

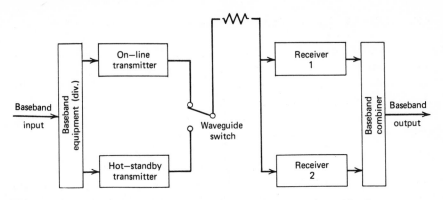

Figure 5.36 One-for-one hot-standby configuration (using a combiner).

There are two approaches to the use of standby equipment. These are called 1-for-1 and 1-for-n. One-for-one provides one full line of standby equipment for each operational system. See Figure 5.36. One-for-n provides only one full line of equipment for n operational lines of equipment, $n \geqq 2$. See Figure 5.37. One-for-one is more expensive but provides a higher order of reliability. Its switching system is comparatively simple. One-for-n is cheaper, with only one line of spare, hot-standby equipment for several operational. It is less reliable (i.e., suppose there were two equipment failures in n), and switching is considerably more complex.

5.13 RADIOLINK REPEATERS

Up to this point the only radiolink repeaters discussed have been baseband repeaters. Such a repeater fully demodulates the incoming RF signal to baseband. In the most simple configuration, this demodulated baseband is used to modulate the transmitter used in the next link section. This type of repeater also lends itself to dropping and inserting voice channels, groups, and supergroups. It may also be desirable to demultiplex the entire baseband down to voice channel for switching, and insert and drop a new arrangement of voice channels for multiplexing. The new baseband would then modulate the transmitter of the next link section. A simplified block diagram of a baseband repeater is shown in Figure 5.38a.

Two other types of repeaters are also available: the IF heterodyne repeater (Figure5.38b), and the RF heterodyne repeater (Figure 5.38c). The IF repeater is attractive for use on long, backbone systems where noise and/or differential phase and gain should be minimized.

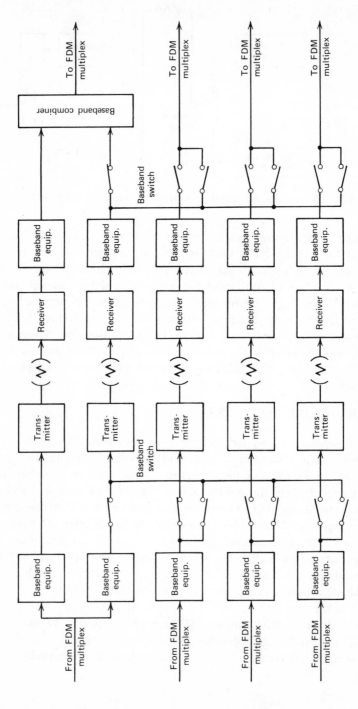

Figure 5.37 One-for-*n* hot-standby configuration (a 1-for-4 configuration is shown).

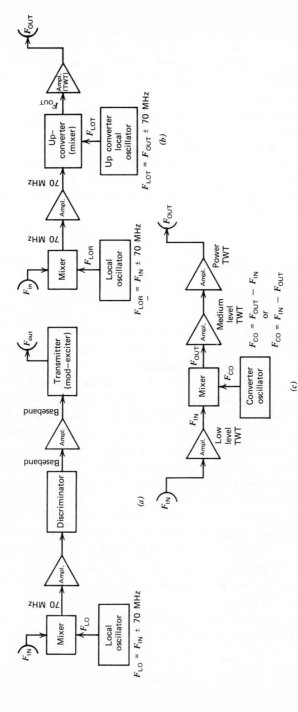

Figure 5.38 Radiolink repeaters. (*a*) Baseband repeater, (*b*) IF heterodyne repeater, (*c*) RF heterodyne repeater. F_{IN}, input frequency to the receiver; F_{OUT}, output frequency of the transmitter; F_{CO}, output frequency of the converter local oscillator; F_{LOR}, frequency of the receiver local oscillator; F_{LOT}, frequency of transmitter local oscillator; F_{LO}, frequency of the local oscillator; TWT, traveling wave tube.

Generally, a system with fewer modulation-demodulation stages or steps is less noisy. The IF repeater eliminates two modulation steps. The repeater simply translates the incoming signal to IF with appropriate local oscillator and a mixer, amplifies the derived IF, and then up-converts it to a new RF frequency. The up-converted frequency may then be amplified by a traveling wave tube amplifier(TWT).

An RF heterodyne repeater is shown in Figure 5.38c. With this type of repeater amplification is carried out directly at RF frequencies. The incoming signal is amplified again, usually by a TWT, and then re-radiated. RF repeaters are troublesome in their design in such things as sufficient selectivity, limiting and automatic gain control, and methods to correct envelope delay. However, some RF repeaters are now available, particularly for operation below 6 GHz.

5.14 FREQUENCY PLANNING (Step *G*)

5.14.1 General

To derive maximum performance from a radiolink system, the systems engineer must set out a frequency usage plan that may or may not have to be approved by the local administration.

The problem has many aspects. First, the useful RF spectrum is limited, from above dc to about 150 GHz. The frequency range of discussion for radiolinks is essentially from the VHF band at 150 MHz (overlapping) to the millimeter region of 15 GHz. Secondly, the spectrum from 150 MHz to 15 GHz must be shared with other services such as radar, navigational aids, research (i.e., space), meteorological, broadcast, etc. For point-to-point communications, we are limited by international agreement to those bands shown in Table 5.1.

Although many of the allocated bands are wide, some up to 500 MHz in width, FM by its very nature is a wide band form of emission. It is not uncommon to have $B_{rf}* = 25$ or 30 MHz for just one emission. Guard bands must also be provided. These are a function of frequency drift of transmitters as well as "splatter" or out-of-band emission, which in some areas is not well specified.

Occupied bandwidth has been specified in Section 5.4.2, Carson's Rule. This same rule is followed by the U.S. regulatory agency, FCC.

---

* B_{rf} = RF bandwidth.

5.14.2 Spurious Emission

The United States Sandards Institute (now the American National Standards Institute) in its memorandum No. 40 to the U.S. Defense Communication Agency quotes the following at the "RF interface":

> The spurious emissions, referred to the RF carrier frequency, shall not exceed the following values, excluding oscillator leak:
>
> $$F_0 \pm 70 \text{ MHz}, -24 \text{ dBm}$$
> $$F_0 \pm 140 \text{ MHz}, -54 \text{ dBm}$$
>
> Harmonics of carrier frequency, -24 dBm

Local Oscillator Leakage

The amount of local oscillator power as measured at the branching network / transmission line point, shall not exceed -60 dBm.

Bandwidth

Each radio channel bandpass filter shall have a bandwidth not less than 40 MHz (at the 3 dB points relative to band center), and the transmission loss shall not vary by more than 0.1 dB over the central 10 MHz.

5.14.3 Radio Frequency Interference (RFI)

On planning a new radiolink system or on adding RF carriers to an existing installation, careful consideration must be given to RFI of the existing (or planned) emitters in the area. Usually the governmental authorizing agency has information on these and their stated radiation limits. Typical limits have been given above. Equally important as those limits is that of antenna directivity and side lobe radiation. Not only must the radiation of other emitters be examined from this point of view but also the capability of the planned antenna to reject unwanted signals. The radiation pattern of all licensed emitters should be known. Convert the lobe level in the direction of the planned installation and convert to EIRP in dBW. This should be done for all interference candidates within interference frequency range. For each emitter's EIRP, run a path loss to the planned installation to determine interference. Such a study could well affect a frequency plan or antenna design.

Nonlicensed emitters should also be looked into. Many such emitters may be classified as industrial noise sources such as heating devices, electronic ovens, electric motors, unwanted radiation from your own and other microwave installations (i.e., radar harmonics). In the 6-GHz band a coordination contour should be carried out to verify interference from earth stations (see CCIR Rep. 448 and Rec. 359-2). For general discussion on the techniques for calculating interference noise in radiolink systems see CCIR Rep. 388-1.

5.14.4 Overshoot

Overshoot interference may occur when radiolink hops in tandem are in a straight line. Consider stations A, B, C, and D in a straight line, or that a straight line on a map drawn between A and C also passes through B and D. Link A-B has frequency F_1 from A to B. F_1 is reused in direction C to D. Care must be taken that some of the emission F_1 on the A-B hop does not spill into the receiver at D. Reuse may even occur on an A, B, and C combination, so F_1 at A to B may spill into a receiver at C tuned to F_1. This can be avoided provided stations are removed from the straight line. In this case the station at B should be moved to the north of a line A to C, for example.

5.14.5 Transmit-Receive Separation

If a transmitter and receiver are operated on the same frequency at a radiolink station, the loss between them must be at least 120 dB. One way to assure the 120 figure is to place all "go" channels in one-half of an assigned band and all "return" channels in the other. The terms "go" and "return" are used to distinguish between the two directions of transmission.

5.14.6 Basis of Frequency Assignment

"Go" and "return" channels are assigned as in the preceding section. For adjacent RF channels in the same half of the band, horizontal and vertical polarizations are used alternately. To carry this out we may assign, as an example, horizontal polarization (H) to the odd numbered channels in both directions on a given section and vertical polarization (V) to the even numbered channels. The order of isolation between polarization is on the order of 20 dB, but often specified as 26 dB or more.

Table 5.8 Frequency Frogging at Radiolink Repeaters

	Minimum Separation (MHz)	
Number of Voice Channels	2000–4000 MHz	6000–8000 MHz
120 or less	120	161
300 or more	213	252˙

Source: *USAF Pub. TO 31R5-1-9.*

In order to prevent interference between antennas at repeaters between receivers on one side and transmitters in the same chain on the other side of the station, each channel shall be shifted in frequency (called frequency "frogging") as it passes through the repeater station. Recommended shifts of frequency are shown in Table 5.8.

5.14.7 IF Interference

Care must be taken when assigning frequencies of transmitter and receiver local oscillators, as to whether these are placed above or below the desired operating frequency. Avoid frequencies that emit F_R, the received channel frequency, or $F_R \pm 70$ MHz for equipment with 70-MHz IF's or $F_R \pm 140$ MHz when 140-MHz IF's are used. Often plots of all station frequencies are made on graph paper to assure that forbidden combinations do not exist. When close frequency stacking is desired, and/or nonstandard IF's are to be used, the system designer must establish rules as to minimum adjacent channel spacing and receive/transmit channel spacing. A listing of CCIR recommendations regarding channel spacing is given in the next section.

5.14.8 CCIR Recommendations

Regarding frequency assignment, the system designer should consider the following CCIR recommendations:

Rec. No.	Description
279-1	300-channel systems using FDM in the 2- and 4-GHz band.
283-2	60–120- and 300-channel telephony systems operating in the 2-GHz band.

Rec. No.	Description
382-2	Telephony and equivalent TV systems, RF channel arrangements for systems for 600–1800 voice channels, or equivalent operating in the 2- and 4-GHz bands.
383-1	Telephone and equivalent TV systems for systems with a capacity of 1800 voice channels operating in the 6-GHz band.
384-1	Telephone and equivalent TV systems for systems with a capacity of either 2700 or 960 telephone channels or equivalent operating in the 6-GHz band.
385	60–120 and 300 FDM channel systems operating in the 7-GHz band.
386-1	Telephone and equivalent TV systems for systems with a capacity of 960 FDM voice channels or equivalent operating in the 8-GHz band.
387-1	Telephony and equivalent TV systems with 600–1800 FDM channel capacity or equivalent operating in the 11-GHz bands.
389-1	Auxiliary radio systems, for telephony or TV transmission, for systems operating in the 2-, 4-, 6-, or 11-GHz bands.

5.15 ALARM AND SUPERVISORY SYSTEMS

5.15.1 General

Many radiolink sites are unattended and many others remain unattended for weeks or months. To assure improved system reliability, it is desirable to know the status of unattended sites at a central, manned location. This is accomplished by means of a fault reporting system. Such fault alarms are called status reports. Those radiolink sites originating status reports are defined as reporting locations. A site which receives and displays such reports is defined as a supervisory location. This is the standard terminology of the industry. Normally supervisory locations are those terminals terminating a radiolink section. Status reports may also be required to be extended over a wire circuit to a remote location.

5.15.2 Monitored Functions

The following functions at a radiolink site which is a reporting location (unmanned) may be desirable for status reports:

Equipment Alarms

Loss of receive signal
Loss of pilot (receiver)
High noise (receiver)
Power supply failure
Loss of modulating signal
TWT overcurrent
Low transmitter output
Off-frequency operation
Hot-standby actuation

Site Alarms

Illegal entry
Commercial power failure
Low fuel supply
Standby power unit failure
Standby power unit on-line
Tower light status

Often alarms are categorized into "major" (urgent) and "minor" (non-urgent) in accordance with their importance. A major alarm may be audible as well as visible on the status panel. A minor alarm may then show only as an indication on the status panel.

The design intent is to make all faults binary: a light is either on or off, the receive signal level has dropped below -100 dBw or the noise is above so many picowatts; the power is below -3 dB of its proper level, etc. By keeping all functions binary using relay closure (or open) or equivalent transistor circuitry, the job of coding alarms for transmission is much easier. Thus all alarms are of a "go, no-go" nature.

5.15.3 Transmission of Fault Information

Common practice today is to transmit fault information in a voice channel associated with the service channel groupings of voice channels (see Section 5.5.1). Binary information is transmitted by voice frequency telegraph equipment using FSK or tone-on, tone-off as described in Section 8.13 or on-off tone equipment. Depending on the system used, 16, 18, or 24 tone channels may occupy the voice channel assigned. A tone channel is assigned to each reporting location (i.e., each reporting location will have a tone transmitter operating on specific tone frequency assigned to it). The supervisory location will have a tone receiver for each reporting (unmanned site under its supervision) location.

At each reporting location the points described in Section 5.15.2 are scanned every so many seconds (between 2 and 8) and the information from each monitor or scan point is time division multiplexed in a simple code. The pulse output from each tone receiver at the supervisory location represents a series of reporting information on each remote, unmanned site. The coded sequence is demultiplexed and displayed on a status panel.

A simpler method is the tone-on or tone-off method. Here the presence of a tone indicates a fault in a particular time slot; in the other method, it is indicated by the absence of a tone. A device called a fault-interrupter panel is used to code the faults so that different faults may be reported on the same tone frequency.

5.15.4 Remote Control

Through a similar system operating in the opposite direction the supervisory station can control certain functions at reporting locations via a VFTG tone link, with a tone frequency assigned to each separate reporting location. In this case the VFTG is almost always FSK. If only one function is to be controlled as turning on tower lights, then a mark condition would represent lights on, and a space, lights off. If more than one condition to be controlled, then coded sequences are used to energize or deenergize the proper function at the remote reporting location.

5.16 ANTENNA TOWERS AND MASTS

5.16.1 General

Two types of towers are used for radiolink systems: guyed and self-supporting. However, other natural or man-made structures should also be considered, or at least taken advantage of. The radiolink engineer should consider mountains, hills, and ridges so that tower heights may be reduced. He should also consider office buildings, hotels, grain elevators, high-rise apartment houses, and other steel structures (e.g., the sharing of a TV broadcast tower) for direct antenna mounting. For tower heights of 30–60 ft, wooden masts are often used.

One of the most desirable construction materials for a tower is hot dipped galvanized steel. Guyed towers are usually preferred because of overall economy and versatility. Although guyed towers have the advantage that they can be placed closer to a shelter or building than self-supporting types, the fact that they need a larger site may be a disadvantage where land values are high. The larger site is needed because additional space is required for installing guy anchors. Table 5.9 shows approximate land areas needed for several tower heights.

Tower foundations should be reinforced concrete with anchor bolts firmly embedded. Economy or cost versus height tradeoffs usually limit tower heights to no more than 300 ft (188 m). Soil bearing pressure is a major consideration in tower construction. Increasing the foundation area increases soil bearing capability or equivalent design pressure. Wind loading under no-ice (i.e., normal) conditions is usually taken as 30 lb/ft^2 for flat surfaces. A design guide (EIA RS-222-A) indicates that standard tower foundations and anchors for self-supporting and guyed towers should be designed for a soil pressure of 4000 lb/tf^2 acting normal to any bearing area under specified loading.

5.16.2   Tower Twist and Sway

As any other structure, a radiolink tower tends to twist and sway due to wind loads and other natural forces. Considering the narrow beamwidths referred to in Section 5.5.4 (Table 5.5), with only a little imagination we can see that only a very small deflection of a tower and/or antenna will cause a radio ray beam to fall out of the reflection face of an antenna on the receive side or move the beam out on the far-end transmit side of a link.

Twist and sway, therefore, must be limited. Table 5.10 sets certain limits. The table has been taken from *EIA RS-222A*. From the table we can see that angular deflection and tower movement is a function of wind velocity. It should also be noted that the larger the antenna, the smaller the beamwidth, besides the fact that the sail area is larger. Thus the larger the antenna (and the higher the frequency of operation), the more we must limit the deflection.

Table 5.9 Minimum Land Area Required for Guyed Towers

Tower Height (ft)	Area Required[a] (ft)				
	80% Guyed	75% Guyed	70% Guyed	65% Guyed	60% Guyed
60	87 × 100	83 × 96	78 × 90	74 × 86	69 × 80
80	111 × 128	105 × 122	99 × 114	93 × 108	87 × 102
100	135 × 156	128 × 148	120 × 140	113 × 130	105 × 122
120	159 × 184	150 × 174	141 × 164	132 × 154	123 × 142
140	183 × 212	178 × 200	162 × 188	152 × 176	141 × 164
160	207 × 240	195 × 226	183 × 212	171 × 198	159 × 184
180	231 × 268	218 × 252	204 × 236	191 × 220	177 × 204
200	255 × 296	240 × 278	225 × 260	210 × 244	195 × 226
210	267 × 304	252 × 291	236 × 272	220 × 264	204 × 236
220	279 × 322	263 × 304	246 × 284	230 × 266	213 × 246
240	303 × 350	285 × 330	267 × 308	249 × 288	231 × 268
250	315 × 364	296 × 342	278 × 320	254 × 282	240 × 277
260	327 × 378	308 × 356	288 × 334	269 × 310	249 × 288
280	351 × 406	330 × 382	309 × 358	288 × 332	267 × 308
300	375 × 434	353 × 408	330 × 382	308 × 356	285 × 330
320	399 × 462	375 × 434	351 × 406	327 × 376	303 × 350
340	423 × 488	398 × 460	372 × 430	347 × 400	321 × 372
350	435 × 502	409 × 472	383 × 442	356 × 411	330 × 381
360	447 × 516	420 × 486	393 × 454	366 × 424	339 × 392
380	471 × 544	443 × 512	414 × 478	386 × 446	357 × 412
400	495 × 572	465 × 536	425 × 502	405 × 468	375 × 434
420	519 × 599	488 × 563	456 × 527	425 × 490	393 × 454
440	543 × 627	510 × 589	477 × 551	444 × 513	411 × 475

Tower Height (ft)	Area Required (acre)				
	80% Guyed	75% Guyed	70% Guyed	65% Guyed	60% Guyed
60	0.23	0.21	0.19	0.17	0.15
80	0.38	0.34	0.30	0.26	0.23
100	0.56	0.50	0.44	0.39	0.34
120	0.77	0.69	0.61	0.53	0.46
140	1.03	0.91	0.80	0.70	0.61
160	1.31	1.16	1.03	0.90	0.77
180	1.63	1.45	1.27	1.11	0.96
200	1.99	1.76	1.55	1.35	1.16
210	2.18	1.93	1.70	1.48	1.27
220	2.38	2.11	1.85	1.61	1.39

Table 5.9   (*Continued*)

Tower Height (ft)	Area Required (acre)				
	80% Guyed	75% Guyed	70% Guyed	75% Guyed	60% Guyed
240	2.81	2.49	2.18	1.90	1.63
250	3.04	2.69	2.36	2.05	1.76
260	3.27	2.89	2.54	2.21	1.90
280	3.77	3.33	2.92	2.54	2.18
300	4.30	3.80	3.33	2.89	2.49
320	4.87	4.30	3.77	3.27	2.81
340	5.48	4.84	4.24	3.65	3.15
350	5.79	5.11	4.48	4.88	3.33
360	6.12	5.40	4.73	4.10	3.52
380	6.79	5.99	5.25	4.55	3.90
400	7.50	6.62	5.79	5.02	4.30
420	8.24	7.27	6.36	5.52	4.73
440	9.03	7.96	6.96	6.03	5.17

[a] Preferred area is a square using the larger dimesnion of minimum area. This will permit orienting tower in any desired position.

To reduce twist and sway, tower rigidity must be improved. One generality we can make is that towers that are designed to meet required wind load or ice load specifications are sufficiently rigid to meet twist and sway tolerances. One way to increase rigidity is to increase the number of guys, particularly at the top of the tower. This is often done by doubling the number of guys from three to six.

5.17   PLANE REFLECTORS AS PASSIVE REPEATERS

A plane reflector as a passive repeater offers some unique advantages. Suppose we wish to provide multichannel telephone service to a town in a valley and a mountain is nearby with poor access to its top. The radiolink engineer should consider the use of a passive repeater as an economic alternative. A prime requirement is that the plane reflector be within line-of-sight of the terminal antenna in town as well as line-of-sight of the

Table 5.10 Nominal Twist and Sway Values for Microwave Tower—Antenna—Reflector Systems[a]

A	Tower Mounted Antenna		Tower Mounted Passive Reflector		
Total Beamwidth of Antenna or Passive Reflector between Half-Power Points (°)	B Limits of Movement of Antenna Beam with Respect to Tower (±°)	C Limits of Tower Twist or Sway at Antenna Mounting Point (±°)	D Limits of Movements of Passive Reflector with Respect to Tower (±°)	E Limits of Tower Twist at Passive Reflector Mounting Point (±°)	F Limits of Tower Sway at Passive Reflector Mounting Point (±°)
14	0.75	4.5	0.2	4.5	4.5
13	0.75	4.5	0.2	4.5	4.3
12	0.75	4.5	0.2	4.5	3.9
11	0.75	4.5	0.2	4.5	4.6
10	0.75	4.5	0.2	4.5	3.3
9	0.75	4.5	0.2	4.5	3.9
8	0.75	4.2	0.2	4.5	2.6
7	0.6	4.1	0.2	4.5	2.3
6	0.5	4.0	0.2	4.3	2.1
5	0.4	3.4	0.2	3.7	1.8
4	0.3	3.1	0.2	3.3	1.6
3.5	0.3	2.9	0.2	2.9	1.4
3.0	0.3	2.3	0.2	2.5	1.2
2.5	0.2	1.9	0.1	2.1	1.0
2.0	0.2	1.5	0.1	1.7	0.9
1.5	0.2	1.1	0.1	1.2	0.6

1.0	0.1	0.9	0.1	0.9	0.5
0.75 [b]	0.1	0.7	0.1	0.7	0.4
0.5 [b]	0.1	0.4	0.1	0.4	0.2

Source: *EIA RS-222A*.

[a] The values are tabulated as a guide for systems design and are based on values that have been found satisfactory in the operational experience of the industry. These data are listed for reference only.

[b] These deflections are extrapolated and are not based on experience of the industry.

Notes

1. Half-power beamwidth of the antenna to be provided by the purchaser of the tower.

2. a. The limits of beam movement resulting from an antenna mounted on the tower are the sum of the appropriate figures in columns *B* and *C*.

   b. The limits of beam movement resulting from twist when passive reflectors are empolyed are the sum of the appropriate figures in columns *D* and *E*.

   c. The limits of the beam movement resulting from sway when passive reflectors are employed are twice the sum of the appropriate figures in columns *D* and *F*.

   d. The tabulated values in columns *D*, *E*, and *F* are based on a vertical orientation of the antenna beam.

3. The maximum tower movement shown above (4.5°) will generally be in excess of that actually experienced under conditions of 20 lb/ft² wind loading.

4. The problem of linear horizontal movement of a reflector-parabola combination has been considered. It is felt that in a large majority of cases, this will present no problem. According to tower manufacturers, no tower will be displaced horizontally at any point on its structure more than 0.5 ft/100 ft of height under its designed wind load.

5. The values shown correspond to 10 db gain degradation under the worst combination of wind forces at 20 lb/ft². This table is meant for use with standard antenna-reflector configurations.

6. Twist and sway limits apply to 20 lb/ft² wind load only, regardless of survival or operating specifications. If there is a requirement for these limits to be met under wind loads greater than 20 lb/ft², such requirements must be specified by the user.

255

distant radiolink station. Such a passive repeater installation may look like the following example where $a =$ the net path loss in decibels:

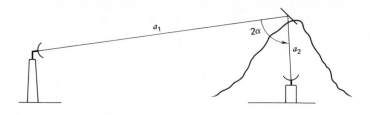

$$a = G_t + G_r + G_A - a_1 - a_2 \qquad (5.23)$$

where $a_1 =$ path loss (dB) on path 1
$\quad\quad\; a_2 =$ path loss (dB) on path 2
$\quad\quad\; G_t =$ transmitting antenna gain
$\quad\quad\; G_r =$ receiving antenna gain
$\quad\quad\; G_A =$ passive reflector gain

Let us concern ourselves with G_A for the moment. The gain of a passive reflector results from the capture of a ray beam of RF energy from a distant antenna emitter and the redirection of it toward a distant receiving antenna. The gain of the passive is divided into two parts: (1) incoming energy, and (2) redirected or reflected energy. The gain for (1) and (2) is

$$G_A = 20 \log \frac{4\pi A \cos \alpha}{\lambda^2} \qquad (5.24)$$

where $\alpha = \frac{1}{2}$ the included horizontal angle between incident and reflected wave
$\quad\quad A =$ the surface area (ft^2 or m^2). If A is in square feet, then λ must also be in feet

The reader will find the following relationship useful:

$$\text{wavelength in feet} = \frac{985}{F}$$

where F is in megahertz. Passive reflector path calculations are not difficult. The first step is to determine if the shorter path (a_2 path in the figure above) places the passive reflector in the near field of the nearer parabolic antenna. Remember that when we dealt with periscopic systems, the reflector was always in the near field. See Section 5.5.7. To determine whether near-field or far-field, solve the following formula:

$$\frac{1}{k} = \frac{\pi\lambda d'}{4A}$$

If the ratio $1/k$ is less than 2.5, a near-field condition exists; if $1/k$ is greater than 2.5, a far-field condition exists. d' is the length of the path in question (i.e., the shorter distance).

For the far-field condition: consider path 1 and path 2 as separate paths and sum their free-space path losses. Determine the gain of the passive plane reflector using formula 5.24. Sum this gain with the two free-space path losses algebraically to obtain the net path loss.

For the near-field condition: (where $1/k$ is less than 2.5): the free-space path loss will be that of the longer hop (a_1 above, for instance); algebraically add the repeater gain (or loss) which is determined as follows.

Compute the parabola/reflector coupling factor l:

$$l = D'\sqrt{\pi/4A}$$

where D' = diameter of parabolic antenna (ft)

$\quad\quad A$ = effective area of passive (ft^2)

Figure 5.39 is now used to determine near-field gain or loss. The $1/k$ value is on the abscissa and l, the family of curves shown.

Example. *Far-field:* a plane passive reflector 10×16 ft, or 160 ft^2, is erected 21 mi from one active site and only 1 mi from the other. $2\alpha = 100°$, $\alpha = 50°$. The operating frequency is 2000 MHz By formula the free-space loss for the longer path is 129.5 dB, and for the shorter, 103 dB.

Calculate gain of passive plane reflector, G_A (formula 5.24):

$$G_A = \frac{20 \log 4 \times 160 \cos 50°}{(985/2000)^2} = 20 \log 5.340$$

$$= 74.6 \text{ dB}$$

Net path loss $= -129.5 - 103 + 74.6 = 157.9$ dB

Example. *Near-field:* The passive reflector selected in this case is 24×30 ft. The operating frequency is 6000 MHz. The long leg is 30 mi and the short leg, 4000 ft. A 10-ft parabolic antenna is associated with the active site on the short leg. (6 GHz is approximately equivalent to 0.164 ft.) Determine $1/k$.

$$\frac{1}{k} = \frac{\pi\lambda d}{4A} = \frac{\pi(0.164)\,(4000)}{4 \times 720} = 0.717$$

Note that this figure is less than 2.5, indicating the near-field condition. Calculate l.

$$l = D'\sqrt{\pi/4A} = 10\sqrt{\pi/4 \times 720} = 0.33$$

Using these two inputs, the value of l and $1/k$, we go to Figure 5.39 and find the net gain of the system is $+0.2$ dB. Net free-space loss is then $+0.2 - 142.3 = 142.1$ dB. The value 142.3 dB is the free-space loss of the 30-mi leg.

5.18 NOISE PLANNING ON RADIOLINKS

CCIR Rec. 393-1 allots 10,000 pW psophometrically weighted noise on a 2500-km reference circuit carrying FDM telephony. Hence total noise accumulation on a per-kilometer basis is

$$\frac{10,000\ \mathrm{pWp}}{2{,}500\ \mathrm{km}} = 4\ \mathrm{pWp/km}$$

Of the 10,000 pW, 2500 is allotted to terminal equipment (i.e., the multiplex equipment; see Chapter 3) and 7500 pW to line equipment. In the case of radiolinks, the line equipment is the radio equipment. Thus

$$\frac{7500\ \mathrm{pWp}}{2500\ \mathrm{km}} = 3\ \mathrm{pWp/km}$$

This is the maximum noise accumulation permitted on a per-kilometer basis. (See CCIR Rec. 395-1 as well.)

Note. To obtain pWp, when "flat" noise is given in a 3.1-kHz channel, reduce the figure in picowatts by 2.5 dB. See Section 1.9.6.

For this discussion we are dealing with thermal noise and intermodulation noise which may be considered additive. For basic discussions on noise, refer to Chapter sections 1.6.7 and 5.6.5.

Here we assume that a system is loaded as indicated in Section 5.6.2 such that NPR is optimized, that is, loaded (i.e., the baseband input level) with sufficient deviation to minimize the effects of thermal noise, but not loaded to such an extent that intermodulation noise is excessive. The curves in Figure 5.30 provide data on this sort of loading.

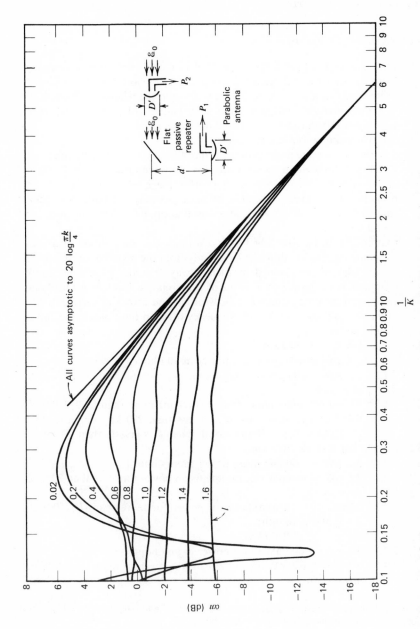

Figure 5.39 Antenna-reflector efficiency curves. Courtesy Microflect Inc., Salem, Ore.

259

For noise engineering real circuits, CCIR Rec. 395-1 offers the following guidelines (quoted in part):

CCIR UNANIMOUSLY RECOMMENDS

1. that, in circuits established over real links which do not differ appreciably from the hypothetical reference circuit, the psophometrically weighted* noise power at a point of zero relative level in the telephone channels of frequency-division multiplex radio-relay systems of length L, where L is between 280 and 2500 km, should not exceed:
1.1 $3 L$ pW mean power in any hour†;
1.2 $3 L$ pW one-minute mean power for more than 20% of any month;
1.3 47,500 pW one-minute mean power for more than $(L/2500) \times 0.1\%$ of any month; it is recognized that the performance achieved for very short periods of time is very difficult to measure precisely and that, in a circuit carried over a real link it may, after installation, differ from the planning objective;
2. that circuits to be established over real links, the composition of which, for planning reasons, differs substantially from the hypothetical reference circuit, should be planned in such a way that the psophometrically weighted noise power at a point of zero relative level in a telephone channel of length L, where L is between 50 and 2500 km, carried in one or more baseband sections of frequency-division multiplex radio links, should not exceed:
2.1 for 50 km $\leqq L \leqq$ 840 km:
　2.1.1 $3 L$ pW + 200 pW mean power in any hour,†
　2.1.2 $3 L$ pW + 200 pW one-minute mean power for more than 20% of any month,
　2.1.3 47,500 pW one-minute mean power for more than $(280/2500) \times 0.1\%$ of any month when L is less than 280 km, or more than $(L/2500) \times 0.1 \times$ of any month when L is greater than 280 km;
2.2 for 840 km $< L \leqq$ 1670 km:
　2.2.1 $3 L$ pW + 400 pW mean power in any hour,†
　2.2.2 $3 L$ pW + 400 pW one-minute mean power for more than 20% of any month,
　2.2.3 47,500 pW one-minute mean power for more than $(L/2500) \times 0.1\%$ of any month;
2.3 for 1670 km $< L \leqq$ 2500 km:
　2.3.1 $3 L$ pW + 600 pW mean power in any hour,†
　2.3.2 $3 L$ pW + 600 pW one-minute mean power for more than 20% of any month,

---

* The level of uniform-spectrum noise power in a 3.1 kHz band must be reduced by 2.5 dB to obtain the psophometrically wieghted noise power.

† The hourly mean noise power objective and its subdivision are at present under study (see Recommendation 393-1).

2.3.3   47,500 pW one-minute mean power for more than $(L/2500) \times$ 0.1 % of any month;

3.   that the following Notes should be regarded as part of the Recommendation:

Note 1. Noise in the frequency-division multiplex equipment is excluded. On a 2500 km hypothetical reference circuit the CCITT allows 2500 pw mean value for this noise in any hour.

Considering that the hypothetical reference circuit described in CCIR Rec. 393-1 is divided into nine homogeneous sections alloting 840 pW per section (approximately), then each section may be treated as in Table 5.11 (taken from Table 4, Section B.IV.4 of the ITU publication *Economics and Technical Aspects of the Choice of Transmission Systems*).

Table 5.11   Noise on Radiolinks

Radiolink, FM

1 modulation section consisting of . . . . hops

		Lower	Center	Upper	kHz
Number of modulation sections	(1)		1		
Number of hops	(2)		6		
Measuring channel	(3)	Lower	Center	Upper	kHz
Basic noise					
Thermal receiver noise	(4)		−72.2[a]	−71.2[a]	dBm0p
	(5)		60	76[a]	pW0p
Basic noise of the RF equipment	(6)		130[a]	120[a]	pW0p
Basic noise of the modem equipment	(7)		65[a]	60[a]	pW0p
Total thermal noise under no-fading	(8)		255	256	pW0p
condition (sum of 5 + 6 + 7)	(9)		−65.9	−65.9	dBm0p
Intermodulation noise					
RF and IF equipment	(10)		389[a]	360[a]	pW0p
Modem	(11)		65[a]	60[a]	pW0p
Total intermodulation noise	(12)		454	420	pW0p
(sum of 10 plus 11)	(13)		−63.4	−63.8	dBm0p
Total noise under no-fading con-	(14)		709	676	pW0p
ditions (sum of 8 plus 12)	(15)		−61.6	−61.7	dBm0p
Permissible total noise	(16)		840	840	pW0p
	(17)		−60.8	−60.8	dBm0p

Table 5.11   (*Continued*)

Permissible rise in receiver thermal noise (difference between 16 and 14)	(18)	131	164	pW0p
Permissible thermal receiver noise (sum of 18 plus 5)	(19)	191	240	pW0p
	(20)	−67.2	−66.2	dBm0p
Average fading margin (20 minus 4)	(21)	5	5	dB

Courtesy International Telecommunications Union.
a Values taken to demonstrate the procedure of calculation.

It should be noted that row 21, the 5-dB fade margin, is academic. As we established previously in Sections 5.8 and 5.9, each system must be examined under real conditions. Besides, the margin of fade in Table 5.10 is for six hops in tandem and assumes noncoincident fading in the individual hops. Nonetheless, the table offers a good approach to noise engineering of radiolink systems.

REFERENCES AND BIBLIOGRAPHY

1. *Reference Data for Radio Engineers*, 5th ed., Howard W. Sams & Co., Indianapolis.
2. *Microwave Path Engineering Considerations*, Lenkurt Electric Co., San Carlos, Calif., Sept. 1961.
3. CCIR, New Delhi, 1970, Study Group 9, Vol. IV.
4. CCITT White Books, Mar del Plata, 1968, Vol. III, G. Recommendations.
5. D. M. Hamsher, *Communication System Engineering Handbook*, McGraw-Hill, New York, 1967.
6. *Collins CEL 19* and *22*, Collins Radio Co., Dallas, Texas.
7. *Microwave Radio Relay Systems*, USAF T.O. 31R5-1-9, 1 April 1965, U.S. Department of Defense, Washington, D.C.
8. W. Oliver, *White Noise Loading of Multichannel Communication Systems*, Marconi Instruments, Sept. 1964.
9. *Electrical Communications Systems Engineering—Radio*, U.S. Army TM 11-486-6, U.S. Department of Defense, Washington, D.C.
10. *Transmission Systems for Communications*, 4th ed., Bell Telephone Laboratories.
11. B. D. Holbrook and J. T. Dixon, "Load Rating Theory for Multichannel Amplifiers," *Bell Syst. Tech. J.*, **18**, 624–644 (Oct. 1939).
12. *Technical Note 100*, National Bureau of Standards, Boulder, Colo., Jan. 1962.
13. *Technical Note 103*, National Bureau of Standards, Boulder, Colo., Jan. 1963.
14. *United States Standards Institute Memorandum No. 40*, Microwave Line-of-sight Systems to Participating Companies, Sept. 1966.
15. F. E. Terman, *Radio Engineering*, 4th ed., McGraw-Hill, New York.

16. D. C. Livingston, *The Physics of Microwave Propagation*, GTE Monograph, General Telephone and Electronics, 1967.

17. *Microflect Passive Repeater Engineering Manual No. 161*, Microflect Inc., Salem, Ore., 1962.

18. *Jerrold Path Calculations*, Jerrold Electronics Corp., Philadelphia, Pa., 1967.

19. Electronics Industry Association (EIA) Standards: RS-195-A, RS-203, RS-222A, and RS-250A.

20. K. Bullington, "Radio Propagation Fundamentals," *Bell Syst. Tech. J.*, May 1957.

21. J. Jasik, *Antenna Engineering Handbook*, McGraw-Hill, New York.

22. R. L. Marks et al., *Microwave and Troposcatter Systems*, Rome Air Development Center, USAF (AD 617 686), April 1965.

23. J. Fagot and P. Magne, *Frequency Modulation Theory*, Pergamon Press, London, 1961

24. *Andrew Catalog 27*, Andrew Corporation, Orland Park, Ill., 1971.

25. *Collins Telecommunication Equipment Catalog No. 2*, Collins Radio Co., Dallas, Texas.

26. *A Survey of Microwave Fading Mechanisms, Remedies and Applications*, ESSA Technical Report ERL 69-WPL 4, Boulder, Colo., March 1968.

27. Philip F. Panter, *Communication Systems Design—Line-of-Sight and Tropo-scatter Systems*, McGraw-Hill, New York, 1972.

28. Robert F. White, *Reliability in Microwave Communication Systems—Prediction and Practice*, Lenkurt Electric Corp., San Carlos, Calif., 1970.

29. A. P. Barsis et al., *Analysis of Propagation Measurements over Irregular Terrain in 76 to 9200 MHz Range*, ESSA Technical Report ERL 114-ITS 82, Boulder, Colo., March 1969.

30. K. W. Pearson, "Method for the Prediction of the Fading Performance of the Multisection Microwave Link," *Proc. IEE*, **112**(7), July 1965.

31. R. G. Medhurst, "Rainfall Attenuation of Centimeter Waves: Comparison of Theory and Measurement," *IEEE Trans. Antennas Propag.*, July 1965.

6 | TROPOSPHERIC SCATTER

6.1 INTRODUCTION

Tropospheric scatter is one method of propagating microwave energy beyond line-of-sight or "over-the-horizon." Communication systems utilizing the tropospheric scatter phenomena handle from 12 to 240 frequency division multiplex telephone channels. Well-planned tropospheric scatter links may have propagation reliabilities on the order of 99.9% or better. These reliabilities are comparable to those of radiolink systems (LOS microwave) discussed in the preceding chapter. In fact, the discussion of tropo (tropospheric scatter) is a natural extension of Chapter 5.

Tropo takes advantage of the refraction and reflection phenomena in in a section of the earth's atmosphere called the troposphere. This is the lower portion of the atmosphere from sea level to a height of about 11 km (35,000 ft). UHF signals are scattered in such a way as to follow reliable communications on hops up to 640 km (400 mi). Long distances of many thousands of kilometers may be covered by operating a number of hops in tandem. The North Atlantic Radio System (NARS) of the United States Air Force is an example of a lengthy tandem system. It extends from Canada to Great Britain via Greenland, Iceland, the Faeroes, and Scotland. A mix of radiolinks (LOS microwave) and tropo is becoming fairly common. The Canadian National Telephone Company (CNT) operates such a system in the Northwest Territories. The Bahama Islands are interconnected for communications by a mix of radiolinks, tropo, and HF.

Tropospheric scatter systems generally use transmitter power outputs of 1 or 10 kW, parabolic type of antennas with diameters of 4.5 (15), 9 (30), or 18 (60 ft), and sensitive (uncooled) broadband FM receivers with front-end noise figures on the order of 2.5–5 dB. A tropo installation is obviously a bigger financial investment than a radiolink (LOS micro-

264

wave) installation. Tropo, however, has many advantages for commercial application that could well outweigh the issue of high cost. These advantages are summarized as follows:

1. Reduces the number of stations required to cover a given large distance when compared to radiolinks. Tropo may require from one-third to one-tenth the number of stations as a radiolink system over the same path.
2. Provides reliable multichannel communication across large stretches of water (e.g., over inland lakes, to offshore islands, between islands) or between areas separated by inaccessible terrain.
3. May be ideally suited to meet toll connecting requirements of areas of low population density.
4. Useful when radio waves must cross territories of another political administration.
5. Requires less maintenance staff per route-kilometer than conventional radiolink systems over the same route.
6. Allows multichannel communication with isolated areas, especially. when intervening territory limits or prevents the use of repeaters.

6.2 THE PHENOMENON OF TROPOSPHERIC SCATTER

There are a number of theories explaining over-the-horizon communications by tropo. One theory postulates atmospheric air turbulence, irregularities in the refractive index, or similar homogeneous discontinuities capable of diverting a small fraction of the transmitted radio energy toward a receiving station. This theory accounts for the scattering of radio energy in a way much as fog or moisture seems to scatter a searchlight on a dark night. Another theory is that the air is stratified into discrete layers of varying thickness in the troposphere. The boundaries between these layers become partially reflecting surfaces for radio waves and thereby scatter the waves downward over the horizon.

Figure 6.1 is a simple diagram of a tropo link showing two important propagation concepts. These are as follows:

- *Scatter angle,* which may be defined as either of two acute angles formed by the intersection of the two portions of the tropo beam (lower boundaries) tangent to the earth's surface. Keeping the angle small effectively reduces the overall path attenuation.
- *Scatter volume* or *"common volume"* is the common enclosed area where the two beams intercept.

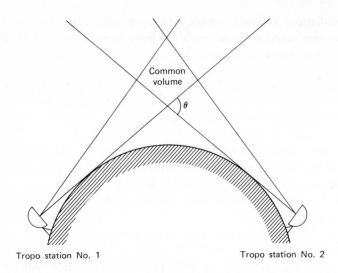

Figure 6.1   Tropospheric scatter model. $\theta =$ scatter angle.

6.3   TROPOSPHERIC SCATTER FADING

Fading is characteristic of tropo. It is handy to break fading in tropospheric scatter systems into two types, slow and fast fading. Expressed another way, these are long-term (slow) and short-term variations in the received signal level.

When referring to tropo received signal levels, we usually use the median received level as reference. In general the hourly median and minute median are the same. In Chapter 5, the reference level was the unfaded signal level, which turned out to approximate sufficiently the calculated level under no-fade conditions. Such a straightforward reference signal level is impossible in tropo because a tropo signal is in a constant condition of fade. Thus for path calculations and path loss, we refer to the long-term median, usually extended over the whole year.

At any one moment a received tropo signal will be affected by both slow and fast fading. It is believed that fast fading is due to the effects of multipath (i.e., due to a phase incoherence at various scatter angles). Fast fading is treated statistically "within the hour" and has a Rayleigh distribution with a sampling time of 1–7 min, although in some circumstances it has been noted up to 1 hr. The fading rate depends on both frequency and distance or length of hop.

Table 6.1   Time Block Assignments

Time Block number	Month	Hours
1	Nov.–April	0600–1300
2	Nov.–April	1300–1800
3	Nov.–April	1800–2400
4	May–Oct.	0600–1300
5	May–Oct.	1300–1800
6	May–Oct.	1800–2400
7	May–Oct.	0000–0600
8	Nov.–April	000–00600

The (U.S.) National Bureau of Standards (NBS) describes long-term variations in signal level as variations of *hourly median* values of transmission loss. This is the level of transmission loss which is exceeded for a total of one-half of a given hour. A distribution of hourly medians gives a measure of long-term fading. Where these hourly medians are considered over a period of 1 month or more, the distribution is log normal.

In studying variations in tropo transmission path loss (fading), we have had to depend on empirical information. Signal level varies with time of day, season of year, and latitude, among other variables. To assist in analysis and prediction of long-term signal variation, the hours of the year have been broken down into eight time blocks, given in Table 6.1.

Most commonly we refer to time block 2 for a specific median path loss. Time block 2 may be thought of as an average winter afternoon in the temperate zone of the northern hemisphere.

It should be noted that signal levels average 10 dB lower in winter than in summer, and that morning or evening signals are at least 5 dB higher than midafternoon. Slow fading is believed due to changes in path conditions such as atmospheric changes, for example, a change in the index of refraction of the atmosphere.

6.4   PATH LOSS CALCULATIONS

Tropospheric scatter paths have large losses when compared to typical radiolink (LOS microwave) paths. In the case of tropo, median path loss may be estimated by several methods. Two such methods are reviewed here. The first we shall discuss is called the "Yeh"* method. Here median path loss,

$$L_{mp} = L_{fs} + L_{sl} - 0.2(N_s - 310) \tag{6.1}$$

Let us consider the three terms separately. The first term is L_{fs}, the free-space loss in decibels discussed in Section 5.3.1, or

$$L_{fs} = 36.6 + 20 \log_{10} D_{mi} + 20 \log_{10} F_{MHz}$$

where D_{mi} is the distance between transmitter at A and receiver at B in statute miles, and F is the operating frequency in megahertz. D_{mi} in this case is the great-circle distance. *Note.* Great-circle caclulations are covered in Section 4.10.

The second term of the equation L_{sl}, is the scatter loss in dB.

$$L_{sl} = 57 + 10(\theta - 1) + 10 \log_{10} \frac{F_{MHz}}{400} \tag{6.2}$$

This is valid for $\theta > 1°$, where $\theta =$ the scatter angle in degrees; and $F_{MHz} =$ carrier frequency (operating frequency) in megahertz. The reader will note that at scatter angles of about 1° at 400 MHz, the scatter loss approaches 57 dB. He will also note that scatter loss is dependent on the scatter angle, increasing about 10 dB for each degree of increase in the scatter angle. Scatter loss is also frequency dependent.

The third term of the equation, $0.2(N_s - 310)$, corrects for variations of the mean yearly refractive index, N_s. Dr. Yeh uses a 0.2-dB increase in scatter loss per unit increase in N_s. Refer to CCIR Rep. 233-1, Vol. II, Oslo, 1966 (Ref. 17). The 0.2 dB is a good figure for temperature zones.

The second method is a modification of the NBS (National Bureau of Standards) method as used in Ref. 3. The median path loss

$$L_{mp2} = 30 \log F - 20 \log d + F(\theta d) + A_a \tag{6.3}$$

where F = frequency (carrier) (MHz)

d = path length (great circle) (km)

θ = scatter angle (*radians*)

$F(\theta d) =$ scatter loss, taken from Figure 6.2

θd = product of the scatter angle (rad) and d, the path length (km)

A_a = atmospheric absorption term and may be disregarded for carrier frequencies below 500 MHz. A value for A_a may be taken from Figure 6.3

---

* Dr. Luang P. Yeh (see Ref. 1).

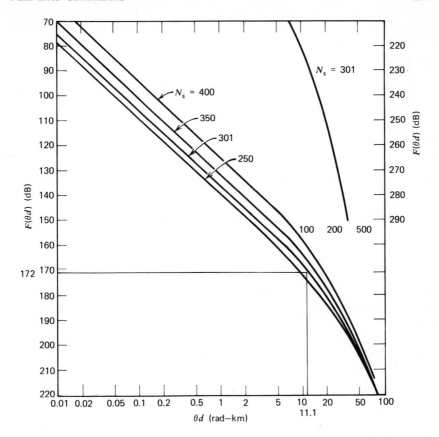

Figure 6.2 The attenuation function for the determination of scatter loss. Courtesy Institute of Telecommunication Sciences, Office of Telecommunications, U.S. Department of Commerce.

The reader should refer to *NBS Tech. Note 101* (revised), Vol. 1, for more details on the NBS method.

Example of Method 1. Given a tropo path 200 mi long, a carrier frequency of 900 MHz, a scatter angle of 2°, and an index of refraction of 300, compute the median path loss.

$L_{fs} = 36.6 + 20 \log 200 + 20 \log 900$
$\quad = 36.6 + 46 + 59.1$
$\quad = 141.7$ dB
$L_{sl} = 57 + 10(2 - 1) + 10 \log 400/900$
$\quad = 57 + 10 + 10 \ (\log 2.25)$
$\quad = 70.5$ dB

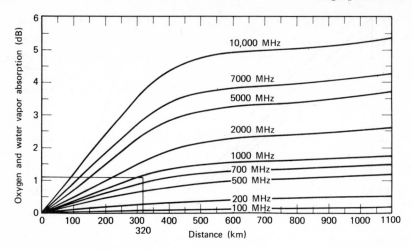

Figure 6.3 Determination of median oxygen and water vapor absorption in decibels, given path length for various operating frequencies (for August, Washington, D.C.). Courtesy Institute of Telecommunication Sciences, Office of Telecommunications, U.S. Department of Commerce.

Correcting for refractive index,

$$-0.2(300-310) = 2 \text{ dB}$$

Thus

$$L_{\text{mp}} = 141.7 + 70.5 + 2$$
$$= 214.2 \text{ dB}$$

Example of Method No. 2. Use the same input data as in method 1. The 200-mi path is equivalent to 320 km.

$$L_{\text{bsr}} = 30 \log F - 20 \log d + F(\theta d) + A_{\text{a}}$$
$$= 30 \log 900 - 20 \log 320 + F(2 \times 1.74 \times 10^{-2} \times 320) + 1.1$$
$$= 30 \times 2.95 - 20 \times 2.50 + F(11.14) + 1.1$$
$$= 88.5 - 50 + 172 + 1.1$$
$$= 211.6 \text{ dB}$$

6.5 APERTURE-TO-MEDIUM COUPLING LOSS

Some tropo link designers include aperture-to-medium coupling loss as another factor in the path loss equation, and others prefer to include it as another loss in the path calculation portion of link design as if it were a waveguide loss or similar. In any event this loss must be included somewhere.

Aperture-to-medium coupling loss has sometimes been called the "antenna gain degradation." It occurs because of the very nature of tropo in that the antennas used are not doing the job we would expect them to do. This is evident if we use the same antenna on a line-of-sight (LOS or radiolink) path. The problem stems from the concept of the common volume. High gain parabolic antennas used on tropo paths have very narrow beamwidths (see Section 6.9.3). The tropo path loss calculations consider a larger common volume than would be formed by these beamwidths. As the beam becomes more narrow due to the higher gain antennas, the received signal level does not increase in the same proportion as it would under free-space (LOS) propagation conditions. The difference between the free-space expected gain and its measured gain on a tropo hop is called the antenna-to-medium coupling loss. This loss is proportional to the scatter angle, θ, and the beamwidth Ω. The beamwidth may be calculated from the formula:

$$\Omega = \frac{7.3 \times 10^4}{F \times D_r} \tag{6.4}$$

where F = carrier frequency (MHz)
D_r = antenna reflector diameter (ft)

The ratio θ/Ω is computed and from this ratio the aperture-to-medium coupling loss may be derived from Table 6.2.

Example of calculation of aperture-to-medium coupling loss. As inputs, use a 30-ft antenna at each end of the path with a 2° scatter angle and a 900-MHz operating frequency.

$$\Omega = \frac{7.3 \times 10^4}{30 \times 900}$$

= 2.7° equivalent to 46 mrad
= 2° is equivalent to 35 mrads, the scatter angle

Thus the ratio $\theta/\Omega = 35/46$ or approximately 0.75. From Table 6.2 this is equivalent to a loss of 1.2 dB.

6.6 TAKEOFF ANGLE

The takeoff angle is probably the most important factor under control of the engineer selecting a tropo site in actual path design. The takeoff angle is the angle between a horizontal ray extending from the radiation

Table 6.2   Antenna-to-Medium Coupling Loss

Coupling Loss (dB)	Antenna Beam-width Ratio, θ/Ω	Coupling Loss (dB)	Antenna Beam-width Ratio, θ/Ω
0.18	0.3	2.95	1.4
0.40	0.4	3.22	1.5
0.60	0.5	3.55	1.6
0.90	0.6	3.80	1.7
1.10	0.7	4.10	1.8
1.20	0.75	4.25	1.9
1.40	0.8	4.63	2.0
1.70	0.9	4.90	2.1
1.95	1.0	5.20	2.2
2.2	1.1	5.48	2.3
2.42	1.2	5.70	2.4
2.75	1.3	6.00	2.5

Source: Ref. 11.

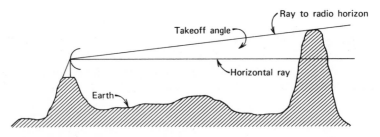

Figure 6.4   Definition of takeoff angle.

center of an antenna and a ray extending from the radiation center of the antenna to the radio horizon. Figure 6.4 illustrates the definition.

The takeoff angle is computed by means of path profiling several miles out from the candidate site location. It then can be verified by means of a transit siting. Path profiling is described in Section 5.3.3.

Figure 6.5 is a graph showing the effect of takeoff angle on transmission loss. As takeoff angle is increased about 12 dB of loss is added for each degree increase in takeoff angle. This loss shows up in the scatter loss term of the equation for computing median path loss (see methods 1 and 2 above). This approximation is valid at 0° in the range of + 10 to − 10°.

The advantage of siting a tropo station on as high a site as possible is obvious. The idea is to minimize obstructions to the horizon in the direc-

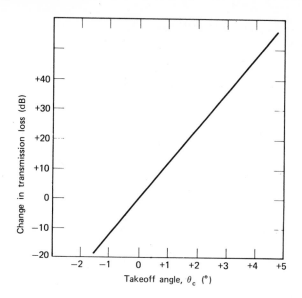

Figure 6.5 Effect of takeoff angle on transmission loss.

tion of the "shot." As we shall see later, every decibel saved in median path loss may represent a savings of many thousands of dollars. Thus the more we can minimize the takeoff angle, the better. Negative takeoff angles are very desirable. Figure 6.6 illustrates this criterion.

6.7 OTHER SITING CONSIDERATIONS

6.7.1 Antenna Height

Increasing antenna height decreases takeoff angle, in addition to the small advantage of getting the antenna up and over surrounding obstacles. Raising an antenna from 20 ft above the ground to 100 ft above the ground provides something on the order of less than 3-dB improvement in median path loss at 400 MHz and about 1 dB at 900 MHz (Ref. 4).

6.7.2 Distance to Radio Horizon

The radio horizon may be considered one more obstacle which the tropo ray beam must get over. Varying the distance to the horizon varies the takeoff angle. If we maintain a constant takeoff angle, distance to the

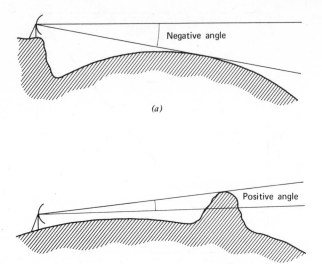

Figure 6.6 (*a*) A more desirable site regarding takeoff angle; (*b*) a less desirable site regarding takeoff angle.

horizon can vary widely with insignificant effect on overall transmission loss.

6.7.3 Other Considerations

If we vary path length with constant takeoff angle, the median path loss varies about 0.1 dB/mi. The primary effect of increasing path length is to change the takeoff angle, which will notably affect total median path loss. This is graphically shown in Figure 6.5.

6.8 PATH CALCULATIONS

This paragraph provides information to assist in determining the basic transmission parameters of a specific tropo hop to meet a set of particular transmission objectives, usually related to an overall system plan. Such objectives may be found in the CCIR, namely, CCIR Recs. 395-1, 396-1, and 397-2. Another objective used by the U.S. Department of Defense is that found in *DCA Circ. 370-175-1,* paragraph 3.2.2.4.1.1.3. These performance objectives deal with noise in the voice channel stating that it should not exceed a particular level during a particular percentage of time. It is recommended that the reader review Section 5.6.5. Here we showed how carrier-to-noise ratio (C/N) can be related to signal-to-noise ratio (S/N). If carrier-to-noise ratio can be related to the desired time frame, then as a consequence, signal-to-noise ratio may be related to the same time percentage.

Path calculations here are based on the same criterion and starting point as in Chapter 5, namely, the thermal noise in the far-end receiver. Thus we start with -204 dBW as the thermal noise absolute floor value for the perfect receiver at room temperature with 1 Hz of bandwidth. The receiver thermal noise threshold is then calculated by algebraically adding the receiver noise figure (dB) and 10 log B_{if}, IF bandwidth in hertz (see Section 1.9.6).

Let us now consider an example. Given:

A 200-mi hop (320 km)
Median path loss 212 dB
Operating frequency 1000 MHz
60 VF channels to be transmitted (i.e., highest modulating frequency
 of 300 kHz); thus $B_{if} = 2.5$ MHz
Receiver noise figure 3 dB
Path (propagation) reliability 99.7%

From this information we can calculate the receiver noise threshold (refer to Section 5.4.2):

Noise threshold $= -204 + 10$ log $B_{if} + 3$
$= -204 + 10$ log $2.5 \times 10^6 + 3$
$= -204 + 10 \times 6.398 + 3$
$= -137$ dBW

(i.e., where the carrier-to-noise ratio, C/N, exactly equals the thermal noise of the receiver; by definition, the noise threshold).

CCIR Rec. 395-1 recommends a thermal plus intermodulation noise accumulation of noise at 3 pW/km. Our path is 320 km long; therefore we may accumulate 320×3 pW of noise or 960 pWp of noise. It should be noted that CCIR refers to noise psophometrically weighted; thus all figures are in pWp.

960 pWp equates to 24 dBa (Section 1.9.6) for a signal-to-noise ratio (S/N) of 58 dB. This conversion is made using figure 5.31.

From a given S/N we can derive a carrier-to-noise ratio (C/N) as follows:

$$\frac{S}{N} = \frac{C}{N} + FM_{dB} + D_{im} - L_f + P_{im} \qquad (6.5)$$

where FM_{dB} = FM improvement factor (dB)

$\quad D_{im}$ = diversity improvement factor (dB)

$\quad L_f$ = NLR = noise load factor (dB) derived in Section 5.6

$\quad P_{im}$ = preemphasis improvement factor (dB) taken from Figure 5.33 and discussed in Section 5.5.6

For the present assume FM_{dB} to be the traditional 20 dB implied in Section 5.4.2. Let

$$D_{im} = 7.2 \text{ dB (refer to Section 5.5.6) (Ref. 5)}$$

L_f = NLR = $- 1 + 4 \log N$ (we use CCIR loading)

$\quad\quad = - 1 + 4 \log 60$

$\quad\quad = 6.12$ dB

$P_{im} = 2.8$ dB (from Figure 5.33), using 300 kHz as the highest modulating frequency. Thus

$$\frac{S}{N} = 58 = \frac{C}{N} + 20 + 7.2 - 6.12 + 2.8 \text{ (all in dB)}$$

$$58 = \frac{C}{N} + 23.88$$

$$\frac{C}{N} = 34.12 \text{ dB}$$

This C/N of 34.12 dB must be maintained 99.7% of the time. Thus to this figure we must add a fade margin suitable for the path conditions. This fade margin may be taken from Figure 6.7. Assuming quadruple diversity, we require a fade margin of 18 dB, or the C/N must be 18 dB higher than the median received signal level, or

$$\frac{C}{N} = 34.12 + 18 \text{ dB} = 52.12 \text{ dB for this particular path}$$

Therefore the signal level entering the receiver must be 52.12 dB higher than the noise threshold of the receiver, or

$$-137 \text{ dBW} + 52.12 \text{ dB} = -84.88 \text{ dBW}$$

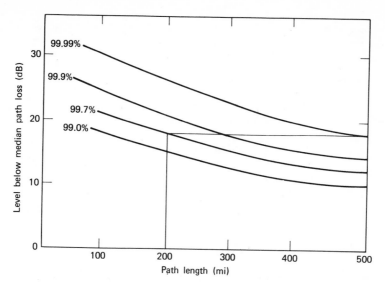

Figure 6.7 Fade margin in decibels related to median path loss (quadruple diversity).

Figure 6.8 Network analogy of tropo path analysis.

To achieve this level, allowing 4 dB total for line losses (2 dB at each end), we must select appropriate antenna sizes for the transmit and receive antennas and a transmitter output. We can reduce the problem to a set of networks in series as shown in Figure 6.8.

From the network in series in Figure 6.8 a simple formula may be derived as follows:

$$-X \text{ dBW} + 2 \text{ dB} - Y \text{ dB} + 212 \text{ dB} - Y \text{ dB} + 2 = -84.88 \text{ dBW}$$

or

$$-X \text{ dBW} - 2Y \text{ dB} + 216 \text{ dB} = -84.88 \text{ dBW}$$
$$X \text{ dBW} + 2Y \text{ dB} = 131.12 \text{ (dBW)}$$

From this we can build a table of values for X and Y to meet the required equivalent EIRP of $+131$ dBW. Table 6.3 shows various combinations of standard tropo transmitter power outputs, and the antenna gains at each

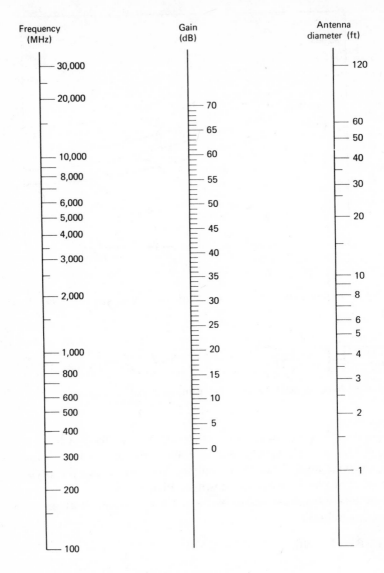

Figure 6.9 Antenna gain nomogram.

Table 6.3   Equipment Selection Table

Transmitter Power (kW)	$-X$ (dBW)	Antenna Reflector Diameters, Transmit and Receive (ft)	Y-dB Antenna Gain for 1 Antenna (dB)	$2Y$-dB Gains Both Antennas (dB)	Sum of Gains to Attain Level (dBW)
1	+30	120	48.5	97	+127[a]
10	+40	90	45.5	91	+131
20	+43	60	44	88	+131
50	+47	53	42	84	+131

[a] Does not meet required equivalent EIRP, thus is discarded.

end to achieve the desired receiver input level. Obviously the 1-kW selection must be discarded. The antenna gains versus reflector diameter are taken from Figure 6.9, the antenna gain nomogram. As we proceed in the discussion on tropo, we can see how to make the most economical selection. For example, one possibility, which would ease the path gain requirement, would be to adopt a method of threshold extension to reduce overall costs.

The sample path shown used very tight requirements regarding noise (i.e., CCIR Rec. 395-1). Tropo paths often are engineered to considerably reduced requirements such as described in CCIR Rec. 397-2, which allows deeper fades (Ref. 18).

Note that if the path reliability (Figure 6.8) were increased to 99.9%, the fade margin would have to be increased about 2.5 dB correspondingly. It can be seen from Figure 6.7 that fade margin in tropo tends to be relaxed (reduced) as path length increases. This tendency is just the reverse of radiolink design, where fade margin increases as path length.

There are several approaches to predicting the proper fade margin. The reader should consult Refs. 3, 5, and 14 for more information.

6.9   EQUIPMENT CONFIGURATIONS

6.9.1   General

As indicated in Section 6.8, tropo equipment must be configured in such a way as to (a) meet path requirements and (b) be an economically viable installation. All tropo installations use some form of diversity (see

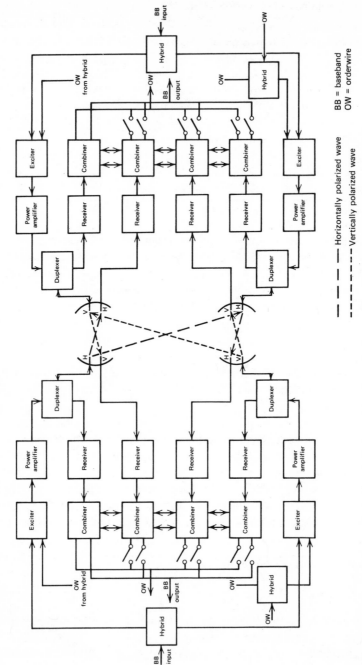

Figure 6.10 Simplified functional block diagram of a quadruple diversity tropospheric scatter configuration.

——— Horizontally polarized wave

------- Vertically polarized wave

BB = baseband

OW = orderwire

Section 5.5.6). Except for some military transportable tropo systems, quadruple diversity is the rule in nearly every case. A typical quadruple diversity tropospheric scatter system layout is shown in Figure 6.10. It is made up of identifiable sections as follows:

1. Antennas, duplexer, transmission lines.
2. Modulator-exciters and power amplifiers.
3. Receivers, preselectors, and threshold extension devices.
4. Diversity and diversity combiners.

Through proper site selection and system layout involving the four categories above, realistic tropo systems can be set up on paths up to 250 statute mi (400 km) in length.

6.9.2 Tropo Operational Frequency Bands

Tropospheric scatter installations commonly operate in the following frequency bands:

350– 450 MHz	755– 985 MHz
1700–2400 MHz	4400–5000 MHz

The reader should also consult CCIR Rec. 388 and CCIR Rep. 285-2 and 286.

6.9.3 Antennas, Transmission Lines, Duplexer, and Other Related Transmission Line Devices

ANTENNAS

The antennas used in tropo installations are broadband, high gain, parabolic reflector devices. The antennas covered here are similar in many respects to those discussed in Section 5.5.4, but have higher gain and therefore larger and considerably more expensive. As we discussed in Section 5.5.4, the gain of this type of antenna is a function of the reflector diameter. Table 6.4 gives some typical gains in dB for several frequency bands and several standardized reflector diameters. Figure 6.9 is a nomogram from which gain in decibels can be derived given the operating frequency and the diameter of the parabolic reflector in feet. A 55% efficiency is assumed for the antenna. It should be noted that the tendency

Table 6.4   Some Typical Antenna Gains

Reflector Diameter (ft)	Frequency (GHz)	Gain (dB)
15	0.4	23
	1.0	31
	2.0	37
	4.0	43
30	0.4	29
	1.0	37
	2.0	43
	4.0	49
60	0.4	35
	1.0	43
	2.0	49
	4.0	55

today is to improve feed methods, particularly where "decibels are so expensive," such as in the case of tropo and earth station installations. Improved feeds illuminate the reflector more uniformly and reduce spillover with the consequent improvement of antenna efficiency. For example, for a 30-ft reflector operating at 2 GHz, improving the efficiency from about 55% to 61% will increase the gain of the antenna about 0.5 dB.

It is desirable, but not always practical, to have the two antennas (as shown in Figure 6.10) spaced not less than 100 wavelengths apart to assure proper space diversity operation. Antenna spillover (i.e., radiated energy in side lobes and back lobes) must be reduced to improve radiation efficiency and to minimize interference with simultaneous receiver operation and with other services.

The first side lobe should be down (attenuated) at least 23 dB and the rest of the unwanted lobes down at least 40 dB from the main lobe. Antenna alignment is extremely important because of the narrow beamwidths. These beamwidths are usually less than 2° and often less than 1° at the half-power points. (See Section 5.5.4.)

A good VSWR is also important, not only from the standpoint of improving system efficiency, but also because the resulting reflected power with a poor VSWR may damage components further back in the transmission system. Often load isolators are required to minimize the damaging effects of reflected waves. In high power tropo systems these devices may even require a cooling system.

A load isolator is a ferrite device with approximately 0.5 dB insertion loss. The forward wave (the energy radiated toward the antenna) is attenuated 0.5 dB; the reflected wave (the energy reflected back from the antenna) is attenuated more than 20 dB.

Another important consideration in planning a tropo antenna system is polarization. See Figure 6.10. For a common antenna the transmit wave should be orthogonal to the receive wave. This means that if the transmitted signal is horizontally polarized, the receive signal should be vertically polarized. The polarization is established by the feeding device, usually a feed horn. The primary reason for using opposite polarizations is to improve isolation, although the correlation of fading on diversity paths may be reduced. A figure commonly encountered for isolation between polarization on a common antenna is 26 dB. However, improved figures may be expected in the future.

TRANSMISSION LINES

In selecting and laying out transmission lines for tropospheric scatter installations, it should be kept in mind that losses must be kept to a minimum. That additional fraction of a decibel is much more costly in tropo than in radiolink installations. The tendency, therefore, is to use waveguide on most tropo installations because of its lower losses rather than coaxial cable. Waveguide is universally used above 1.7 GHz.

Transmission line runs should be less than 200 ft (60 m). The attenuation of the line should be kept under 1 dB from the transmitter to the antenna feed and from the antenna feed to the receiver, respectively. To minimize reflective losses, the VSWR of the line should be 1.05:1 or better when terminated in its characteristic impedance. Figures 5.25 and 5.26 show several types of transmission lines commercially available.

THE DUPLEXER

The duplexer is a transmission line device which permits the use of a single antenna for simultaneous transmission and reception. For tropo application a duplexer is a three-port device (see Figure 6.11) so tuned that the receiver leg appears to have an admittance approaching (ideally) zero at the transmitting frequency. At the same time the transmitter leg has an admittance approaching zero at the receiving frequency. To establish this, sufficient separation in frequency is required between the transmitted and received frequencies. Figure 6.11 is a simplified block diagram of a duplexer. The insertion loss of a duplexer in each direction should be less

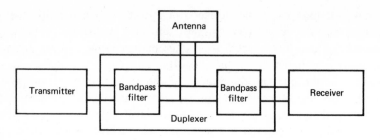

Figure 6.11   Simplified block diagram of a duplexer.

than 0.5 dB. Isolation between the transmitter port and the receiver port should be better than 30 dB. High power duplexers are usually factory-tuned. It should be noted that some textbooks call the duplexer a diplexer.

6.9.4   Modulator-Exciter and Power Amplifier

The type of modulation used on tropospheric scatter transmission systems is nearly exclusively FM. As our discussion develops, keep in mind that tropo systems are high gain, low noise extensions of the radiolink systems discussed in Chapter 5.

The tropo transmitter is made up of a modulator-exciter and a power amplifier (see Figure 6.10). The power outputs are fairly well standardized at 1, 2, 10, 20, and 50 kW. For most commercial applications the 50-kW installation is not feasible from an economic point of view. Installations that are 2 kW or below are usually air-cooled. Those above 2 kW are liquid-cooled, usually with a glycol-water solution using a heat exchanger. If klystron power amplifiers are used, such tubes are about 33% efficient. Thus a 10-kW klystron will require at least 20 kW of heat exchange capacity.

The transmitter frequency stability (long-term) should be $\pm 0.001\%$. Spurious emission should be down better than 80 dB below the carrier output level. Preemphasis is used as described in Section 5.5.3 and depends on the highest modulating frequency of the applied baseband.

The baseband configuration of the modulating signal, depending on the number of channels to be transmitted, is selected in the spectrum 60–552 kHz (CCITT Supergroups 1 and 2—see Section 3.3.6). However, CCITT Subgroup A, 12–60 kHz, is often used as well. For longer route tropo systems, the link design engineer may tend to limit the number of voice channels, selecting a baseband configuration that lowers the highest modulating frequency to be transmitted as much as possible. This tends to increase equivalent overall system gain by reducing B_{if} (the bandwidth of the IF) (equation 6.6), which is equivalent to reducing the RF bandwidth.

The modulator injects an RF pilot tone which is used for alarms at both ends as well as to control far-end combiners. 60 kHz is common in U.S. military systems. CCIR recommends 116 or 119 kHz for 24-channel systems, 304 or 331 kHz for 60-channel systems, and 607 (or 304) kHz for 120-channel systems (CCIR Rec. 401-2).

The modulator also has a service channel input. This is covered in CCIR Rec. 400-2. It recommends the use of the band 300–3400 Hz. U.S. military systems often multiplex more than one service orderwire in the band 300–12,000 Hz. Often one of these channels may be used to transmit fault and alarm information.

The power amplifier should come equipped with a low-pass filter to attenuate second harmonic output by at least 40 dB and third harmonic output by at least 50 dB.

6.9.5 The FM Receiver Group

The receiver group in tropo installations usually consists of two or four identical receivers in dual or quadruple diversity configurations, respectively. Receiver baseband outputs are combined in maximal ratio square combiners. See Section 5.5.6 for a discussion of combiners and the function of the ratio-square type of combiner. A simplified functional block diagram of a typical quadruple diversity receiving system is shown in Figure 6.12.

Receiver noise threshold may be computed as follows:

$$\text{Noise threshold}_{dBW} = 10 \log kTB_{if} + NF \qquad (6.6)$$

where K = Boltzmann's constant, 1.38×10^{-23} J°K
  T = 290°K
  B_{if} = IF bandwidth (Hz)
  NF = the receiver noise (dB)

Typical receiver front-end noise figures are given in Table 6.5. Table 6.6, which provides maximum IF bandwidths (B_{if}) for several voice channel configurations, will be helpful in calculating some receiver noise thresholds when receiver front end noise figures are given. For our dis-

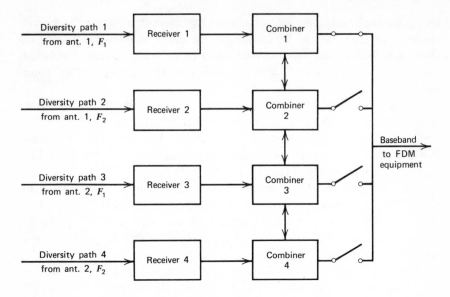

Figure 6.12 Simplified functional block diagram of a quadruple diversity tropo receiving system. F_1, frequency 1; F_2, frequency 2.

Table 6.5   Typical Receiver Front-End Noise Figures

Frequency Band (MHz)	Noise Figure, NF (dB)		
	Tunnel Diode	Parametric Amplifier	Transistor Amplifier
350–450	3.5	2.5	5
775–985	4.0	3.0	6
1700–2400	4.5	3.5	7
4400–5000	5.0	4.0	8

cussion, the receiver front-end noise figure will be the noise figure for the entire receiver. The figures in Table 6.6 were taken from *DCA Circ. 330–175-1*.

Having B_{if} and NF, we can now calculate the noise threshold for each of the three voice channel configurations given using Tables 6.5 and 6.6. Table 6.7 tabulates the three bandwidths for each value of NF given in Table 6.5 giving the equivalent noise threshold. The derived formula

Table 6.6   Maximum IF Bandwidths for Several Voice Channel Configurations

Number of Voice Channels in FDM Bandband	Maximum IF Bandwidth, B_{if} (Hz)
36	3×10^6
60	6×10^6
120	10×10^6

(i.e., -204 dBW $+ NF + 10 \log B_{if}$) assumes uncooled (room temperature) receivers.

From Table 6.7 it is obvious that to achieve FM improvement threshold, the receiver must have an input carrier-to-noise ratio equivalent to adding $+ 10$ dB algebraically to the noise threshold. For instance, if a receiver has a 3-dB noise figure, a B_{if} of 3 MHz, its noise threshold is $- 136.2$ dBW and its FM improvement is $- 126.2$ dBW.

Another method of improving the equivalent equipment gain on a path is to use threshold extension techniques. The FM improvement threshold of a receiver can be "extended" by using a more complex and costly demodulator called a threshold extension demodulator. The amount of improvement that can be expected using threshold extension over conventional receivers is on the order of 7 dB.* Thus for the above example, where the FM improvement threshold was -126.2 dBW without extension, with extension it would be -133.2 dBW.

Threshold extension works on an FM feedback principle which reduces the equivalent instantaneous deviation, thereby reducing the required bandwidth (B_{if}), which in turn effectively lowers the receiver noise threshold. A typical receiver with a threshold extension module may employ a tracking filter which instantaneously tracks the deviation with a steerable bandpass filter having a 3-dB bandwidth of approximately 4 times the top baseband frequency. The control voltage for the filter is derived by making a phase comparison between the feedback signal and the IF input signal.

6.9.6   Diversity and Diversity Combiners

Some form of diversity is mandatory in tropo. Most present-day operational systems employ quadruple diversity. There are several ways of

---

* Assuming a modulation index of 3.

TABLE 6.7   Noise and FM Thresholds for Several Receiver Figures at Three IF Bandwidths

Noise Figure (dB)	Bandwidth (MHz)	Noise Threshold (dBW)	FM Improvement Threshold (dBW)
2.5	3	−136.7	−126.7
	6	−133.7	−123.7
	10	−131.5	−121.5
3.0	3	−136.2	−126.2
	6	−133.3	−123.2
	10	−131.0	−121.0
3.5	3	−135.7	−125.7
	6	−132.7	−122.7
	10	−130.5	−120.5
4.0	3	−135.2	−125.2
	6	−132.2	−122.0
	10	−130.0	−120.0
4.5	3	−134.7	−124.7
	6	−131.7	−121.7
	10	−129.5	−119.5
5.0	3	−134.2	−124.2
	6	−131.2	−121.2
	10	−129.0	−119.0
6.0	3	−133.0	−123.2
	6	−130.2	−120.2
	10	−128.0	−118.0
7.0	3	−132.2	−122.2
	6	−129.2	−119.2
	10	−127.0	−117.0
8.0	3	−131.2	−121.2
	6	−128.2	−118.2
	10	−126.0	−116.0

obtaining some form of quadruple diversity. One of the most desirable is shown in Figure 6.10 (and 6.12), where both frequency and space diversity are utilized. For the frequency diversity section, the system designer must consider the aspects of frequency separation illustrated in Figure 5.2. Space diversity is almost universally used, but the physical separation of antennas is normally in the horizontal plane with a separation distance greater than 100λ, and preferably 150λ.

Frequency diversity, although very desirable, often may not be permitted owing to RFI considerations. Another form of quadruple diversity, perhaps better defined as quasi-quadruple diversity, involves polarization, or what some engineers call *polarization diversity*. This is actually another form of space diversity and has been found not to provide a complete additional order of diversity. However, it often will make do when the additional frequencies are not available to implement frequency diversity.

Polarization diversity is usually used in conjunction with conventional space diversity. The four space paths are achieved by transmitting signals in the horizontal plane from one antenna and in the vertical plane from a second antenna. On the receiving end two antennas are used, each antenna having dual polarized feed horns for receiving signals in both planes of polarization. The net effect is to produce four signal paths that are relatively independent.

A discussion of diversity combiners is covered in Section 5.5.6. There the feasibility of the maximal ratio square combiner is demonstrated, and consequently, it is the most commonly used combiner on tropospheric scatter communication systems.

6.10 ISOLATION

An important factor in tropo installation design is the isolation between the emitted transmit signal and the receiver input. Normally we refer to the receiver sharing a common antenna feed with the transmitter. *DCA Circ. 330-175-1* (Ref. 6) is quoted:

> The transmitter and receiver shall be electrically isolated to the extent that the transmitter signal shall be attenuated by 100 dB when measured at the output of the receiver part of the duplexer at a frequency removed 100 MHz from the transmitted frequency.

To achieve overall isolation such that the transmitted signal interferes in no way with receiver operation when operating simultaneously, the following items aid the required isolation when there is sufficient frequency separation between transmitter and receiver.

1. Polarization.
2. Duplexer.
3. Receiver preselector.
4. Transmit filters.
5. Normal isolation from receiver conversion to IF.

6.11 INTERMODULATION

Noise power ratio (NPR, see Section 5.6.2) measurements are a good indication of operational intermodulation distortion capabilities of tropo equipment. When the NPR is measured on a back-to-back basis with 120-channel loading, we could expect an NPR as high as 55 dB. Once the same equipment is placed in operation on an active path, the NPR from the near-end transmitter to the far-end receiver may drop as low as 47 dB. The deterioration of the NPR is due to intermodulation noise that can be traced to the intervening medium. It is just this intermodulation distortion brought about in the medium that limits useful transmitted bandwidths in tropo systems.

The bandwidth that a tropo system can transmit without excessive distortion is related to the multipath delays experienced. These delays depend on the size of the scatter volume. The common volume is determined by antenna size and scattering characteristics.

6.12 MAXIMUM FEASIBLE MEDIAN PATH LOSS

A figure in decibels can be derived for a maximum median path loss that is feasible by considering the following problem.

Given a path with quadruple diversity reception

120-ft dishes

50-kW power amplifiers

12 FDM/SSB voice channels transmitted with the highest modulating frequency of 60 kHz

Receiver front-end noise figure 2.5 dB

Index of modulation 3

Aperture-to-medium coupling loss 2 dB

Line losses 2 dB at each end

Thus the peak deviation is 60×3 or 180 kHz. From this we calculate

$$B_{if} = 2(FM_p + 2F_{bb}) = 2(180 + 2 \times 60) = 600 \text{ kHz}$$

"twice the peak deviation plus four times the highest modulating frequency"
From this we calculate the receiver noise threshold:

$$-204 + 10 \log 600 \times 10^3 + 2.5 = -143.7 \text{ dBW}$$
$$(\text{from equation 6.6})$$

Assume FM improvement threshold is 10 dB above this figure, or -133.7 dBW. To this figure we add a 20-dB fade margin so that the median input level of the carrier to the receiver is -113.7 dBW.

Calculate the EIRP from the transmit antenna:

$$\text{Transmitter power output} = 10 \log 50 \times 10^3 \, \text{dBW} = \quad 47 \, \text{dBW}$$
$$\text{gain of 120-ft dish} \quad + \underline{\ 55 \, \text{dB at 2 GHz}}$$
$$+ 102 \, \text{dBW}$$
$$\text{minus line and other insertion losses} \quad - \underline{\ \ 2 \, \text{dB}}$$
$$+ 100 \, \text{dBW}$$

Calculate receiver gain equivalent o⸗ antenna gain at receiver minus line losses and aperture-to-medium coupling loss, or $55 - 2 - 2 = 51$ dB.

Consider the following networks in series to assist us in solving the problem.

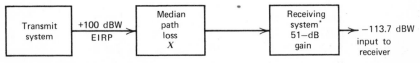

*Antenna and transmission line only.

Thus the problem boils down to the following, assigning X as the maximum feasible median path loss:

$$+100 \, \text{dBW} + X + 51 \, \text{db} = -113.7 \, \text{dBW}, \text{ the medium carrier input}$$
$$\text{level to the receiver}$$
$$X = 264.7 \, \text{dB}$$

50-kW transmitters and 120-ft parabolic dishes are not feasible on a commercial basis. By reducing the installation size to 60-ft dishes and 10-kW amplifiers, an equivalent figure can be calculated which is indicative of a commercially feasible maximum.

Thus the transmit system has an EIRP of $+ 40$ dBW $+ 49$ dB $- 2$ dB (transmitter output plus antenna gain minus lines losses)$= + 87$ dBW. The receiving system equivalent gain will be reduced by 6 dB from the previous example, or 45 dB (i.e., $51 - 6$).

Thus we have 87 dBW $+ X + 45$ dB $= -113.7$ dBW

and $X = 245.7$ dB (see Table 6.9)

6.13 TYPICAL TROPOSPHERIC SCATTER PARAMETERS

Table 6.8 presents some of the more important path parameters for several operational tropospheric scatter paths. Note that takeoff angles are not given nor is the median path loss. Now consult Table 6.9, which

Table 6.8   Some Typical Troposcatter System Parameters

Path Distance (km)	Frequency Band (MHz)	Transmitter Power (kW)	Antenna Diameter (m)[a]	Diversity	Channel Capacity	Comments
150–250	1000	1	9	4	72	
	1000	2	9	2	84	
	1000	2	14	4	108	
	1000	10	18	4	132	Parametric amplifier
	4000	5	9	4	24	
250–320	1000	10	18	2	36	
	1000	10	18	4	72–240	
	2000	10	9	4	72	
	2000	10	18	4	36	
320–420	1000	0.5	18	2	6	
	1000	10	18	2	36	
	1000	10	18	4	24–48	
	2000	10	18	4	36	
320–500	1000	10	18	4	24	Parametric amplifier
500–600	1000	10	37	4	24	Parametric amplifier
700–900	1000	50	37	4	24	
	1000	57	37	4	24[b]	

[a] Nominal diameter of parabolic reflector.
[b] Can support only 12 operational channels with a bit error rate on data dystems in excess of 1×10^{-3} indicative of noise hits due to fades.

supports the preceding section. This table compares measured median path loss with calculated values using three of the accepted median path loss calculation methods.

6.14   FREQUENCY ASSIGNMENT

The problem of frequency assignment in tropospheric scatter systems is similar to that of radiolink systems (see section 5.14). The problem with tropo becomes more complex because of the following:

- Radiated power is much greater (on the order of 30–60 dB greater).
- Nearly all installations are quadruple diversity.
- Receivers are more sensitive, with front-end noise figures of about 3 dB versus about 10 dB for radiolink receivers.

Table 6.9   Median Path Loss—Measured Values Versus Calculated Values Using Three Calculation Methods

World Area	Path Length (km)	Fre-quency (MHz)	Predicted Median Path Loss (dB)			Meas-ured Loss
			NBS 101	CCIR	Yeh	
NE USA	275	460	191.8	192.4	200.9	196.0
UK	275	3480	218.7	220.3	221.6	220.9
Japan	300	1310	204.1	206.0	207.2	211.0
SE Asia	595	1900	261.0	254.8	243.9	260.9
S Caribbean	314	900	184.7	186.0	186.8	189.0
Canada-NY State	465	468	218.2	218.7	223.9	220.6
S Asia	738	1840	255.0	254.5	236.8	245.7

Source: Ref. 16.

Further splatter must be controlled so as not to affect other nearby services. The splatter may be a result of side lobe radiation or from radiation on unwanted frequencies.

For CCIR references, the reader may wish to consult CCIR Rec. 283-2. CCIR Reps. 285-2 and 286 offer some guidance on frequency arrangement.

To reduce splatter from harmonics *DCA Circ. 330-175-1* recommends a transmitter low-pass filter attenuating second harmonic output at least 40 dB and third harmonics at least 50 dB. *DCA Circ. 330-175-1* also specifies the following:

- The minimum separation between transmit and receive carrier frequency of the same polarization on the same antenna shall be 120 MHz.
- The minimum separation between transmit and receive carrier frequency at a single station shall be 50 MHz, but in any case an integral multiple of 0.8 MHz.
- To avoid interference within a single station, separation of the transmit and receive frequencies shall not be near the first IF of the receiver.
- The minimum separation of transmit (receive) carrier frequencies is 5.6 MHz on systems with 36 FDM voice channels or less, 11.2 MHz for 60-voice channels, and 16.8 MHz for 120 voice channels (B_{if} assumed as in Table 6.6).

This document further states that frequency channels shall be assigned on a hop-by-hop basis such that the median value of an unwanted signal in the receiver shall be at least 10 dB below the inherent noise of the receiver (i.e., noise threshold) when using the same or adjacent frequency channels in two relay sections.

REFERENCES AND BIBLIOGRAPHY

1. Luang P. Yeh, "Tropospheric Scatter Communication Systems," presented to ITU World Planning Committee Meeting, Mexico City, 1967.

2. Roger L. Freeman, "Multichannel Transmission by Tropospheric Scatter," *Telecommun. J.* (ITU), Geneva, June 1969.

3. P. L. Rice et al., *Transmission Loss Predictions for Tropospheric Scatter Communication Circuits*, NBS Techn. Note 101 as revised, Jan. 1967.

4. K. O. Hornberg, *Siting Criteria for Tropospheric Scatter Propagation Communication Circuits*, NBS Memo. Rept. PM-85-15, April 1959.

5. *Telecommunication Performance Standards*, Chapter V, "Tropospheric Scatter Systems," USAF T.O. 31Z1-10, U.S. Department of Defense, 1 Dec. 1962.

6. *DCA Circ.* 330-175-1, through Change 9, U.S. Department of Defense, Washington, D.C.

7. A. P. Barghausen et al., *Equipment Characteristics and their Relationship to Performance for Tropospheric Scatter Communication Circuits*, NBS Tech. Note 103, National Bureau of Standards, Boulder, Colo.

8. R. L. Marks et al., *Some Aspects of Design for FM Line-of-Sight and Troposcatter Systems*, USAF Rome Air Development Center, N.Y., AD 617-686.

9. E. F. Florman and J. J. Tory, *Required Signal-to-Noise Ratios, RF Signal Power and Bandwidth for Multichannel Radio Communication Systems*, NBS Tech. Note 100, National Bureau of Standards, Boulder, Colo.

10. A. P. Barsis et al., *Predicting the Performance of Long Distance Tropospheric Communication Circuits*, NBS Rep. 6032, National Bureau of Standards, Boulder, Colo., Dec. 1958.

11. *Forward Propagation Tropospheric Scatter Communications Systems*, Handbook for Planning and Siting, USAF T.O. 31R5-1-11, U.S. Department of Defense, Washington, D.C., as revised 30 Nov. 1959.

12. E. D. Sunde, "Digital Troposcatter Transmission and Modulation Theory," *Bell Syst. Tech. J.*, **43**(1), Jan. 1964.

13. E. D. Sunde, "Intermodulation Distortion in Analog FM Tropospheric Scatter Systems," *Bell Syst. Tech, J.*, Jan. 1964.

14. P. Panter, *Communication Systems Design—Line-of-Sight and Tropo-scatter Systems*, McGraw-Hill, New York, 1972.

15. *Naval Shore Electronics Criteria—Line-of-Sight and Tropospheric Scatter Communication Systems*, Department of the Navy, Washington, D.C., May 1972, Navelex 0101.112.

16. R. Larsen, "A Comparison of Some Troposcatter Prediction Methods," conference paper, IEE Conference on Tropospheric Radio Wave Propagation, Sept.-Oct. 1968.

17. CCIR Rep. 233-1, Vol. II, CCIR, Oslo, 1966.

18. CCIR Rep. 397-2, Vol. IV, CCIR, Oslo, 1966.

7 | EARTH STATION TECHNOLOGY

7.1 INTRODUCTION

The U.S. Federal Communications Commission and a number of world bodies have come to accept the term "earth station" as a radio facility located on the earth's surface that communicates with satellites (or other space vehicles, for that matter). A "terrestrial station" is a radio facility on the earth's surface operating directly with other similar facilities on the earth's surface. Today the term "earth station" has come more to mean a radio station operating with other stations on the earth via an orbiting satellite relay.

This chapter deals with the design of those earth stations that operate with synchronous earth satellites. Such satellites orbit the earth with a period of 24 h. Thus they appear to be stationary over a particular geographic location on earth. The altitude of a synchronous satellite is 22,300 mi (35,900 km) above the surface of the earth.

7.2 THE SATELLITE

A communication satellite is an RF repeater. It may be represented in its most simple configuration as that shown in Figure 7.1. Theoretically three such satellites properly placed in synchronous orbit could provide 100% earth coverage from one earth station located anywhere on the earth's surface. This concept is shown in Figure 7.2.

7.3 EARTH-SPACE WINDOW

In selecting a band of frequencies that is optimum for earth-space communication, two important phenomena must be taken into account: atmos-

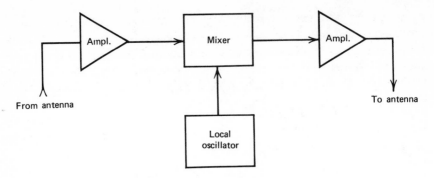

Figure 7.1 Simplified functional block diagram of the radio relay portion of a typical communication satellite.

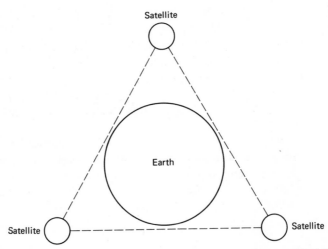

Figure 7.2 Three synchronous satellites properly placed can provide 100% earth coverage.

pheric absorption and noise, both galactic and man-made. Thus the problem is to select a band of frequencies that will permit broadband communication, where the earth-space signal will suffer minimum attenuation due to absorption and where the inherent noise level is least. Regarding absorption, an opening exists between 10 and 10,000 MHz. Such an opening we more commonly call a "window." This same band is limited on the lower end by noise to frequencies above 1000 MHz. Therefore the optimum earth-space window is between 1000 and 10,000 MHz. This by no means indicates that earth-space communication cannot be and is not carried out on frequencies outside this region. The noise effect is shown in Figure 7.3 and the absorption effect in Figure 7.4*a*. Figure 7.4*b* shows the variation of attenuation due to precipitation and adds emphasis to the desirability of operating below 10 GHz. This point is delved into in Section 10.7.

In the 1000–10,000 MHz "window," four bands, each 500 MHz wide, have been assigned for use with communication satellites. These bands are as follows:

> 3700–4200 MHz (satellite-earth)
> 5925–6425 MHz (earth-satellite)
> 7250–7750 MHz (satellite-earth)
> 7900–8400 MHz (earth-satellite)

The first two bands shown above are used for commercial service, and are the two that this chapter essentially covers.

Frequency sharing exists with terrestrial services for all four of the bands. CCIR discusses ways in which one service can coexist with the other in its Recs. 382-2 (3700–4200 MHz), 383-1 (5925–6425 MHz), and 386-1. Specific problems arising from frequency sharing and the development of coordination contours are discussed below in Section 7.14.

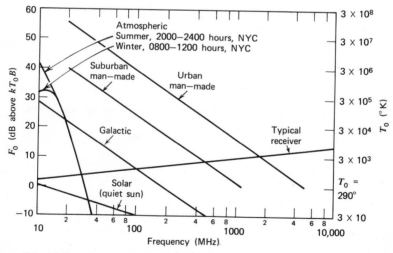

Figure 7.3 Noise window effect—median values of average noise power expected from various sources using omnidirectional antenna near earth's surface (from *Data Handbook for Radio Engineers,* ITT).

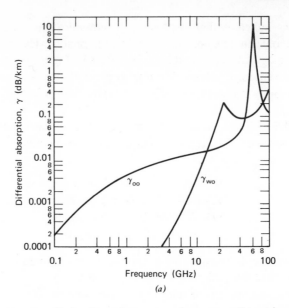

Figure 7.4 (a) Atmospheric absorption versus frequency; (b) attenuation due to precipitation. (From CCIR Rep. 234 (Oslo, 1966) and *Data Handbook for Radio Engineers*, ITT. γ_{oo} = oxygen absorption; γ_{wo} = water vapor absorption.

7.4 PATH LOSS

Section 5.3.1 covered line-of-sight* path loss. The earth-satellite-earth communication link may be considered a radiolink hop to an RF repeater (the satellite) and another hop from the satellite to earth. Path loss must be computed in both cases. Here we use equation 5.1, or

$$L_{dB} = 36.6 + 20 \log D_{mi} + 20 \log F_{MHz}$$

where D is the distance in statute miles and F is the operating frequency in megahertz. For midband, 3950 MHz,

$$L_{dB} = 36.6 + 20 \log (22,300) + 20 \log (3950)$$
$$= 195.498 \text{ dB}$$

For the highest frequency in the 4-GHz band, 4200 MHz,

$$L_{dB} = 36.6 + 20 \log (22,300) + 20 \log (4200)$$
$$= 36.6 + 86.96 + 72.464 = 196.02 \text{ dB}$$

For 6175 MHz (midband),

$$L_{dB} = 36.6 + 20 \log (22,300) + 20 \log (6175)$$
$$= 199.37 \text{ dB}$$

* More properly defined as "free-space" path loss.

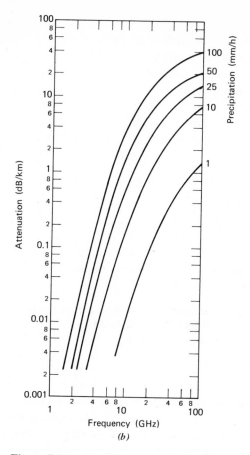

Figure 7.4 (*continued*)

For 6425 MHz, the highest frequency in the 6-GHz band,

$$L_{dB} = 36.6 + 20 \log (22,300) + 20 \log (6425)$$
$$= 199.72 \text{ dB}$$

7.5 SATELLITE-EARTH LINK

By international agreement the maximum "effective isotropically radiated power" (EIRP) of a communication satellite is + 32.1 dBW (Ref. 1) per carrier. This limit has been placed on satellite radiators to minimize interference with shared terrestrial services. The interference effects per se are dealt with in Section 7.14. The problem to be discussed here is how to derive a useful signal at the earth station on the down link.

It will be seen that the down link is the limiting factor on earth station design. The uplink (i.e., the link from earth to satellite) may take advantage of powerful earth station transmitters, high gain, narrow beam antennas with little likelihood of interference with other services. On the other hand, the down link is severely limited in radiated power owing to the possiblity of interfering with terrestrial services. Thus basically we can say that it is the down link around which earth stations are designed.

Let us consider the down link and compute the signal level impinging on an earth station antenna. Use the worst-path loss from Section 7.4; therefore,

$$+32 \text{ dBW} - 196.02 = -164.02 \text{ dBW}$$

Such a broadband signal level is very low indeed, and the earth station must be designed using the most refined radiolink techniques to achieve a proper signal-to-noise ratio at the voice channel output of the demultiplexer.

The reader must remember that earth stations using FDMA frequency division multiple access) are extensions of the technology developed in Chapters 3, 5, and 6. Briefly stated

- The radio signal is frequency modulated by a multi-channel telephone signal made up of FDM/SSB baseband.
- Because the signal traverses so little atmosphere, the received signal level is not affected by fading except under conditions of very heavy rain.

Now let us take the technology of Chapter 5 and, with a little imagination and reasoning, the so-called difficult concepts of earth stations come to light. Use Table 7.1 as the basis for this introductory discussion; it provides parameters of the Intelsat III type of satellites. Intelsat IV will be discussed later in the chapter.

Table 7.1   Intelsat III Characteristics

1. Carrier capacity (number of VF channels)	24	60	132
2. Allocated satellite bandwidth, b_a (MHz)	5	10	20
3. Bottom baseband frequency (excluding service channels, (etc.), f_b (kHz)	12	12	12
4. Top baseband frequency, f_m (kHz)	108	252	52
5. Deviation (rms) for 0 dBm0 test tone, f_r (kHz)	250	410	630
6. Multichannel deviation (rms), f_{mc} (kHz)	420	830	1490
7. Ratio of unmodulated carrier power to maximum carrier power density under full-load conditions, (dB/4 kHz)	24.2	27.2	29.7
8. Occupied bandwidth without guard band, b_s (MHz)	4.0	8.0	14.4

Source: ICSC-37-38E W/1/69.

If -164 dBW impinges on an earth station antenna and, assuming for discussion's sake that the receiving antenna had a zero gain, allowing a 10-dB noise figure for the receiver, a 4-MHz IF bandwidth (B_{if})* for a 24-channel system (item 8, Table 7.1), we can compute the receiver's FM improvement threshold: (equation 5.7):

$$\text{Threshold}_{dBW} = -204 + 10 \log B_{if} + 10 + NF_{dB}$$
$$= -204 + 66 + 10 + 10$$
$$= -118 \text{ dBW}$$

The next step is to compute the minimum gain of a parabolic antenna to bring the nominal signal level of -164 dBW to the FM improvement threshold of the receiver, or

$$-164 \text{ dBW} - 118 \text{ dBW} = 46 \text{ dB}$$

A 46-dB gain requires a reflector diameter of 22 ft (7 m) at 55% efficiency at 4 GHz.

Now open the bandwidth up to 14.4 MHz (Table 7.1, item 8), and it is seen that the antenna system would require nearly 6 dB more gain or about 52 dB gain. (Use the nomogram in Figure 6.10). Thus an antenna of almost 12 m (41 ft) would be necessary.

Let us stretch our imagination a little further by adding a second carrier to our receiver system. A power split would be needed at the antenna feed to allow the addition of a second receiver. This places at least a 3-dB loss in what was a lossless system requiring an increase of the antenna size to something nearer 40 ft (18 m).

The analogy ends here. Let us now consider a modern earth station. Obviously a major change will have to take place in our station design in order to have a station capable of receiving numerous RF carriers that may be assigned anywhere in the 500-MHz down-link bandwidth (i.e., from 3700 to 4200 MHz).

To improve sytem sensitivity, the size of the antenna could be increased and the receiver made more sensitive. Economics have dictated that antennas larger than the nominal 100-ft (30 m) dish have a cost trade-off that is untenable. Large antennas tend to increase in price exponentially as their size increases approximately as the following equation (Ref. 5):

$$C \sim D^{2.8} \tag{7.1}$$

where C is the cost and D is the diameter of the parabolic antenna. Thus commercial earth stations have standardized on the nominal 100-ft (30 m) antenna.

* We make the assumption that $b_s = B_{if}$.

7.5.1 Figure of Merit G/T

G/T, the figure of merit of an earth station, has been introduced into the technology to describe the capability of an earth station to receive a signal from a satellite.

One of the basic publications covering earth stations in the international service, *ICSC-37-38E W/1/69* (Ref. 4) is quoted:

> Approval of an earth station in the category of standard earth stations will only be obtained if the following two minimum conditions are met, these applying at the elevation angle of operation and for the polarization of the satellite concerned, under clear sky conditions, in light wind, and for any frequency in the band 3705 to 3930 and 3970 to 4195 MHz.

$$(a) \quad G/T_{(dB)} = 40.7 \text{ dB} + 20 \log_{10} \frac{f}{4} \qquad (7.2)$$

$$(b) \quad G_{(dB)} = 57 \text{ db} + 20 \log_{10} \frac{f}{4} \qquad (7.3)$$

where G = gain of the antenna at the receiving frequency in question
  T = effective noise temperature of the receiving system in degrees Kelvin
  f = carrier frequency in GHz

At exactly 4 GHz the first formula (a) could be stated in the manner:

$$40.7 \text{ dB} = G_{(dB)} - 10 \log T \qquad (7.4)$$

The earth station system designer must provide a G/T of 40.7 dB or better. The problem is how to achieve it. Equation 7.3 states that the antenna gain at 4 GHz must be 57 dB or better. In order to meet the G/T requirement with a 57-dB antenna, a maser type receiver ($7°K$) is required. It has been found more economical to use cryogenically cooled parametric amplifiers ($20°K$) and an improved antenna with a 60-dB or greater gain. The parametric amplifier system noise temperature coupled with the remainder of the noise contributions provides a system noise temperature not in excess of $85.1°K$. (*Note.* Noise temperature and noise figure are discussed in Section 1.12.)

One may reach this figure by following some simple manipulation of algebra. As stated,

$$G - 10 \log T = 40.7 \text{ dB}$$

Allow the antenna gain at 4 GHz to be 60 dB, which is a conservative figure for a nominal 100-ft (30 m) antenna given the inherent efficiencies when modern feed devices are used (Cassegrain). Now substitute 60 dB in the formula 7.4; thus

$$60 - 10 \log T = 40.7 \text{ dB}$$
$$10 \log T = 60 - 40.7$$
$$= 19.3 \text{ dB}$$

Then

$$T = 81.5°\text{K}$$

Many system designers want a certain field margin to the minimum G/T and design to 41.5 dB. Therefore

$$10 \log T = 60 - 41.5$$
$$= 18.5 \text{ dB}$$

Then

$$T = 70°\text{K}$$

Consider point-to-point (radiolink) microwave at 4 GHz. Allow a 10-dB noise figure for a typical FM receiver operating in that band. The receiver noise temperature turns out to be approximately 2600°K. Noise temperature can be related to noise figure by the formula (equation 1.22)

$$F_{\text{dB}} = 10 \log \left(1 + \frac{T}{290} \right)$$

The T we have been using is more commonly referred to as equivalent noise temperature and is the summation of all noise components in the receiving system. This concept has been with us all the time in Chapters 5 and 6; we have simply shunted it aside because receiver noise temperatures were so large (e.g., the 2600°K) that the other noise components were insignificant. In earth station technology, the receiver noise temperature has been lowered so much that other contributors now become important.

The noise components comprising T may be broken down into four categories, namely:

- Antenna noise, usually taken at the 5° elevation angle of the antenna, which is the worst allowable case. This noise includes antenna "spillover," galactic noise, and atmospheric noise.
- Passive component noise. This is the summation of the equivalent noise from the passive components before the incoming signal reaches the first active component.
- High power amplifier (HPA) leakage noise.
- Sum of the excess noise contributions of the various active amplifying stages of the receiving system.

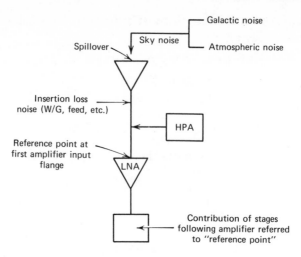

Figure 7.5 Graphical representation of noise contributors. HPA, high power amplifier; LNA, low noise amplifier.

Figure 7.5 represents these noise contributors graphically. Figure 7.6 shows approximate sky noise variation with antenna elevation angle. At an angle of 5° we see that sky noise temperature reaches the order of 25°K. It will also be seen that minimum antenna noise occurs when the antenna is at zenith (i.e., an elevation angle of 90°). *Note.* Elevation angles are referred to the horizon; thus elevation angle would be 0° when the antenna is pointed directly at the horizon.

Antenna spillover refers to radiated energy from the antenna to the ground and scatter off antenna spars. In both cases noise generators are formed and must be considered in a "noise budget." Spars refer here to those metal elements used to support the antenna feeding device. The sum total of antenna noise may reach 39 or 40°K, 25° of which is sky noise.

The next group of noise contributors to the total effective noise temperature (at 4 GHz) is the feed, for which 10°K is a typical noise temperature; directional coupler, 1.45°K; waveguide switch, 0.58°K, and a short piece of flexible waveguide at 2.92°K. Noise temperature of passive, two-port devices (i.e., where G, the active gain, equals 1) in series may be added directly. The noise of these components is a function of their I^2R values. Thus the effective noise temperature of all the passive com-

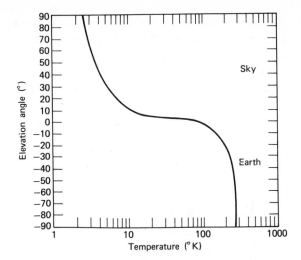

Figure 7.6 Approximate sky noise variation with antenna elevation angle.

ponents (neglecting the insertion loss of the components to the noise energy)

$$T_{ep} = T_{ant} + T_f + T_{dc} + T_{wg} \qquad (7.5)$$
$$= 39 + 10 + 1.45 + 2.92 = 53.37°K$$

Now let us connect this to several stages of a cooled parametric amplifier and a TWT (traveling wave tube) or similar driver. Then the effective noise temperature is

$$T_e = T_{ep} + T_{A1} + \frac{T_{A2}}{G_1} + \frac{T_{A3}}{G_1 G_2} + \frac{T_{twt}}{G_1 G_2 G_3} \qquad (7.6)$$

where T_{ep} = effective noise temperature of passive components
　　　T_{A1} = effective noise temperature of the first stage of the parametric amplifier
　　　T_{A2} = effective noise temperature of the second stage
　　　G_1, G_2, G_3 = gains of the active amplification stages 1, 2, and 3, etc.

If we simplify by considering the parametric amplifier lumped together as one amplifier, then

$$T_e = T_{ep} + T_a + \frac{T_{twt}}{G_a} \qquad (7.7)$$

where T_a = noise temperature of the parametric amplifyer
　　　G_a = gain of the parametric amplifier = 40 dB or 10,000

A good helium-cooled parametric amplifier should display an equivalent noise temperature of no more than 15–20°K at any frequency in the band 3700–4200 MHz. For this discussion 18°K is used. Often the parametric amplifier is followed by a TWT which drives longer lengths of transmission line connecting the antenna facility to the receiving radio equipment (i.e., the down-converters and demodulators). Typically this device may have a 7-dB noise figure or a noise temperature of approximately 1165°K. Thus

$$T_e = 53.37°K + 10°K + \frac{1165°K}{10,000}$$
$$= 53.37 + 18 + 0.11 = 71.48°K$$

Again, assuming G to be 60 dB, then

$$\frac{G}{T} = 60 - 10 \log T_e$$
$$= 60 - 10 \log 71.48$$
$$= 60 - 18.54$$
$$= 41.46 \text{ dB for this example}$$

7.5.2   The Ratio of Carrier-to-Thermal Noise Power (C/T)

The carrier-to-thermal noise power ratio (C/T) provides an absolute measurement of carrier power regardless of bandwidth. It is related to the familiar carrier-to-noise (C/N) ratio in the following way:

$$\frac{C}{T_{dB}} = \frac{C}{N_{dB}} + 10 \log B + 10 \log k$$

where B = the bandwidth (Hz)
k = Boltzmann's constant

This can be derived by letting

$$\frac{C}{N} = \frac{C}{kTB} \qquad (7.8)$$

(This is a simple identity where we let $N = kTB$; see equation 1.11.)

$$\frac{C}{N_{dB}} = \frac{C}{T_{dB}} - 10 \log k - 10 \log B$$

Rearranging terms,

$$-\frac{C}{N} = -\frac{C}{T} + 10 \log k + 10 \log B$$

$$\frac{C}{T} = \frac{C}{N} + 10 \log k + 10 \log B \qquad (7.9)$$

Table 7.2 gives some typical C/T values for stations operating with Intelsat III satellites.

Table 7.2 Typical C/T for Intelsat III

	24	60	132
Voice channel capacity per carrier	24	60	132
Occupied bandwidth (MHz) (from Table 7.1)	4.0	8.0	14.4
Minimum carrier-to-total noise temperature ratio for 50,000 pWp noise performance in the worst channel (dBW/°K)	−158.8	−155.3	−152.5
Carrier-to-total noise temperature ratio at operating point 8400 pWp maximum from RF sources (dBW/°K)	−154.8	−151.3	−148.5

Source: *ICSC-37-38E W /1 /69.*

For an example, let $C/N = 30$ dB, use item 8 of Table 7.1 for 24-channel operation, and let $10 \log k$ be the familiar -228.6 dBW/°K (Section 1.12).

$$\frac{C}{T} = +30 - 228.6 + 66$$
$$= -132.6 \text{ dBW/°K for this example}$$

C/T for tangential noise threshold, of course, is where

$$\frac{C}{N} = 0 \text{ (dB)}$$

$$\frac{C}{T} = -228.6 + 66 \text{ or } -162.6 \text{ dBW/°K for the above example}$$

7.5.3 Relating C/T to G/T

For the down link

$$\frac{C}{T} = \text{EIRP (satellite)} - \text{path loss (4 GHz)}$$
$$+ G/T \text{ (of the earth station)} \qquad (7.10)$$

If $+22.5$ dBW is used for the EIRP of the satellite, -196 dB for path loss, and $+40.7$ dB for the G/T of the earth station, then

$$\frac{C}{T} = +22.5 - 196 + 40.7$$
$$= -132.8 \text{ dBW/°K for this example}$$

Not included here is the EIRP adjustment for satellite backoff of power when more than one carrier is carried on a satellite transponder.

7.5.4 Deriving Signal Input from Illumination Levels

Many standards and specifications for earth stations give illumination levels or flux densities. From this information the earth station system engineer will want to derive a receive level at the feed of the earth station. The flux density is given in a level value per square meter or square foot.

A typical specification may state, "The specification in this exhibit shall apply for all illumination levels over the range of -150 dBW/m^2 to -116 dBW/m^2." If the antenna described uses a 32-ft (10 m) parabolic reflector, what would be the input at the feed?

The first step is to compute the projected area. For the case of a parabolic antenna, this is a circle. For our 32-ft diameter antenna it would be a circle with radius of 16 ft (5 m). For the area of the circle with radius 16 ft we have $\pi(16)^2$ or $156\pi = 489.84$ ft^2 (78.5 m^2). If our antenna were 100% efficient and illuminated with a signal with a flux density of -150 dBW/m^2, we would then multiply 78.5 by -150 dBW/m^2. Thus we have

$$10 \log 78.5 - 150 \text{ dBW/m}^2 = 19.85 - 150 = -131.05 \text{ dBW}$$

for the receive level at the feed for this hypothetical case.

These figures are for an antenna with 100% efficiency. Let us use something more practical, say, an antenna with 60% efficiency. Then we must multiply the projected area by 0.60 to get what is called the effective area:

$$78.5 \text{ m}^2 \times 0.60 = 47.10 \text{ m}^2$$

The input level at the feed will then be

$$+10 \log 47.1 - 150 \text{ dBW/m}^2 = 16.73 - 150 \text{ dBW}$$
$$= -133.27 \text{ dBW}$$

Another example is to use the higher flux density value from the above with the same antenna values; thus

$$+10 \log 47.1 - 116 \text{ dBW/m}^2 = +16.73 - 116 \text{ dBW}$$
$$= -99.27 \text{ dBW}$$

Given the input level at the feed and subtracting the various line losses, the earth station engineer can easily calculate the input level to the receiver front end.

7.5.5 Station Margin

One of the major considerations when designing radiolink and tropo systems is the fade margin. This is the additional signal level that was added in the system calculations to allow for fading. Often this value was on the order of 20–50 dB.

In other words the system in question was overbuilt in receive signal level between 20 and 50 dB to overcome most fading conditions or to ensure that noise would not exceed a certain norm for a fixed time frame.

As we saw in Chapter 5, fading is caused by anomalies in the intervening medium between stations or by the reflected signal causing interference to the direct ray signal. There would be no fading phenomenon on a radio signal being transmitted through a vacuum. Thus satellite earth station signals are subject to fade only during the time they traverse the atmosphere. For this case, most fade, if any, may be attributed to rainfall.

Margin or station margin is the additional design advantage that has been added to the station to compensate for deteriorated propagation conditions or fading. The margin designed into an LOS system is large and is achieved by increasing antenna size, improving the receiver noise figure, or increasing transmitter output power.

The station margin of a satellite earth station in comparison is small, on the order of 4–6 dB. Typical rainfall attenuation exceeding 0.01% of a year may be from 1 to 2 dB without a radome on the antenna, and when the antenna is at a 5° elevation angle. As the antenna elevation increases to zenith, the attenuation notably decreases because the signal passes through less atmosphere. The addition of a radome could increase the attenuation to 6 dB or greater during precipitation. Receive station margin for an earth station, those extra decibels on the down link, is achieved by use of threshold extension demodulation techniques. These were discussed in Section 6.9.5. Up-link margin is provided by using larger transmitters and by increasing power output when necessary. A G/T in excess of 40.7 dB at the 5° elevation angle will also provide margin, but may prove quite expensive to provide and prove out the additional performance.

7.6 UP-LINK CONSIDERATIONS

The up link from the earth station is less critical from the point of view of earth station design. The EIRP (Effective Isotropically Radiated Power) requirements for Intelstat II and III as stated in Attachment A to *ICSC-37-38E-W/1/69* are for maximum EIRP values per carrier during clear sky conditions.

For Intelsat II
    per voice channel   +68 dBW
    for television transmission from +90 to +95 dBW

For Intelsat III
    per voice channel   +61 dBW
    for television transmission   +86 dBW

To determine the EIRP for a specific number of voice channels to be transmitted on a carrier, one takes the required output per voice channel in dBW (the above) and adds logarithmically 10 log N, where N is the number of voice channels to be transmitted.

As an example take the case for Intelstat III for an up-carrier transmitting 60 voice channels; then we would have

$$+61 \text{ dBW} + 10 \log 60 = 61 + 17.78 = +78.78 \text{ dBW}$$

Consider the nominal 100-ft (30-m) antenna to have a gain of 63 dB (at 6 GHz) and losses typically at 3 dB; what transmitter output power, P_t is required?

$$\text{EIRP}_{dBW} = P_t + G_{ant} - \text{line losses}_{dB}$$

where P_t = output power of the transmitter (dBW)
      G_{ant} = antenna gain (dB) (up link)

Then is the example,

$$+78.78 \text{ dBW} = P_t + 63 - 3$$
$$P_t = +18.78 \text{ dBW}$$
$$= 75.6 \text{ W}$$

Typically for television transmission for Intelsat III,

$$+86 \text{ dBW} = P_t + 63 - 3$$
$$P_t = +26 \text{ dBW}$$
$$= 400 \text{ W}$$

The earth station design engineer will probably concern himself more with trade-offs on location of the transmitter power amplifier to minimize line losses, trade-offs on the use of TWTs, or klystrons in the power amplifier(s) considering problems of intermodulation products under multicarrier operation.

7.7 MULTIPLE ACCESS

Our discussion in Sections 7.1 and 7.2 was extremely simplified. We should remain in that sphere of simplicity to describe the term "multiple access," which refers to the manner in which a number of stations may use a repeater (or any other input-output device) simultaneously. There are two methods now in use that permit this simultaneous multiusage; these are more commonly called FDMA and TDMA. The first allows multiple access in the frequency domain, hence the term frequency division multiple access (FDMA). The second permits multiple access in the time domain, hence time division multiple access (TDMA).

FDMA

Allow the satellite repeater a 500-MHz bandwidth. Each earth station is assigned a segment of that usable bandwidth. Sufficient guard band is allocated between segments to assure that one user will not interfere with another, by drifting into or splattering the other user's segment. Assign 2225 MHz as the frequency of the local oscillator in Figure 7.1. The band of frequencies transmitted up to the satellite, representing the input in the figure, is 5925–6425 MHz. Thus a 6000-MHz carrier received by the satellite-repeater is retransmitted in the 4000-MHz band as $6000 - 2225 = 3775$ MHz. Note that in the mixing operation we are taking the difference frequency. To prove the viability of the mixer, check the band-edge frequencies:

$$5925 - 2225 = 3700 \text{ MHz}$$
$$6425 - 2225 = 4200 \text{ MHz}$$

As we already know, 3700–4200 MHz is the band of frequencies transmitted from the satellite to the ground.

Consider the satellite antennas in all examples to be omni-directional. Now let us do a sample exercise with three earth stations within common view of a synchronous satellite that wish to intercommunicate. Assign each earth station a 132-channel configuration and, from Table 7.1, we see that the allocated bandwidth for 132-voice channel configurations is 20 MHz. This includes the guard band. Situate one station in Buenos Aires, another in New York, and the third in Madrid, Spain. (For the sake of simplicity, we will work with only two supergroups, leaving CCITT Subgroup A spare, of the 132-channel configuration.)
Assign the segments as follows:

New York	5930–5950 MHz
Madrid	5990–6010 MHz
Buenos Aires	6220–6240 MHz

for the up-link carriers for this hypothetical case. When converted in the satellite, the down-link carriers will then be

New York	3705–3725 MHz
Madrid	3765–3785 MHz
Buenos Aires	3995–4015 MHz

As we know from Table 7.1, the 20-MHz bandwidth, which includes guard bands, requires at the earth station receivers a $B_{if} = B_{rf}$ of 14.4 MHz. Inside the 14.4-MHz baseband (for Intelsat III operation) are contained two supergroups plus orderwire information. The baseband is transmitted by standard FM techniques, which have been covered in Chapter 5.

Look at the following salient facts for our hypothetical example:

- Each earth station transmitting to the satellite transmits one carrier up to the satellite, which carrier is re-radiated omnidirectionally.
- That re-radiated carrier may be received by any earth station within view (over 5° elevation angle) of the satellite.
- For this hypothetical case information is contained (i.e., in the baseband) on the up-link carrier from the New York facility for both Buenos Aires and Madrid, likewise at Buenos Aires for New York and Madrid, and so forth.

This concept is depicted in Figure 7.7. The baseband assignments are referenced to Table 7.1. The same concept can be expanded showing how groups may be broken out for distinct locations. It should be kept in mind that, although an earth station may transmit only one carrier up

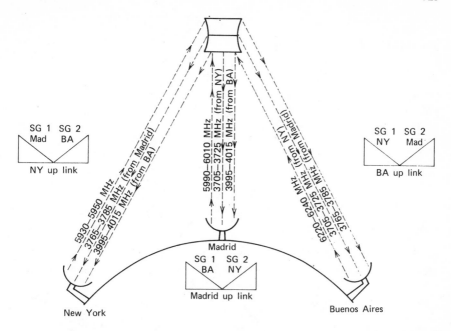

Figure 7.7 A simplified illustration of multiple access in the frequency domain (FDMA) (hypothetical).

to the satellite, it must be prepared to receive at least one carrier for each distant location with which it wishes to communicate. In our hypothetical case, each station would provide one chain of equipment for the up-link and two separate chains of receiving equipments to receive and demodulate the two separate receive carriers from the two distant ends. Some earth stations intercommunicate with 10 or more distant stations, and thus require 10 chains of receiving equipment and transmit to all 10 stations on one up-link carrier. This is just an expansion of the concept depicted in Figure 7.7. For instance Madrid may wish to intercommunicate assigning groups in a 120-channel baseband modulating just one up-link carrier. The up-link baseband assignment is shown in Figure 7.8. Keep in mind that this is just one more hypothetical case.

TDMA

Time division multiple access allows only one carrier to utilize a satellite transponder (repeater) at a time. This has the advantage of limiting the production of intermodulation products due to multiple carrier opera-

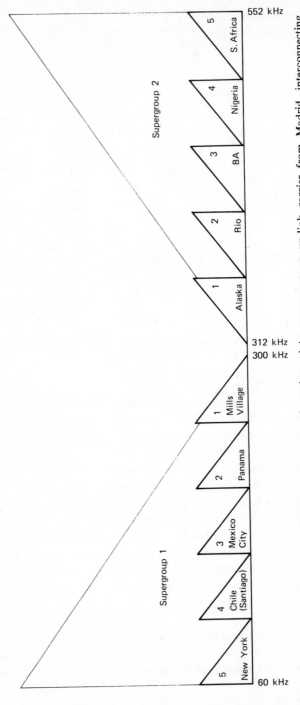

Figure 7.8 As the multiplex (FDM) baseband configuration might appear on an up-link carrier from Madrid, interconnecting with 10 distant points, assigning a 12-channel group to each point (CCITT Subgroup A not assigned).

Supergroup 2

552 kHz

5	S. Africa
4	Nigeria
3	BA
2	Rio
1	Alaska

312 kHz
300 kHz

1	Mills Village
2	Panama
3	Mexico City
4	Chile (Santiago)
5	New York

Supergroup 1

60 kHz

tion. With FDMA several carriers (sometimes up to 10 or more) use a transponder simultaneously. The use of traveling wave tubes (TWT) is universal as power amplifiers in satellite transponders. TWTs are highly susceptible to the production of intermodulation products (IM's) when operating at full excitation. To reduce IM's, the excitation of the TWT (the drive) must be cut back substantially as each additional carrier is added. The cutback is referred to as "backoff." Backoff reduces system efficiency. If only one carrier at a time occupies a transponder, by definition no IM's may be generated. This would allow the TWT to be operated at full power. It should be noted that TDMA in commercial satellite systems is still experimental.

SPADE

SPADE is a third form of multiple access which is really a form of frequency division multiple access. In the SPADE system a single voice channel modulates an RF carrier assigned to a slot 45 kHz wide. A "pool" of 800-voice channel slots are available in a 36-MHz bandwidth such as utilized on an Intelsat IV transponder. SPADE is a far different concept from that of conventional FDMA/FM. Both ends of a telephone may be selected on demand. Thus neither end of a channel is permanently associated with a specific earth station as in the case of FDMA/FM. In the SPADE system voice channels are paired to form a connection as required within the demand-assignment pool.

SPADE offers these advantages:

- Establishes communication links between any two stations whether or not these links are economically justifiable from a viewpoint of traffic density.
- Reduces the number of FDM/FM (FDMA) carriers and the associated reduction of earth station down chains required to service multidestination traffic.
- Reduces the number of revisions of frequency plans to accommodate new users or modifications of present user requirements.
- Permits growth on the basis of traffic at nominal expense on channel-by-channel incremental steps.

Although SPADE appears attractive in every aspect, FDMA/FM will remain the basic method of access between earth stations with comparatively heavy traffic load (i.e., in excess of 6 erlangs* for the busy hour). A more detailed discussion of SPADE is found in Section 7.12.

* See glossary, Appendix D, for definition.

7.8 FUNCTIONAL OPERATION OF A "STANDARD EARTH STATION"

7.8.1 The Communication Subsystem

Figure 7.9 is a simplified functional block diagram of an earth station showing the basic communication subsystem only. We shall use this figure to trace a signal through the station. Figure 7.10 is a more detailed functional block diagram of a typical earth station.

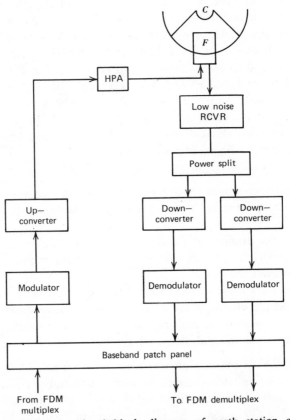

Figure 7.9 Simplified functional block diagram of earth station communication subsystem. F, feed; HPA, high power amplifier; C, Cassegrain subreflector.

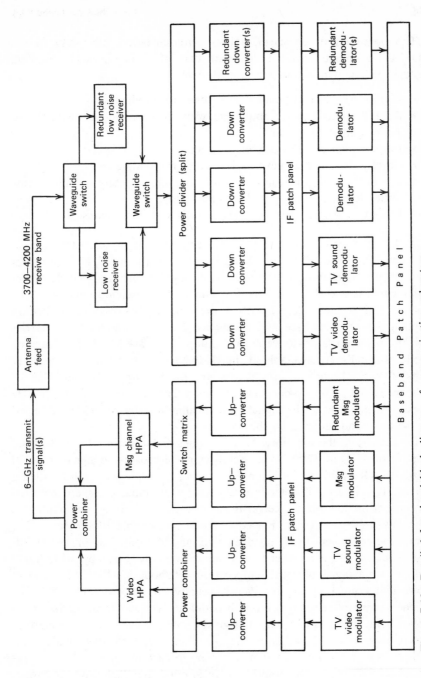

Figure 7.10 Detailed functional block diagram of communication subsystem.

317

The operation of an earth station communication subsystem in the FDMA/FM mode varies little from that of a line-of-sight radiolink system. The variances are essentially these:

- Use of cryogenically cooled low noise preamplifier(s).
- An HPA (high power amplifier) with the capability of from 0.5 to 1 kW output of power.
- Larger, high efficiency antennas, feeds.
- Careful design to achieve as low noise as possible.
- Use of a signal processing technique that allows nearly constant transmitter loading (i.e., spreading waveform).
- Use of threshold extension demodulators in some cases.

Now let us trace a signal through the communication subsystem typical of Figure 7.9. The FDM baseband is fed from the multiplex equipment through the baseband patch facility to the modulator. A spreading waveform signal is added to the very low end of the baseband, and the baseband signal is then shaped with a preemphasis network (described in Section 5.5.2). The baseband so shaped frequency modulates a carrier and the resultant is up-converted or multiplied to a 70-MHz IF. Patching facilities usually are available at the IF to loop back through the receiver subsystem or through a test receiver for local testing. The 70-MHz IF is then passed to an up-converter, which translates the IF to the 6-GHz operating frequency. The signal is then amplified by the HPA, directed to the feed, and radiated by the antenna.

On reception, the low noise receiver looks at the entire 500-MHz band (i.e., from 3700 to 4200 MHz) amplifying the broadband signal from 20 to 40 dB. Often this signal is amplified again by a low level TWT called a driver. This is done if comparatively long waveguide forms are required to connect the low noise receiver to the down-converter and demodulators. The low noise receiver is placed as close as possible to the feed. The remaining equipment in the receive chain may be located in another building. The low noise receiver is cryogenically cooled, usually with liquid helium achieving a physical temperature of about 12–15°K. Equivalent noise temperature of the receiver is usually on the order of 17–20°K.

The comparatively high level, broadband receive signal is fed to a power split. There is one output from the split for every down-converter/demodulator chain. In addition there is often a test receiver available as well as one or several redundant receivers in case of failure of an operational receiver chain. It should be kept in mind that each time the broadband incoming signal is split in two, there is a 3-dB loss due to the split, plus an insertion loss of the splitter. If a splitter has four outputs, the level going into a down-converter is something over 6 dB below the level at the input to the splitter. A splitter with eight outputs will cause a loss of something on the order of 10 dB.

Each down-converter is tuned to its 4-GHz operational frequency and converts it to a 70-MHz IF. Dual conversion is becoming more popular. One equipment on the market converts from 4 GHz to a first IF of 727 MHz and thence to 70 MHz as a second IF. Patching facilities are available for local loop back with the modulator.

The 70-MHz IF is then fed to the demodulator. The resulting demodulated signal, the baseband, is reshaped in the deemphasis network and the spreading waveform signal removed. The baseband output is fed to the baseband patch facilities and thence to the demultiplex equipment. At many earth station facilities, threshold extension demodulators are used in lieu of conventional demodulators to achieve "station margin." As of this writing, threshold extention techniques (see Chapter 6.9.5) are effective on the narrower B_{if}'s used in the so-called *message* service. The terms "message demodulator," "message up-converter," etc., are used to indicate the equipment used to carry telephone traffic. This is in opposition to the equipment used to carry video picture and TV sound traffic.

7.8.2 Antenna Tracking Subsystem

Geostationary earth satellites tend to drift in small suborbits (figure 8's). However, even with improved satellite station-keeping, the narrow beamwidths encountered with "standard" 30-m (100-ft) antennas require precise pointing and subsequent tracking of the earth station antenna to maximize the signal on the satellite and from the satellite. The basic modes of operation to provide these capabilities are

Programmed pointing
Manual pointing
Automatic tracking

Pointing deals with "aiming" the antenna initially on the satellite. Tracking keeps it that way. Programmed pointing may assume both duties. It points the antenna passively to a computed satellite position. With programmed pointing, the antenna is continuously pointed by interpolation between values of a precomputed time-indexed ephemeris. With adequate information as to the actual satellite position and true earth station position, pointing resolutions are on the order of $0.03–0.05°$.

Manual pointing may be effective for initial satellite acquisition or "capture" for later automatic tracking. It is also effective for wider beam, "nonstandard" earth stations when used with satellites with a low inclination and that are keeping station well. Manual pointing is used on certain types of domestic satellite systems, particularly those with smaller antennas, and therefore wider beamwidths.

Automatic tracking is used almost universally on "standard" earth stations. It actively seeks to maximize the received signal power by continuously comparing the relative signal amplitudes or phases of the satellite beacon signal of comparatively narrow RF beamwidth. The automatic tracking systems discussed here require an antenna feed modified for tracking, a sufficiently sensitive tracking receiver, a tracking signal comparator, and a servo unit.

One automatic tracking technique that is used commonly today is called monopulse. Monopulse may be of either the amplitude or phase type. A phase sensing antenna system may be constructed using two or more antenna apertures separated by several wavelengths. When the antenna is off the satellite-antenna axis, there will be a time delay or phase delay between the signals arriving at the different aperture feeds.

An amplitude sensing antenna system is constructed using a single antenna aperture having two or more closely spaced feeds. In an amplitude sensing system each of the closely spaced feeds produces a radiation pattern which is displaced from the antenna-satellite axis. This displacement angle is a function of the separation of the feed-horn phase centers and the focal point of the antenna aperture. The displaced patterns intersect on the antenna-satellite axis. Thus all received signals off-axis will induce unequal signal amplitude in the feeds. When the two signals are subtracted, a null will be produced only for those signals arriving along the axis line.

To achieve the two signal patterns, sum and difference signal envelopes must be produced. This is done in a comparator, which in its simplest form could consist of a folded "T" hybrid. The sum is obtained by vectorial addition in the hybrid sum arm and the difference by vectorial subtraction in the difference or orthogonal arm of the hybrid.

The practical monopulse system uses a more elaborate method for improving the resolution of azimuth and elevation information, but the basic sum and difference concept remains the same.

Figure 7.11 is a simplified functional block diagram of a monopulse tracking subsystem. Key to system operation is the monopulse tracking receiver. It deals with three signals ΔAZ, ΔEL, and Σ (AZ = azimuth,

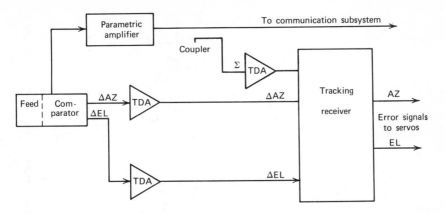

Figure 7.11 Simplified functional block diagram of a monopulse tracking subsystem. TDA, tunnel diode amplifier.

EL = elevation, Σ = sum). The sigma signal derives from coupling some output signal from the operational parametric amplifier. The ΔAZ and ΔEL derive from the monopulse comparator associated with the feed. The main function of the monopulse receiver is to detect the amplitude of the Δ signals relative to the Σ signal. The nature of the desired control signal(s) after detection is a dc null voltage which goes positive for an antenna pointing on one side of the satellite-antenna axis and negative for pointing error on the other side of the axis. The output of the tracking receiver is the input error information for the servo control subsystem for azimuth and elevation control of the antenna. Tracking accuracy has been fairly well standardized at 0.1 antenna beamwidth for winds at steady state of 30 mph with gusts to 45 mph, and 0.15 beamwidth for 45 mph steady-state winds gusting to 60 mph. At 4 GHz for a 60-dB gain antenna, half-power beamwidth is about 0.18°, dictating a tracking accuracy of 0.018° and 0.027°, respectively.

7.8.3 Multiplex, Orderwire, and Terrestrial Link Subsystems

"Standard" earth stations working with Intelsat synchronous satellites connect to a country's national long-distance telephone network via an international four-wire switching center. The switching center is usually located in a center of population, whereas the earth station is usually located in a relatively noise-free area, thus out in the countryside. An

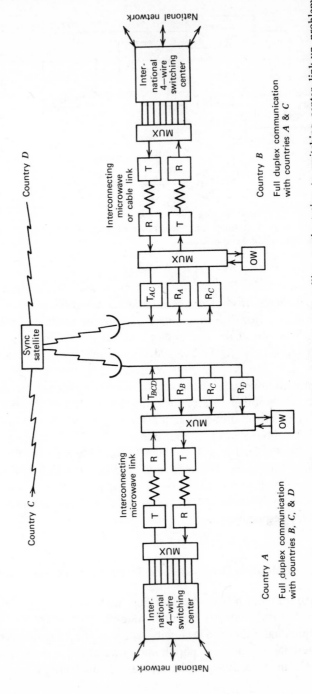

Figure 7.12 Interconnection of national toll networks via a synchronous satellite: earth station to switching center link-up problem. OW, order wire; T, transmitter; R, receiver.

322

interconnection with the switching center is usually over a broad band transmission medium such as a radiolink.

Such a system is shown in Figure 7.12. In the figure a synchronous satellite interconnects four countries: *A, B, C,* and *D.* Consider country A. The up link is simple. The up link baseband is configured at the four-wire switching center in its FDM multiplex equipment, transmitted on one radiolink carrier to the earth station, demodulated to baseband, and the baseband is applied directly to the earth station modulator. This baseband is configured as in Figure 7.13. In our example, earth station *A* receives three "down-link" Carriers: *F*1, *F*2, and *F*3 in the band 3700–4200 MHz. Each carier is demodulated as shown. The problem is to pass the baseband of each of the three incoming carriers on only one RF carrier via a conventional microwave radiolink connecting the earth station to the international switching center.

Figure 7.14 illustrates the problem. In the figure the demodulator "from *B*" contains in its baseband Group 1 of Supergroup 1 the 12 voice channels destined for country *A.* The demodulator "from *C*" (the *F*2 demodulator) contains in its demodulated baseband Supergroup 1 (60 voice channels) destined for country *A;* and the demodulator "from *D*" contains Supergroup 1 (60 voice channels).

Each baseband is processed by the multiplex equipment as shown in Figure 7.14. The output of the multiplex equipment is a single baseband containing Supergroups 1, 2, and 3. It should be noted, in the example shown, that from the country with which there is 12 full-duplex channel communication, only group 1 of Supergroup 1 is used. It has been placed in the Supergroup 3 slot of the terrestrial microwave baseband even though only one group of the supergroup carries active traffic.

The example shown is a very simplified one. As mentioned previously, many earth stations are designed to transmit several carriers for direct communication with 10 or more distinct geographical locations on the earth's surface via one synchronous earth satellite. This would require 10 or more separate receivers to receive 10 distinct carriers. The multiplex

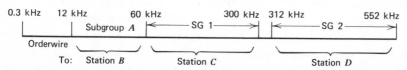

Figure 7.13 Configuration of transmitted baseband (up link) from earth station *A.*

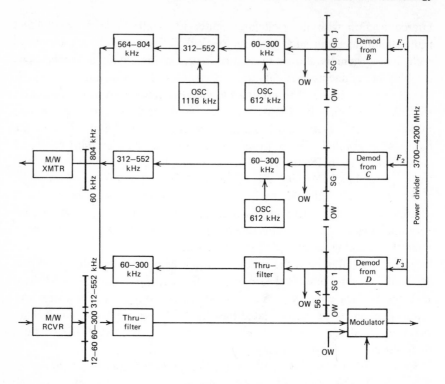

Figure 7.14 Simplified block diagram, reconfiguration of down FDM multiplex basebands for transmission over terrestrial link. OW, orderwire; OSC, oscillator; SG, supergroup; Gp, group M/W, microwave (LOS).

equipment required to place the channels, groups, and supergroups so received on one terrestrial system radio carrier can indeed be complex.

The orderwire facilities are transmitted for message carrier operation in the baseband segment form 300 Hz to 12 kHz as shown in Figure 7.14. The facilities provided are teleprinter using tone telegraph techniques in a "voice plus" arrangement. (Teleprinter, tone telegraph, and voice plus are discussed in Chapter 8.) Two voice channels are provided in each case, denominated "P1" and "P2" (P1 for 4–8 kHz and P2 for 8–12 kHz) and are transmitted inverted. In each channel, voice is transmitted from 300 to 2500 Hz. Five tone telegraph channels are available above 2500 Hz with center frequencies at 2.7, 2.82, 2.94, 3.06, and 3.18 kHz. The voice channels, speech portion, use in-band signaling, 2280 Hz. Video carriers do not transmit orderwire information directly on the carrier, but utilize a portion of aural carrier. The transmission of TV is discussed below.

7.9 INTELSAT IV

A series of four different types of communication satellites have been launched by the ICSC (International Communication Satellite Consortium). Their characteristics are summarized in Table 7.3. As evident in the table, Intelsat IV represents a major advance in communication satellite technology.

Table 7.3 Satellite Characteristics—Intelsat

	I	II	III	IV
Diameter (ft)	2.36	4.67	4.67	7.8
Weight (lb)	85	190	270	1200
Height (ft)	1.94	2.2	3.41	16.9 with ants.
Dc power (W)	33	75	125	570
Number of Transponders	2	1	2	12
Bandwidth per transponder (MHz)	25	130	225	36
Antenna characteristics (denoting "donut" pattern)	11 × 360° centered at +7°	12 × 360° centered on equator	20 × 20°	Two global beam 17°, two spot beam 4.5°
EIRP (W per rptr)	10	35	150	global +22 dBW spot +33.7 dBW
Total full-duplex VF channels	240	240	1200	5000

Figure 7.15 is a simplified functional block diagram of the satellite. The broad band 6-GHz signal is received on a global antenna, downconverted to 4 GHz, and amplified in a common equipment section. A spare section is available in case of failure. The converted and amplified signal is fed to a hybrid dividing the signal into two paths. Six transponders feed from each path. The transponders have a bandwidth of 36 MHz each. Equalization for both delay and amplitude is provided. A variable attenuator provides for the proper drive to the TWT amplifier. The "back-

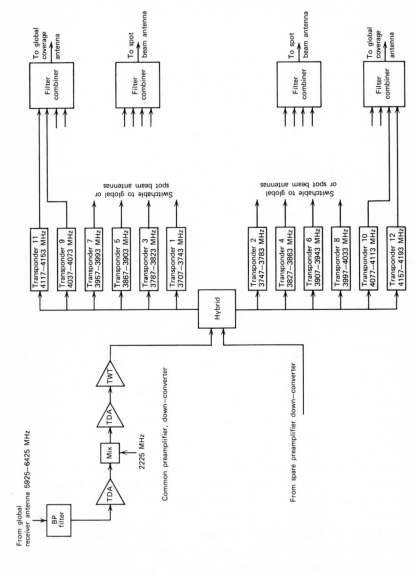

Figure 7.15 Simplified functional block diagram, Intelsat IV communication subsystem.

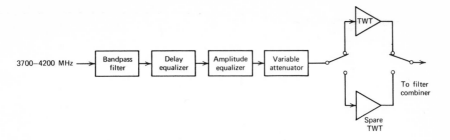

Figure 7.16 Simplified block diagram of Intelsat IV transponder.

off" on drive is a function of the number of carriers to be amplified by the transponder. Figure 7.16 is a simplified functional block diagram of a typical transponder.

The output of each transponder is connected to a filter combiner. There is a filter combiner associated with each transmitting antenna, of which there are four. Two are transmit horn antennas with global coverage and two are front-end fed paraboloids for transmitting the 4.5° spot beams. The spot beam pointing direction can be adjusted on command from the ground. The antennas are connected to their corresponding transponders via the filter combiners as shown in Figure 7.15. It should be noted that transponders 1–8 may be connected to either a spot beam or a global beam antenna. This switching is also done on command from the ground. The overall gain of each transponder is adjusted in 3.5-dB steps by ground command so that the correct TWT operating point can be selected. This refers to the "backoff" for multiple or single carrier operation, global-spot beam antenna usage.

For Intelsat IV polarization is left-hand circular for receive antennas (6 GHz) and right-hand circular for transmit antennas (4 GHz). The satellite beacons transmit right-hand circular on 3947.5 and 3952.5 MHz and are modulated with telemetry data. The transmission of telemetry in no way interferes with earth station tracking functions using the beacon(s) and both may be carried out simultaneously. Beacon EIRP is 0 dBW.

The 36-MHz bandwidth per transponder permits satisfactory TV transmission via global beam antennas and provides single carrier per transponder voice channel capacities of 900 and 1800 in the global and spot beams, respectively. The bandwidth is also suitable for multiple carrier FDM/FM, SPADE, and TDMA modes.

One method of achieving the greater voice channel capacity of Intelsat IV over its predecessors was to reduce guard bands between adjacent

Table 7.4a Intelsat IV—Global Beam

Carrier capacity (number of voice channels)	24	60	96	132	252	432	972
Top baseband frequency (kHz)	108	252	408	552	1052	1796	4028
Allocated satellite bandwidth (MHz)	2.5	5.0	7.0	10.0	15.0	25.0	36.0
Occupied bandwidth (MHz)	2.0	4.0	5.9	7.5	12.4	20.7	36.0
Multichannel deviation (rms) (kHz)	275	546	799	1020	1627	2688	4417
C/T at operating point (8000 pWp from RF sources) dBW/°K	-153.0	-149.9	-148.2	-147.1	-144.1	-141.4	-135.2

Table 7.4b Intelsat IV—Spot Beam

Carrier capacity (Number of voice channels)	60	132	192	252	432	612	792
Top baseband frequency (kHz)	252	552	804	1052	1796	2540	3284
Allocated satellite bandwidth (MHz)	2.5	5.0	7.5	10.0	15.0	20.0	25.0
Occupied bandwidth (MHz)	2.25	4.4	6.4	8.5	13.0	17.8	22.4
Multichannel deviation (MHz)	276	529	758	1009	1479	1996	2494
C/T at operating point (8000 pWp from RF sources) dBW/°K	-144.0	-141.4	-140.6	-139.9	-136.2	-133.8	-132.8

carriers down to about 10%. Thus FM deviation must be controlled carefully. Carrier sizes are multiples of a given bandwidth unit, 2.5 MHz. This permits interchanging one large carrier for two or more smaller carriers and vice versa without causing undue reallocation problems.

Tables 7.4*a* and 7.4*b* give basic transmission characteristics for Intelsat IV for the FDM/FM (FDMA) mode of operation. The reader should consult ICSC document 43-13 for more complete information. Table 7.5 gives required earth station EIRP's for 10° elevation angle for various carrier sizes.

Noise budgets for Intelsat IV operation are as follows (Ref. *ICSC 45-13*):

Thermal up path, thermal down path, and satellite intermodulation	7500 pWp
Multicarrier intermodulation in earth station transmitter	500 pWp
Earth station noise excluding transmit multicarrier intermod and group delay	500 pWp
Noise due to total system group delay after necessary equalization	500 pWp
Interference from terrestrial systems sharing the same bands	1000 pWp
total	10,000 pWp

Table 7.5　Required EIRP's for Various Carrier Sizes for Standard Earth Stations with Elevation Angles of 10°

Carrier Size (Channels)	Required EIRP (dBW) When Elevation Angle = 10°	
	Global Beam	Spot Beam
24	74.7	—
60	77.8	81.4
96	79.5	—
132	80.6	83.4
192	—	84.7
252	82.8	85.4
432	85.1	88.4
612	—	90.1
792	—	91.5
972	90.1	—
1872	—	98.6
TV	88.0	—

7.10   INTELSAT IVA

An improved version of Intelsat IV has been developed and is now operational. It has been called Intelsat IVA. The satellite achieves an increase in communications capability over Intelsat IV by frequency reuse through beam separation. For the Atlantic Region the east beam covers the continents of Europe, Africa, and the Middle East, and the west beam covers South and Central America and the eastern part of North America. Its transponders are similar to Intelsat IV with 36-MHz bandwidths. Thus these twelve 36-MHz frequency bands could be used twice for a total of 24 operational transponders. In practice some global beam coverage is required for television and SPADE operation and for those earth stations not adequately served by the frequency reuse beams. The planned use of Intelsat IVA is to have eight 36-MHz transponders in frequency reuse and four in global beam operation. Thus the satellite will have 20 active transponders.

Each frequency reuse transponder has three output TWT amplifiers, one for the west beam, one for the east beam, and a common spare. The spare amplifier is switched into operation by ground command.

Intelsat IVA is very similar to Intelsat IV such that few changes are required in earth stations that will operate with it. Several recommendations have been suggested, however. These are as follows:

- An improvment in axial ratio from 1.4 to 1.06 for new stations to facilitate possible use of dual polarization on future satellites.
- Earth stations to be required to transmit several carriers since multiple beam satellites require this.
- Allowable EIRP of intermodulation products resulting from multi-carrier operation to be reduced from 26 dBW/4 kHz to 23 dBW/4 kHz.
- 1.25 MHz 12-channel carriers to be included as standard Intelsat IVA carriers.

The capacity of Intelsat IVA has been increased to 12,500 VF channels along with SPADE and television capabilities. The EIRP per transponder is + 29 dBW.

7.11   TELEVISION TRANSMISSION

Standard practice for the transmission of television on the Intelsat series satellites is to transmit the video and audio portions of the TV signal separately. As shown in Table 7.5, the video required EIRP is con-

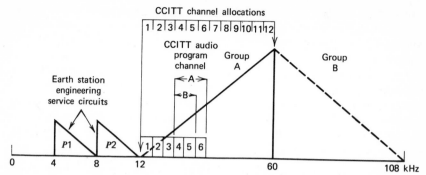

Figure 7.17 Intelsat III baseband allocation for audio program and coordination channels during television service. If additional audio program or coordination channels are required, Group B should be utilized. Channel allocations: (1) Coordination channel. These channels may be used for vision control, technical coordination, program coordination, commentary channel, program return channel, and cue channel. (2) Coordination channel. (3) Coordination channel. (4, 5) Type B audio program channel. (4, 5, 6) Type A audio program channel. (7, 8, 9) Additional audio program channel (when required). (10) Coordination channel. (11) Coordination channel. (12) Coordination channel.

siderably greater than that required for "message" carriers. As mentioned previously, the video portion occupies a separate transponder for Intelsat IV operation.

The aural channel, referred to henceforth as the program channel, with associated orderwire and cuing channels is placed on a standard CCITT group (60–108 kHz), as shown in Figure 7.17. The configurations of program channel and cuing/orderwire channels are handled as standard 12-channel carriers as shown in the figure. These may be CCITT Type A or B program channels (CCITT Rec. J.22). The preferred assembly uses the lower group, Basic Group B, 60–108 kHz inverted, wherein the program channel itself occupies the band 84–96 kHz on a virtual carrier frequency of 95.5 kHz, translated for transmission through the satellite to Basic Group A, 12–60 kHz erect (see Chapter 3).

Standard Intelsat video transmission parameters are shown in Tables 7.6a and 7.6b.

7.12 SPADE (Single Channel Per Carrier Multiple Access Demand Assignment Equipment)

Over the Atlantic basin one transponder of Intelsat IV has been assigned for SPADE operation. The 36-MHz bandwidth of the transponder is divided into 45-kHz segments providing a frequency pool (see Section

Table 7.6a   Intelsat III—Video Transmission Parameters

Allocated satellite bandwidth	40 MHz	
Receiver bandwidth	25 MHz	
C/T total at operating point	-142 dBW/°K	
Television standards	525/60	625/50
Maximum video bandwidth	4 MHz	6 MHz
Peak-to-peak deviation with a 15-kHz test signal		
for preemphasized video	9.4 MHz	7.9 MHz

Table 7.6b   Intelsat IV—Video Transmission Parameters

Allocated bandwidth	36 MHz	
Receiver bandwidth	30 MHz	
C/T total at operating point	-138 dBW/°K	
Television standards	525/60	625/50
Maximum video bandwidth	4.2 MHz	6 MHz
Peak-to-peak deviation with 15 kHz test signal for		
preemphasized video	6.8 MHz	5.1 MHz

7.7) of 800 frequencies. SPADE is another version of FDMA on a per-voice channel basis and each channel is assigned on demand. Figure 7.18 is a simplified block diagram of a SPADE terminal processor. On the left side of the drawing is the interface equipment with the four-wire telephone circuits coming from the international switching center (CT) and on the right-hand side are 70 MHz inputs/outputs of earth station up-converter and down-converter, respectively.

An incoming voice channel has its signaling information removed and is passed to the PCM codec (coder-decoder) where the voice channel is converted into a digital bit stream in PCM format. The format consists of seven-level code and A-law companding (see Chapter 11 for a full discussion of PCM). The output of the transmit side of the codec is a bit stream of 64,000 bps which modulates a four-phase PSK modem (coherent) in a 38-kHz noise bandwidth (the 45 kHz mentioned earlier includes guard bands). The output may be on any one of the 800 frequencies in the "pool." Frequency control is by means of a synthesizer.

For the receive side of a full-duplex voice channel, the above process is reversed. The 70-MHz output of the down-converter is fed to the SPADE terminal input (receive) of the PSK modem. The receive section of the modem, the demodulator, provides an output which is a digital bit

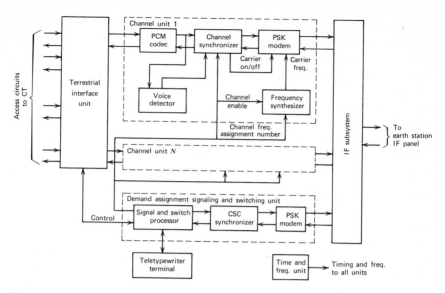

Figure 7.18 Block diagram of a SPADE terminal.

stream, which is fed to the decoder section of the codec where the digital-to-analog processing takes place. The resulting analog voice channel is fed to the terrestrial interface unit and thence to the international switching center after proper signaling information has been restored. The basic technical characteristics of a SPADE voice channel are reviewed in Table 7.7.

How does the SPADE terminal know what frequency slot is vacant at a particular interval of time? This information is provided by the DASS (demand-assignment signaling and switching unit). The terrestrial interface equipment routes signaling information to the DASS unit. The DASS unit

Table 7.7 Technical Characteristics of SPADE (Voice Channel Portion)

Channel encoding	7-bit PCM; A-law companding
Modulation	PSK; four-phase, coherent
Bit rate	64 kb/s
Bandwidth per channel	38 kHz
Channel spacing	45 kHz
Stability requirement	± 2 kHz (with AFC)
Bit error rate at threshold	1×10^{-5} (voice channel receiver)

uses a separate information channel for signaling and provides constant status updating information of busy-idle condition of channels in the "pool," so that busy frequencies become unavailable for assignment of new calls. DASS provides signaling interface on one common RF channel for all voice channels transmitted. This common signaling channel is actually a time division multiple access broadcast channel. It receives status updating information as well. Such a separate channel is required at all SPADE installations and operates on a TDMA basis allowing up to 49 stations plus a reference station to access the channel in 50 successive 1-ms bursts per frame at a bit rate of 128 kbs. The channel modem is different from the voice channel modem. It uses biphase coherent phase shift keying (see Chapter 8 for a discussion of modems). Bit error rate is on the order of 1×10^{-7}. The common signaling and control (CSC) channel operates at an RF level 7 dB higher than the ordinary SPADE voice channel to maintain the lower error rate at the higher bit rate.

7.13   REGIONAL SATELLITE COMMUNICATION SYSTEMS

Regional or "domestic" satellite communication systems are attractive to "regions" of the world with a large community of interest. These regions may be just one country with a large geographical expanse such as the United States, India, or the Soviet Union, or a group of countries with some common interests such as Europe, Spanish-speaking South America, or Oceania. Most important is that the satellite communication system must compete favorably against its terrestrial counterpart, usually microwave radiolinks (Chapter 5) and coaxial cable systems (Chapter 9).

From a cost point of view; the large earth stations operating in the Intelsat series can be justified. They are supported by significantly large toll revenues of transoceanic communications. Such earth stations compete with costly multichannel undersea cable systems, whereas the regional satellite system must compete with relatively inexpensive terrestrial systems.

For a competitive regional satellite system we must look for cost reduction of earth stations without significant performance degradation. The first and most obvious place to look is the antenna. This is the major cost item for the "standard" Intelsat earth station, amounting to perhaps $\frac{1}{3}-\frac{1}{2}$ of the cost.

The nature of regional or domestic satellite systems is that they will cover smaller geographical areas than their Intelsat counterparts. This implies larger minimum elevation angles. Turn to Figure 7.6. The Intelsat "standard" earth station was specified for a minimum elevation angle of

5° from the horizon resulting in approximately 25°K sky noise temperature contribution to the total system noise temperature. If we loosen the specification to a minimum of 10° of elevation angle the sky noise contribution drops to about 10°K (from the figure), or to 20° minimum elevation angle, the contribution reduces to about 7°K.

If we accept 20° as the minimum elevation angle, we can reduce the antenna dish to about 70 ft from 100 ft and still maintain a 40.7 G/T. If we are willing to reduce the G/T to 38 dB, we can reduce the antenna dish diameter to about 58 ft. By doing this, we double the beamwidth. With double the beamwidth, and improved satellite station-keeping, we may be able to remove the auto-track feature and change the type of antenna mount, reducing the antenna costs to well under $100,000.

A tendency in regional satellites such as the European ESRO system is to go to higher frequency bands. For ESRO one possibility is to station a satellite at about 15°E longitude, and with an antenna beamwidth of 7°, coverage can be provided from Iceland to the coastal regions of North Africa and as far east as the Ural Mountains. Reykjavik would have an elevation angle of about 13°. The up path would be nominally 14 GHz and the down path at 11 GHz, and the EIRP of the satellite, + 33 dBW per carrier at beam center. A typical earth station interoperating with this satellite may have the following parameters:

Antenna diameter, 48 ft (15 m)
Cooled parametric amplifier with a noise temperature of 50°K
G/T at 13° elevation angle, 41.4 dB

Another type of earth station with loosened G/T requirements may have the following parameters:

Antenna diameter, 42 ft (13 m)
Uncooled parametric amplifier, 250°K
G/T at 25° elevation angle, 36.1 dB
G/T at 34° elevation angle, 36.3 dB

The principal reason against higher frequency operation is the possibility of rainfall absorption with the consequent signal degradation or complete dropout. The higher elevation angle significantly reduces the distance the signal travels in the atmosphere and hence the distance it travels in rain.

One of the primary applications of regional satellites is television relay. Better than a 54-dB signal-to-noise ratio may be expected at 35° elevation angles for 99% of the time with a 36.3 dB G/T. This will provide a picture better than TASO grade 1. (See Chapter 12 for a discussion of video transmission.)

7.14 THE COORDINATION CONTOUR

The 4- and 6-GHz bands used for earth-satellite-earth communication systems are shared with terrestrial radiolink services. The terms "coordination contour" and "coordination distance" deal with methods to ensure that one service does not interfere with the other. Coordination contour establishes around an earth station an area beyond which the possibility of mutual interference may be considered negligible. The coordination distance is the distance to that boundary at a given azimuth.

Coordination distance is often given in minimum permissible free space path loss (see Section 5.3.1) (basic loss), L_B. For the possibility of interference from an earth station to a terrestrial station CCIR Rep. 382-1 provides this formula:

$$L_B = (P_T + G_T) - F_s - (P_R - G_R) \qquad \text{(dB)}$$

where P_T = power density in any 4-kHz band supplied by an earth station transmitter to the antenna input (dBW)

G_T = isotropic gain of the earth station transmitting antenna in the pertinent direction (dB)

F_s = shielding factor, a function of the elevation angle of the horizon (or obstacle) as seen from the earth station in the pertinent direction

P_R = maximum permissible interfering RF signal power density in any 4-kHz band at the receiver input of the terrestrial station in question (dBW)

G_R = isotropic gain of the receiving antenna of terrestrial station in question less line losses (dB)

The term F_s may be derived from Table 7.8, where θ is the angle of elevation of the obstacle. For main beam obstacles this table is valid only for those obstacles 5–16 km from the earth station. For side lobe obstacles, the shielding factor may be used for correspondingly less distances.

If the terrestrial station receiver has an effective noise temperature of $750°\text{K}$ and $G_R = 42$ dB, assuming -76 dBm0p resulting from the noise temperature in a telephone channel, then

$$L_B \,(0.005\%) = (P_T + G_T) - F_s + 173 \text{ (dB)}$$

Table 7.8 Site Shielding Factor

Elevation Angle of Horizon, θ	Site Shielding Factor, F_s (dB)
<1°	0
1–2°	10
2–3°	17
3–4°	23
≥4°	25

Source: CCIR Rep. 382-1.

The approach to be used for a terrestrial station to an earth station for computing coordination distance follows the same general reasoning. The reader is urged to review CCIR Rep. 382-1 (New Delhi, 1970, Vol. 4, pp. 349–266). Recs. 355-1 and 357-1 should also be consulted.

REFERENCES AND BIBLIOGRAPHY

1. Extraordinary Administrative Radio Conference (EARC), Geneva 1963, ITU (Final Acts).
2. *Reference Data for Radio Engineers*, 5th ed., Howard W. Sams & Co., Indianapolis.
3. CCITT Green Books New Delhi, 1970, Vol. IV, Recs. 382-2, 383-1, and 386-1.
4. *ICSC*-37-38*E W*/1/69, Communication Satellite Corp., Washington, D.C.
5. F. D. Doidge, "Antenna Gain as it Applies to Satellite Communication Earth Stations," *Edited Lectures, U.S. Seminar on Communication Satellite Earth Station Technology*, May 1966, ComSat, Washington, D.C.
6. M. Hoffman, "Antenna Noise Temperature," *Edited Lectures, U.S. Seminar on Communication Satellite Earth Station Station Technology*, May 1966, ComSat, Washington, D.C.
7. Roger L. Freeman, "An Approach to Earth Station Technology," *Telecommun. J.*, (ITU), June 1971.
8. B. Cooper, "Station Margin," *Edited Lectures, U.S. Seminar on Communication Satellite Earth Station Technology*, May 1966, ComSat, Washington, D.C.
9. *A Satellite Time Division Multiple Access Experiment*, ICSC/T-17-6E W/1/67, Communication Satellite Corporation, Washington, D.C.
10. *Spade System Specifications*, ICSC 47-14, May 1970, Communication Satellite Corporation, Washington, D.C.
11. A. M. Werth, "Spade: A PCM FDMA Demand Assignment System for Satellite Communications," INTELSAT/IEE Symposium on Digital Satellite Communications, London, November 1969.

12. Cohen and Steinmetz, "Amplitude and Phase Sensing Monopulse Systems Parameters," *Microwave J.*, October 1959.

13. *INTELSAT IV System Specifications*, ICSC 45-13, Communication Satellite Corp., Washington, D.C.

14. Stamininger and Jeffers, "Transmission System Planning for INTELSAT IV," Conference on Earth Station Technology, London 1970.

15. R. R. Willett, "Some Basic Criteria for a Regional Communication Satellite System Operating Above 10 GHz," Conference on Earth Station Technology, London, 1970.

8 | THE TRANSMISSION OF DIGITAL DATA

8.1 INTRODUCTION

Up to this point in our discussion we have been dealing with analog transmission. An analog transmission system has an output at the far end which is a continuously variable quantity representative of the input. With analog transmission there is continuity (of waveform); with digital transmission there is discreteness.

The simplest form of digital transmission is binary, where an information element is assigned one of two possibilities. There are many binary situations in real life where only one of two possible values can exist; for example, a light may be either on or off, an engine is running or not, and a person is alive or dead.

An entire number system has been based on two values, which by convention have been assigned the symbols 1 and 0. This is the binary system, and its number base is 2. Our everyday number system has a base of 10 and is called the decimal system. Still another system has a base of 8 and is called the octal system.

The basic information element of the binary system is called the bit, which is an acronym for *binary digit*. The bit, as we know, may have the values 1 or 0.

A number of discrete bits can identify a larger piece of information and we may call this larger piece a character. A code is defined by the IEEE as "A plan for representing each of a finite number of values or symbols as a particular arrangement or sequence of discrete conditions or events."

Binary coding of written information and its subsequent transmission have been with us for a long time. An example is teleprinter service (i.e., the transmission of a telegram).

The greater number of computers now in operation operate in binary languages; thus binary transmission fits in well for computer-to-computer communication and the transmission of data.

Although communication networks have been advocated that are built exclusively for digital transmission, and at least one such network now exists, most data are transmitted over the existing telephone network facilities or other analog facilities that are based on the nominal 4-kHz telephone channel as the basic building block.

This chapter introduces the reader to the transmission of binary information or data over telephone networks. It considers data on an end-to-end basis and the effects of variability of transmission characteristics on the final data output. There is a review of the basics of the makeup of digital data signals and their application. Therefore the chapter endeavors to cover the entire field of digital data and its transmission over telephone network facilities. It includes a discussion of the nature of digital signals, coding, information theory, constraints of the telephone channel, modulation techniques, and the dc nature of data transmission.

No distinction is made between data and telegraph transmission. Both are binary and often the codes used for one serve equally for the other. Likewise the transmission problems of data apply equally to telegraph. Telegraph communication is in message format, but not always) transmitted at rates less than 110 wpm, and is asynchronous. Data transmission is most often synchronous (but not necessarily so) and usually is an alphanumeric mix of information, much of the time destined for computers (EDP, electronic data processing). Data often are transmitted at rates in excess of 110 wpm.

8.2 THE BIT AND BINARY CONVENTION

In a binary transmission system the smallest unit of information is the bit. As we know, either one or two conditions may exist, the 1 or the 0. We call one state a mark, the other a space. These conditions may be indicated electrically by a condition of current flow and no current flow. Unless some rules are established, an ambiguous situation would exist. Is the "1" condition a mark or a space? Does the "no-current" condition mean a "0" is transmitted, or a "1"? To avoid confusion and to establish a positive identity to binary conditions, CCITT Rec. V.1 recommends equivalent binary designations. These are shown in Table 8.1. If the table is adhered to universally, no confusion will exist as to which is a mark, which is a space, which is the active condition, which is the passive condition, which is "1," and which is "0." It defines the *sense* of transmission so that the mark and space, the "1" and "0" will not be inverted. Data transmission engineers often refer to such a table as a table of *mark-space convention*.

Table 8.1   Equivalent Binary Designations

Active Condition	Passive Condition
Mark or marking	Space or spacing
Current on	Current off
+ voltage	− voltage
Hole (in paper tape)	No hole (in paper tape)
Binary "1"	Binary "0"
Condition Z	Condition A
Tone-on (amplitude modulation)	Tone-off
Low frequency (frequency shift keying)	High frequency
No phase inversion (differential phase shift keying)	Inversion of phase
Reference phase	Opposite to reference phase

Source: CCITT Rec. V.1.

8.3   CODING

8.3.1   Introduction to Binary Coding Techniques

Written information must be coded before it can be transmitted over a digital system. The discussion of coding below covers only binary codes. But before launching into coding itself, the term *entropy* is introduced.

Operational telecommunications systems transmit information. We can say that information has the property of reducing uncertainty of a situation. The measurement of uncertainty is called entropy. If entropy is large, then a large amount of information is required to clarify a situation; if entropy is small, then only a small amount of information is required for clarification. Noise in a communications channel is a principal cause of uncertainty. From this we now can state Shannon's Noisy Channel Coding Theorem (stated approximately):

> If an information source has an entropy H and a noisy channel capacity C, then provided $H < C$, the output from the source can be transmitted over the channel and recovered with an arbitrarily small probability of error. If $H > C$, it is not possible to transmit and recover information with an arbitrarily small probability of error [Ref. 1, 38-12].

Entropy is a major consideration in the development of modern codes. Coding can be such as to reduce transmission errors (uncertainties) due to the transmission medium and even correct the errors at the far end. This is done by reducing the entropy per bit (by adding redundancy).

We shall discuss errors and their detection in greater detail in Section 8.4. Channel capacity is discussed in Section 8.10.

Now the question arises, how big a binary code? The answer involves yet another question, how much information is to be transmitted?

One binary digit (bit) carries little information; it has only two possibilities. If two binary digits are transmitted in sequence, there are four possibilities,

$$00 \qquad 10$$
$$01 \qquad 11$$

or four pieces of information. Suppose 3 bits are transmitted in sequence. Now there are eight possibilities:

$$000 \qquad 100$$
$$001 \qquad 101$$
$$010 \qquad 110$$
$$011 \qquad 111$$

We can now see that for a binary code, the number of distinct information characters available is equal to 2 raised to a power equal to the number of elements or bits per character. For instance, the last example was based on a three-element code giving eight possibilities or information characters.

Another, more practical example is the Baudot teleprinter code. It has 5 bits or information elements per character. Therefore, the different or distinct graphics or characters available are $2^5 = 32$. The American Standard Code for Information Interchange (ASCII) has 7 information elements per character or $2^7 = 128$; so it has 128 distinct combinations of marks and spaces that are available for assignment as characters or graphics.

The number of distinct characters for a specific code may be extended by establishing a code sequence (a special character assignment) to shift the system or machine to uppercase (as is done with a conventional typewriter). Uppercase is a new character grouping. A second distinct code sequence is then assigned to revert to lowercase. As an example, the CCITT ITA No. 2 code (Figure 8.1) is a 5-unit code with 58 letters, numbers, graphics, and operator sequences. The additional characters, etc., (additional above $2^5 = 32$) come from the use of uppercase. Operator sequences appear on a keyboard as "space" (spacing bar), "figures" (uppercase), "letters" (lowercase), "carriage return," "line feed" (spacing vertically), etc.

Characters				Code Elements[a]						
Letters Case	Communi-cations	Weather	CCITT #2[b]	START	1	2	3	4	5	STOP
A	—	↑			■	■				■
B	?	⊕			■			■	■	■
C	:	○				■	■	■		■
D	$	⚡	WRU		■			■		■
E	3	3			■					■
F	1	⟶	Unassigned		■		■	■		■
G	&	↘	Unassigned			■		■	■	■
H	STOP[c]	↓	Unassigned				■		■	■
I	8	8				■	■			■
J	′	↯	Audible signal		■	■		■		■
K	(	←			■	■	■	■		■
L	)	↖				■			■	■
M	.	.					■	■	■	■
N	,	◍					■	■		■
O	9	9						■	■	■
P	θ	θ				■	■		■	■
Q	1	1			■	■	■		■	■
R	4	4				■		■		■
S	BELL	BELL	,		■		■			■
T	5	5							■	■
U	7	7			■	■	■			■
V	;	◎	=			■	■	■	■	■
W	2	2			■	■			■	■
X	/	/			■		■	■	■	■
Y	6	6			■		■		■	■
Z	″	+	+		■				■	■
BLANK		—								■
SPACE							■			■
CAR. RET.								■		■
LINE FEED						■				■
FIGURE					■	■		■	■	■
LETTERS					■	■	■	■	■	■

[a] Blank, spacing element; crosshatched, marking element.

[b] This column shows only those characters which differ from the American "communications" version.

[c] Figures case H(COMM) may be stop or +.

Figure 8.1 Communications and weather codes, CCITT International Telegraph Alphabet No. 2.

When we refer to a 5-unit, 6-unit, 7-unit, or 12-unit code, we refer to the number of information units or elements that make up a single character or symbol. That is, we refer to those elements assigned to each character that carry information and that make it distinct from all other characters or symbols of the code.

8.3.2 Some Specific Binary Codes for Information Interchange

In addition to the ITA No. 2 code, some of the more commonly used codes are Field Data Code, IBM Data Transceiver Code (Figure 8.2), the American Standard Code for Information Interchange (ASCII) (Figure 8.3), CCITT No. 5 Code (Figure 8.4), the EBCDIC Code (Figure 8.5), the Hollerith Code (Figure 8.6), and the BCD Code.

The Field Data Code was adopted by the U.S. Army in 1969 and is now being replaced by ASCII. The Field Data Code is an 8-bit code consisting of 7 information bits and a control bit. With the 8 bits, 256 (2^8) bit patterns or permutations are available. Of these, 128 are available for assignment to characters. These 128 bit patterns are subdivided into 64 each. The groups are distinguished one from the other by the value or state of a particular bit known as a control bit. Owing to this arrangement of bits and bit patterns, the Field Data Code has parity check capability.

Parity checks are one way to determine if a character contains an error after transmission. We speak of even parity and odd parity. On a system using an odd parity check, the total count of "1"s or marks has to be an odd number per character (or block) (e.g., it carries 1, 3, 5, or 7 marks or "1"s). Some systems, such as the Field Data Code, use even parity (i.e., the total number of marks must be an even number, e.g., 2, 4, 6, or 8). The code is used by the U.S. Army for communication in a common language between computing, input/output, and terminal equipments of the Field Data Equipment family.

To explain parity and parity checks a little more clearly, let us look at some examples. Consider a seven-level code with an extra parity bit. By system convention, even parity has been established; suppose a character is transmitted as 1111111. There are seven marks, so to maintain even parity we would need an even number of marks. Thus an eighth bit is added and must be a mark (1). Look at another bit pattern, 1011111. Here there are six marks, even; then the eighth (parity) bit must be a space. Still another example would be 0001000. To get even parity a mark must be added on transmission and the character transmitted would be 00010001, maintaining even parity. Suppose that, owing to some sort of signal interference, one signal element was changed on reception. No matter which element was changed, the receiver would indicate an error because we would no longer have even parity. If two elements were changed, though, the error could be masked. This would happen in the case of even or odd parity if two marks were substituted for two spaces or vice versa at any element location in the character.

The IBM Data Transceiver Code (see Figure 8.2) is used for the transfer of digital data information recorded on perforated cards. It is an 8-bit

Bit Number	Code Assignments							
	0	0	0	0	1	1	1	1
	0	0	1	1	0	0	1	1
	0	1	0	1	0	1	0	1
X O N R 7 4 2 1 ↓↓↓↓ ↓↓↓↓								
0 0 0 0 0								
0 0 0 0 1								TPH/TGR
0 0 0 1 0								@
0 0 0 1 1				(NA)		(NA)	Space	
0 0 1 0 0								#
0 0 1 0 1				(NA)		(NA)	9	
0 0 1 1 0				(NA)		(NA)	8	
0 0 1 1 1		G	P		X			
0 1 0 0 0								(NA)
0 1 0 0 1				(NA)		(NA)	6	
0 1 0 1 0				(NA)		(NA)	5	
0 1 0 1 1		D	M		U			
0 1 1 0 0				(NA)		(NA)	3	
0 1 1 0 1		B	K		S			
0 1 1 1 0		A	J		/			
0 1 1 1 1	0							
1 0 0 0 0								Restart
1 0 0 0 1				SOC/EOC		$\overset{+}{0}$	EOT	
1 0 0 1 0				*		%		
1 0 0 1 1		&	—		Ø			
1 0 1 0 0				$		,	.	
1 0 1 0 1		I	R		Z			
1 0 1 1 0		H	Q		Y			
1 0 1 1 1	7							
1 1 0 0 0				(NA)		(NA)	(NA)	
1 1 0 0 1		F	O		W			
1 1 0 1 0		E	N		V			
1 1 0 1 1	4							
1 1 1 0 0		C	L		T			
1 1 1 0 1	2							
1 1 1 1 0	1							
1 1 1 1 1								

Figure 8.2 IBM Data Transceiver Code.

Transmission Order: Bit X → Bit 1.
Legend: TPH/TGH Telephone/Telegraph $\overset{+}{0}$ Plus zero
       SOC/EOC Start or end of card $\overline{0}$ Minus zero
       EOT   End of transmission   "Lozenge"
       (NA)   Valid but not assigned   (special symbol)

code providing a total of 256 mark-space combinations. Only those combinations or patterns having a "fixed count" of 4 ones (marks) and 4 zeros (spaces) are made available for assignment as characters. Thus only 70 bit patterns satisfy the fixed count condition and the remaining, 186 combinations are invalid. The parity (fixed count) or error checking advantage is obvious. Of the 70 valid characters, 54 are assigned to alphanumerics and a limited number of punctuation signs and other symbols. In addition, the code includes special bit patterns assigned to control functions peculiar to the transmission of cards such as "start card" and "end card."

The American Standard Code for Information Interchange (see Figure 8.3), better known as ASCII, is the latest effort on the part of the U.S.

Bit Number									
		0	0	0	0	1	1	1	1
		0	0	1	1	0	0	1	1
		0	1	0	1	0	1	0	1
$b_7\,b_6\,b_5\,b_4\,b_3\,b_2\,b_1$	Row \ Column	0	1	2	3	4	5	6	7
0 0 0 0	0	NUL	DLE	SP	Ø	`	P	@	p
0 0 0 1	1	SOH	DC1	!	1	A	Q	a	q
0 0 1 0	2	STX	DC2	"	2	B	R	b	r
0 0 1 1	3	ETX	DC3	#	3	C	S	c	s
0 1 0 0	4	EOT	DC4	$	4	D	T	d	t
0 1 0 1	5	ENQ	NAK	%	5	E	U	e	u
0 1 1 0	6	ACK	SYN	&	6	F	V	f	v
0 1 1 1	7	BEL	ETB	'	7	G	W	g	w
1 0 0 0	8	BS	CAN	(	8	H	X	h	x
1 0 0 1	9	HT	EM	)	9	I	Y	i	y
1 0 1 0	10	LF	SS	*	:	J	Z	j	z
1 0 1 1	11	VT	ESC	+	;	K	[	k	{
1 1 0 0	12	FF	FS	,	<	L	~	l	⌐
1 1 0 1	13	CR	GS	–	=	M	⌐	m	}
1 1 1 0	14	SO	RS	.	>	N	∧	n	|
1 1 1 1	15	SI	US	/	?	O	–	o	DEL

Figure 8.3 American Standard Code for Information Interchange.

Notes

1. Columns 2, 3, 4, and 5 indicate the printable characters in DCS Autodin (U.S. Defense Communication System) Automatic Digital Network.

2. Columns 6 and 7 fold over into columns 4 and 5, respectively, except DEL.

GENERAL DEFINITIONS

Communication Control: a function character intended to control or facilitate transmission of information over communication networks.

Format Effector: a functional character which controls the layout or positioning of information in printing or display devices.

Information Separator: a character used to separate and qualify information in a logical sense. There is a group of four such characters, which are to be used in a hierarchical order. In order rank, highest to lowest, they appear as follows: FS, file separator; GS, group separator; RS, record separator; and US, unit separator.

SPECIFIC CHARACTERS

NUL: the all-zeros character.

SOH, Start of Heading: a communication control character used at the beginning of a sequence of characters which constitute a machine-sensible address or routing information. Such a sequence is referred to as the "heading." An STX character has the effect of terminating a heading.

STX, Start of Text: a communication control character which precedes a sequence of characters that is to be treated as an entity and thus transmitted through to the ultimate destination. Such a sequence is referred to as "text." STX may be used to terminate a sequence of characters started by SOH.

ETX, End of Text: a communication control character used to terminate a sequence of characters started with STX and transmitted as an entity.

EOT, End of Transmission: a communication control character used to indicate the conclusion of a transmission, which may have contained one or more texts and any associated headings.

ENQ, Enquiry: a communication control character used in data communication systems as a request for a response from a remote station.

ACK, Acknowledge: a communication control character transmitted by a receiver as an affirmative response to a sender.

BEL: a character used when there is a need to call for human attention. It may trigger an alarm or other attention devices.

BS, Backspace: a format effector which controls the movement of the printing position one printing space backward on the same printing line.

HT, Horizontal Tabulation: a format effector which controls the movement of the printing position to the next in a series of predetermined positions along the printing line.

LF, Line Feed: a format effector which controls the movement of the printing position to the next printing line.

VT, Vertical Tabulation: a format effector which controls the movement of the printing position to the next in a series of predetermined printing lines.

FF, Form Feed: a format effector which controls the movement of the printing position to the first predetermined printing line on the next form or page.

CR, Carriage Return: a format effector which controls the movement of the printing position to the first printing position on the same printing line.

SO, Shift Out: a control character indicating that the code combinations which follow shall be interpreted as outside of the character set of the standard code table until a Shift In (SI) character(s) is (are) reached.

SI, Shift In: a control character indicating that the code combinations which follow shall be interpreted according to the standard code table.

DLE, Data Link Excape: a communication control character which will change the meaning of a limited number of contiguously following characters. It is used exclusively to provide supplementary controls in data communication networks. DLE is usually terminated by a Shift In character(s).

DC1, DC2, DC3, DC4, Device Controls: characters for the control of ancillary devices associated with data processing or telecommunication systems, especially for switching devices on or off.

NAK, Negative Acknowledgment: a communication control character transmitted by a receiver as a negative response to the sender.

SYN, Synchronous Idle: a communication control character used by a synchronous transmission system in the absence of any other character to provide a signal from which synchronism may be achieved or retained.

ETB, End of Transmission Block: a communication control character used to indicate the end of a block of data for communication purposes.

CAN, Cancel: a control character used to indicate that the data with which it is sent is in error or is to be disregarded.

EM, End of Medium: a control character associated with the sent data which may be used to identify the physical end of the medium, or the end of the used or wanted portion of information recorded on a medium.

SS, Start of Special Sequence: a control character used to indicate the start of a variable length sequence of characters which have special significance or which are to receive special handling. SS is usually terminated by a Shift In (SI) character(s).

ESC, Escape: a control character intended to provide code extension (supplementary characters) in general information interchange. The Escape character itself is a prefix affecting the interpretation of a limited number of contiguously following characters. ESC is usually terminated by a Shift In (SI) character(s).

DEL, Delete: this character is used primarily to erase or obliterate erroneous or unwanted characters in perforated tape.

SP, Space: normally a nonprinting graphic character used to separate words. It is also a format effector which controls the movement of the printing position, one printing position forward.

"Diamond": a noncoded graphic which is printed by a printing device to denote the sensing of an error when such an indication is required.

"Heart": a noncoded graphic which may be printed by a printing device in lieu of the symbols for the control characters shown in columns 0 and 1.

industry and common carrier systems, backed by the United States Standards Institute (now the American National Standards Institute), to produce a universal common language code. ASCII is a 7-unit code with all 128 combinations available for assignment. Here again the 128 bit patterns are divided into two groups of 64. One of the groups is assigned to a subset of graphic printing characters. The second subset of 64 is assigned to control characters. An eighth bit is added to each character for parity check. ASCII is widely used in North America and has received considerable acceptance in Europe and Latin America.

CCITT Rec. V.3 offers a seven-level code as an international standard for information interchange. It is not intended as a substitute for CCITT No. 2 code. CCITT No. 5, or the new alphabet No. 5, as the seven-level code is more commonly referred to, is basically intended for data transmission.

Although CCITT No. 5 is considered a seven-level code, CCITT Rec. V.4 advises that an eighth bit may be added for parity. Under certain circumstances odd parity is recommended; on others, even parity.

Figure 8.4 shows the CCITT No. 5 code, b_1 is the first signal element in serial transmission and b_7 is the last element of a character. Like the ASCII code, CCITT No. 5 does not normally need to shift out (i.e., uppercase, lowercase as in CCITT No. 2). However, like ASCII, it is provided with an escape, 1101100. Eight footnotes explaining peculiar character usage are provided. Few differences exist between the ASCII and CCITT No. 5 codes.

Bit Number									
	→ 0	0	0	0	1	1	1	1
	→ 0	0	1	1	0	0	1	1
	→ 0	1	0	1	0	1	0	1
	Column								
$b_7 b_6 b_5 b_4 b_3 b_2 b_1$ Row	0	1	2	3	4	5	6	7
0 0 0 0 0	NUL	(TC$_r$)DLE	SP	0	(@)[3]	P	' [4]	p
0 0 0 1 1	(TC$_1$)SOH	DC$_1$	!	1	A	Q	a	q
0 0 1 0 2	(TC$_2$)STX	DC$_2$	" [6]	2	B	R	b	r
0 0 1 1 3	(TC$_3$)ETX	DC$_3$	£ [3][7]	3	C	S	c	s
0 1 0 0 4	(TC$_4$)EOT	DC$_4$	$ [3][7]	4	D	T	d	t
0 1 0 1 5	(TC$_5$)ENQ	(TC$_8$)NAK	%	5	E	U	e	u
0 1 1 0 6	(TC$_6$)ACK	(TC$_9$)SYN	&	6	F	V	f	v
0 1 1 1 7	BEL	(TC$_{10}$)ETB	' [6]	7	G	W	g	w
1 0 0 0 8	FE$_0$(BS)	CAN	(	8	H	X	h	x
1 0 0 1 9	FE$_1$(HT)	EM	)	9	I	Y	i	y
1 0 1 0 10	FE$_2$(LF)[1]	SUB	*	: [8]	J	Z	j	z
1 0 1 1 11	FE$_3$(VT)	ESC	+	; [8]	K	([) [3]	k	[3]
1 1 0 0 12	FE$_4$(FF)	IS$_4$(FS)	,	<	L	[9]	l	[3]
1 1 0 1 13	FE$_5$(CR)[1]	IS$_3$(GS)	—	=	M	(]) [3]	m	[3]
1 1 1 0 14	SO	IS$_2$(RS)	.	>	N	∧ [4][6]	n	− [4][6]
1 1 1 1 15	SI	IS$_1$(US)	/	?	O	—	o	DEL

Figure 8.4 CCITT No. 5 Code for Information Interchange (see CCITT Rec. V.3).

Notes

1. The controls CR and LF are intended for printer equipment which requires separate combinations to return the carriage and to feed a line.

For equipment which uses a single control for a combined carriage return and line feed operation, the function FE$_2$ will have the meaning of "new line" (NL).

These substitutions agreement between the sender and the recipient of the data.

The use of this function NL is not allowed for international transmission on general switched telecommunication networks (telegraph and telephone networks).

2. For international information interchange, $ and £ symbols do not designate the currency of a given country. The use of these symbols combined with other graphic symbols to designate national currencies may be the subject to other Recommendations.

3. Reserved for national use. These positions are intended primarily for alphabetic extensions. If they are not required for that purpose, they may be used for symbols and a recommended choice is shown in parentheses in some cases.

4. Positions 5/14, 6/0, and 7/14 of the 7-bit set table normally are provided for the diacritical signs "cricumflex," "grave accent," and "overline." However, these positions may be used for other graphical symbols, when it is necessary to have 8, 9, or 10 positions for national use.

5. For international information interchange, position 7/14 is used for the graphical symbol – (overline), the graphical representation of which may vary according to national use to represent ∼ (tilde) or another diacritical sign provided that there is no risk of confusion with another graphical symbol included in the table.

6. The graphics in positions 2/2, 2/7, 5/14 have respectively the significance of "quotation mark," "apostrophe," and "upwards arrow"; however, these characters take on the significance of the diacritical signs "diaeresis," "acute accent," and "circumflex accent" when they precede or follow the "backspace" character.

7. For international information interchange, position 2/3 of the 7-bit code table has the significance of the symbol £ and position 2/4 has the significance of the symbol $.

By agreement between the countries concerned where there is no requirement for the symbol £, the symbol "number sign" (#) may be used in position 2/3. Likewise, where there is no requirement for the symbol $, the symbol "currency sign" (   ) may be used in position 2/4.

8. If 10 and 11 as single characters are needed (for example, for Sterling currency subdivision), they should take the place of "colon" (:) and "semicolon" (;), respectively. These substitutions require agreement between the sender and the recipient of the data. On the general telecommunication networks, the characters "colon" and "semicolon" are the only ones authorized for international transmission.

The reader's attention is called to what are known as computable codes such as ASCII and CCITT No. 5 codes. Computable codes have the letters of the alphabet plus all other characters and graphics assigned values in continuous binary sequence. Thus these codes are in the native binary language of today's common digital computers. The CCITT No. 2 (ITA No. 2) is not, and when used with a computer often requires special processing.

The EBCDIC (Extended Binary Coded Decimal Interchange Code) is similar to ASCII but is a true 8-bit code. The eighth bit is used as an added bit to "extend" the code, providing 256 distinct code combinations for assignment. Figure 8.5 illustrates the EBCDIC code.

The Hollerith Code was specifically designed for use with perforated (punched) cards. It has attained wide acceptance in the business machine and computer fields. Hollerith is a 12-unit character code in that a character is represented on a card by one or more holes perforated in one

Bit Levels (first group, bit 0 = 0)

0→	0	0	0	0	0	0	0	0
1→	0	0	0	0	1	1	1	1
2→	0	0	1	1	0	0	1	1
3→	0	1	0	1	0	1	0	1

4567 ↓	0000	0001	0010	0011	0100	0101	0110	0111
0000	NULL		DS		SP	&	−(Minus)	
0001			SOS				/(Slash)	
0011			FS					
0100	PF	RES	BYP	PN				
0101	HT	NL	LF	RS				
0110	LC	BS	EOB	US				
0111	DEL	IL	PRE	EOT				
1000								
1001								
1010			SM			!(Exclam. pt.)		:(Colon)
1011					.(Period)	&	,(Comma)	#
1100					<	*	%	
1101						)	−(Undesc.)	'(Prime)
1110					(	;(Semicolon)	>	=(Equals)
1111					(Vert. Bar)	(not)	?	"(Quote)

Bit Levels (second group, bit 0 = 1)

0→	1	1	1	1	1	1	1	1
1→	0	0	0	1	1	1	1	1
2→	0	0	1	0	0	0	1	1
3→	0	1	0	0	1	0	0	1

4567 ↓	1000	1001	1010	1011	1100	1101	1110	1111
0000					PZ	MZ		0
0001	a	j			A	J		1
0010	b	k	s		B	K	S	2
0011	c	l	t		C	L	T	3
0100	d	m	u		D	M	U	4
0101	e	n	v		E	N	V	5
0110	f	o	w		F	O	W	6
0111	g	p	x		G	P	X	7
1000	h	q	y		H	Q	Y	8
1001	i	'	z		I	R	Z	9
1010								
1011								
1100								
1101								
1110								
1111								

Figure 8.5 Extended Binary Coded Decimal Interchange Code (EBCDIC).

column having 12 potential hole positions. It is most commonly used with the standard 80-column punched cards.

The theoretical capacity of a 12-unit binary code is very great and by our definition, using all hole patterns available, is 2^{12} or 4096 bit combinations. In the modern version of the Hollerith code only 64 of these, none using more than 3 holes, are assigned to graphic characters.

Because of its unwieldiness, the Hollerith code is seldom used directly for transmission. Most often it is converted to one of the more conventional transmission codes such as ASCII, CCITT No. 5, the BCD interchange code, or EBCDIC. Figure 8.6 shows the Hollerith Code and its BCD interchange equivalents.

BCD Code Bit Nos: B A 8 4 2 1	Graphics	Hollerith Code Rows Punched	BCD Code Bit Nos: B A 8 4 2 1	Graphics	Hollerith Code Rows Punched
0 0 0 0 0 0	Blank	No holes	1 0 0 0 0 0	–	11
0 0 0 0 0 1	1	1	1 0 0 0 0 1	J	11–1
0 0 0 0 1 0	2	2	1 0 0 0 1 0	K	11–2
0 0 0 0 1 1	3	3	1 0 0 0 1 1	L	11–3
0 0 0 1 0 0	4	4	1 0 0 1 0 0	M	11–4
0 0 0 1 0 1	5	5	1 0 0 1 0 1	N	11–5
0 0 0 1 1 0	6	6	1 0 0 1 1 0	O	11–6
0 0 0 1 1 1	7	7	1 0 0 1 1 1	P	11–7
0 0 1 0 0 0	8	8	1 0 1 0 0 0	Q	11–8
0 0 1 0 0 1	9	9	1 0 1 0 0 1	R	11–9
0 0 1 0 1 0	0	0	1 0 10 1 0	!	11–0
0 0 1 0 1 1	① # or =	3–8	1 0 1 0 1 1	$	11–3–8
0 0 1 1 0 0	① @ or '	4–8	1 0 1 1 0 0	*	11–4–8
0 0 1 1 0 1	:	5–8	1 0 1 1 0 1	]	11–5–8
0 0 1 1 1 0	>	6–8	1 0 1 1 1 0	;	11–6–8
0 0 1 1 1 1	√	7–8	1 0 1 1 1 1	△	11–7–8
0 1 0 0 0 0	ƀ	2–8	1 1 0 0 0 0	① & or +	12
0 1 0 0 0 1	/	0–1	1 1 0 0 0 1	A	12–1
0 1 0 0 1 0	S	0–2	1 1 0 0 1 0	B	12–2
0 1 0 0 1 1	T	0–3	1 1 0 0 1 1	C	12–3
0 1 0 1 0 0	U	0–4	1 1 0 1 0 0	D	12–4
0 1 0 1 0 1	V	0–5	1 1 0 1 0 1	E	12–5
0 1 0 1 1 0	W	0–6	1 1 0 1 1 0	F	12–6
0 1 0 1 1 1	X	0–7	1 1 0 1 1 1	G	12–7
0 1 1 0 0 0	Y	0–8	1 1 1 0 0 0	H	12–8
0 1 1 0 0 1	Z	0–9	1 1 1 0 0 1	I	12–9
0 1 1 0 1 0	‡	0–2–8	1 1 1 0 1 0	?	12–0
0 1 1 0 1 1	,	0–3–8	1 1 1 0 1 1	.	12–3–8
0 1 1 1 0 0	① % or (	0–4–8	1 1 1 1 0 0	① ⌑ or)	12–4–8
0 1 1 1 0 1	∨	0–5–8	1 1 1 1 0 1	[	12–5–8
0 1 1 1 1 0	/	0–6–8	1 1 1 1 1 0	<	12–6–8
0 1 1 1 1 1	₦	0–7–8	1 1 1 1 1 1	‡	12–7–8

Figure 8.6 Hollerith Card Code and BCD Interchange Code.

① "Arrangement A" (for reports) or "Arrangement H" (for programming).

Control Characters: √ TM (tape mark)   ₦ SM (segment mark)
     ƀ SB (substitute blank)   △ MC (mode change)
     ‡ RM (record mark)   ‡ GM (group mark)
     ∨ WS (word separator)   ⌑ CW (clear word mark)

Special Characters: ! MZ (minus zero)   ? PZ (plus zero)

The BCD (binary coded decimal) code is a compromise code assigning binary numbers to the digits between 0 and 9 simplifying the straight and cumbersome binary code. These equivalents appear as follows:

Decimal Digit	BCD Digit	Decimal Digit	BCD Digit
0	1010	5	0101
1	0001	6	0110
2	0010	7	0111
3	0011	8	1000
4	0100	9	1001

To cite some examples, consider the number 16. It is broken into 1 and 6; thus its BCD equivalent is 0001 0110. If it were written in straight binary, it would appear as 10000. 25 in BCD combines the digits 2 and 5 above as 0010 0101.

8.4 ERROR DETECTION AND ERROR CORRECTION

In the transmission of data the most important goal in design is to minimize the error rate. Error rate may be defined as the ratio of the number of bits incorrectly received to the total number of bits transmitted. On many data circuits the design objective is an error rate no poorer than one error in 1×10^5 and on telegraph circuits, one error in 1×10^4.

One method to minimize the error rate is to provide a "perfect" transmission channel, one that will introduce no errors in the transmitted information by the receiver. The engineer designing a data transmission system can never achieve that perfect channel. Besides improvement of the channel transmission parameters themselves, error rate can be reduced by forms of systematic redundancy. In old-time Morse on a bad circuit words often were sent twice; this is redundancy in its simplest form. Of course, it took twice as long to send a message; that is not very economical if useful words per minute received is compared to channel occupancy.

This brings up the point of channel efficiency. Redundancy can be increased such that the error rate could approach zero. Meanwhile, the information transfer or "throughput" across the channel also approaches zero. Thus unsystematic redundancy is wasteful and merely lowers the rate of useful communication. Maximum efficiency or throughput could be obtained in a digital transmission system if all redundancy and other code elements, such as start and stop elements, were removed from the code and, in addition, if advantage were taken of the statistical phenomenon of our written language by making high usage letters short in code length, such as E, T, and A, and low usage letters longer, such as Q and X.

As discussed previously isolated errors can be detected in a binary sequence by adding redundant elements. Odd and even parity have been explained previously. In general it now can be said that it is possible to detect the existence of a single error in a block* of data symbols if all single errors lead to sequences which have not been assigned meanings. The eighth bit assigned for parity in ASCII and CCITT No. 5 codes essentially did this. M out of N ratio codes detect errors in this manner. One common M out of N code is the 3 out of 7 binary code. As we know, a 7-unit code will have 2^7 distinct character sequences. Now allow only those sequences that have 4 zeros and 3 ones to be meaningful characters. Any single error would change the ratio. Add a one and it must replace a zero, so the results is 3 zeros and 4 ones—an error. Similar reasoning follows when a zero replaces a one. Note that the code will detect odd errors with certainty, but not all even errors because in the case of two errors a one could replace a zero and a zero a one. This would be a case of an undetected error. The loss of code efficiency is considerable because the number of admissible 7-element words is reduced from 128 to 35.

One must differentiate between error detection and error correction. Error detection identifies that a symbol has been received in error. Parity discussed above is primarily used for error detection. Parity bits add redundancy and decrease channel efficiency.

Error correction corrects the detected error. Basically, there are two types of error correction techniques: forward acting and two-way error correction. The latter system uses a return channel (backward channel). When an error is detected, the receiver signals this fact to the transmitter over the backward channel and the block of information containing the error is transmitted again. Error detection in this case is in the form of parity, and this time it is called block parity.

The retransmission form of error correction may be a very efficient method of error control in channels where interference occurs only infrequently, but then destroys a large numbers of symbols at a time when the error rate is high. It has several weaknesses. One is evident when a transmitter sends to several receivers over different paths. In such an event error correction by retransmission becomes cumbersome and inefficient because an error in any one link requires that the block be retrans-

---

* A block is a discrete grouping of data characters or symbols. Data blocks are usually of fixed length.

mitted to all. Another weakness shows up when there is a "timeliness" regarding the data, such as in the case of telemetry or radar warning. The storage and retransmission time detracts from the "timeliness."

Forward acting error correction uses codes which are devised so that they are, up to a certain point, self-correcting. Such codes greatly increase redundancy. One type of error correcting code is called a geometric code, and it embodies a generalization of parity checks in more than one dimension.

Forward acting error correction loses its attractiveness to the engineer when he tries to implement it. In most instances, it is no more than a laboratory item, and in few cases has it been applied in the field. Secondly, except in very special circumstances, the expense will be prohibitive. A third drawback is the great increase in redundancy required, cutting channel efficiency by half or even more. However, it has found some application with communication satellites.

Two way error correction has been implemented and is used comparatively widely. One such system has particular application on HF radio and any other similar noisy and fading transmission media. It is called ARQ, derived from an old Morse and telegraph signal, Automatic Repeat Request.

The ARQ equipment is designed to operate in conjunction with regular 5-unit teletype code. At the transmitting end of the circuit this code is translated into the 7-unit Moore ARQ Code, Figure 8.7. Of the 128 combinations available with a 7-unit code, only those having three mark digits and four space digits, 35 in number, are used for the transmission of characters (an M out of N code). At the receiving end a check is made to see if each character has the right number of mark and space digits. If it does, the character is printed. If some other combination is received, the printing is held up while a request for a repeat is automatically sent back to the transmitting station. At the transmitting station enough characters are stored (generally, three) so that by the time the repeat request is received, the required character is still available. On receipt of a correct signal, the transmission continues until another error is detected.

The improvement obtained in undetected character errors by the use of ARQ equipment is in the area of two orders of magnitude. In other words if the error rate on a particular circuit without ARQ was one error in 1×10^3, with ARQ it would improve to one error in 1×10^5.

The circuit efficiency when using the ARQ equipment is lower than without the ARQ equipment, if by efficiency is meant the ability to transmit the greatest number of characters in a given time without regard to

Five-Unit TTY Code Bit Numbers 5 4 3 2 1					Code Assignments Letters Case	Figures Case	Moore ARQ Code Bit Numbers 7 6 5 4 3 2 1						
0	0	0	0	0	Blank	Blank	1	1	1	0	0	0	0
0	0	0	0	1	E	3	0	0	0	1	1	1	0
0	0	0	1	0	Line feed	Line feed	0	0	0	1	1	0	1
0	0	0	1	1	A	–	0	1	0	1	1	0	0
0	0	1	0	0	Space	Space	0	0	0	1	0	1	1
0	0	1	0	1	S	' (Apos.)	0	1	0	1	0	1	0
0	0	1	1	0	I	8	0	0	0	0	1	1	1
0	0	1	1	1	U	7	0	1	0	0	1	1	0
0	1	0	0	0	Carr. ret.	Carr. ret.	1	1	0	0	0	0	1
0	1	0	0	1	D	✪	0	0	1	1	1	0	0
0	1	0	1	0	R	4	0	0	1	0	0	1	1
0	1	0	1	1	J	Bell	1	1	0	0	0	1	0
0	1	1	0	0	N	, (Comma)	0	0	1	0	1	0	1
0	1	1	0	1	F	□	1	1	0	0	1	0	0
0	1	1	1	0	C	:	0	0	1	1	0	0	1
0	1	1	1	1	K	(	1	1	0	1	0	0	0
1	0	0	0	0	T	5	1	0	1	0	0	0	1
1	0	0	0	1	Z	+	1	0	0	0	1	1	0
1	0	0	1	0	L	)	0	1	0	0	0	1	1
1	0	0	1	1	W	2	1	0	1	0	0	1	0
1	0	1	0	0	H	▨	0	1	0	0	1	0	1
1	0	1	0	1	Y	6	1	0	1	0	1	0	0
1	0	1	1	0	P	0	0	1	0	1	0	0	1
1	0	1	1	1	Q	1	1	0	1	1	0	0	0
1	1	0	0	0	O	9	0	1	1	0	0	0	1
1	1	0	0	1	B	?	1	0	0	1	1	0	0
1	1	0	1	0	G	⊟	1	0	0	0	0	1	1
1	1	0	1	1	Figures	Figures	0	1	1	0	0	1	0
1	1	1	0	0	M	.	1	0	0	0	1	0	1
1	1	1	0	1	X	/	0	1	1	0	1	0	0
1	1	1	1	0	V	=	1	0	0	1	0	0	1
1	1	1	1	1	Letters	Letters	0	1	1	1	0	0	0
					Signal I	Signal I	0	0	1	0	1	1	0
					Idle α	Idle α	1	0	0	1	0	1	0
					Idle β	Idle β	0	0	1	1	0	1	0

Figure 8.7 Moore "ARQ" Code. Transmission order: Bit 1 → Bit 7.

the number of errors received. If the pulse width were kept the same, the 7-unit code would be only 70–80% as fast as the 5-unit code, depending upon whether an allowance is made for stop-start digits or not. In practice the pulse width is shortened with the 7-unit code so as to keep the word speed the same as with the 5-unit code.

When repetitions are required, circuit time in the reverse direction is used as well as time for the repetition and for some additional characters in the forward direction. As a result of this, the net number of words that can be transmitted in the two directions drops as the number of mutilated characters increases. It should be noted that the reduction in efficiency is hardly significant until the error rate exceeds 1 character/100.

Another form of two-way error correction is finding more and more favor in data transmission. This is block transmission. A number of such systems are available commercially. One of these that is widely used has a block size that is convenient to standard punched cards. Such a block is 80 characters long and is preceded and followed by two unique control characters, called framing characters. Therefore the line block is actually 84 characters long. The framing characters serve two purposes. First, they identify the type of block (i.e., whether it is a multiblock message) and the block number as well as the type of message (card or tape). The second function is parity, and in this case it is block parity. It corresponds to the even longitudinal parity count of bits in each of the eight levels in the code (ASCII with parity) excluding the first framing character.

In this block transmission system, as in most, each block must be correctly received and acknowledged before the next block is sent, and if an error is indicated the block is retransmitted. Two blocks are stored at each end of the system and alternate acknowledgment codes are used to identify uniquely to which of the two consecutively transmitted line blocks the acknowledgment pertains. The backward channel for this system is the return traffic channel of the full-duplex data system.

8.5 THE DC NATURE OF DATA TRANSMISSION

8.5.1 Loops

Binary data is transmitted on a dc loop. More correctly the binary data end instrument delivers to the line and receives from the line one or several dc loops. In its most basic form a dc loop consists of a switch, a dc voltage, and a termination. A pair of wires interconnects the switch and termination. The voltage source in data and telegraph work is called the battery although the device is usually electronic, deriving the dc voltage from an ac power line source. The battery is placed in the line such that it provides voltage(s) consistent with the type of transmission desired. A simplified dc loop is shown in Figure 8.8.

8.5.2 Neutral and Polar Dc Transmission Systems

Nearly all dc data and telegraph systems functioning today are operated in either a neutral or a polar mode. The words "neutral" and "polar" describe the manner in which battery is applied to the dc loop. On a "neutral" loop, following the mark-space convention in Table 8.1, battery is applied during marking (1) conditions and is switched off during spacing (0). Current therefore flows in the loop when a mark is sent and the loop is closed. Spacing is indicated on the loop by a condition of no current. Thus we have the two conditions for binary transmission, an open loop (no current flowing) and a closed loop (current flows). Keep in mind that we could reverse this, change the convention (Table 8.1) and, say, assign spacing to a condition of current flowing or closed loop and

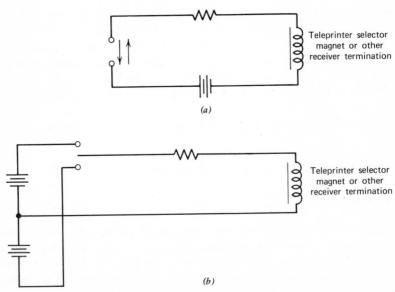

(a)

(b)

Figure 8.8 Simplified diagram of a dc loop with (a) neutral and (b) polar keying.

marking to a condition of no current or an open loop. This is sometimes done in practice and is called changing the sense. Either way, a neutral loop is a dc loop circuit where one binary condition is represented by the presence of voltage, flow of current, and the other by the absence of voltage/current.

Polar transmission approaches the problem a little differently. Two batteries are provided. One is called negative battery and the other positive. During a condition of marking, a positive battery is applied to the loop, following the convention of Table 8.1, and a negative battery is applied to the loop during the spacing condition. In a polar loop, current is always flowing. For a mark or binary "one," it flows in one direction and for a space or binary "zero" it flows in the opposite direction. Figure 8.8b shows a simplified polar loop.

8.6 BINARY TRANSMISSION AND THE CONCEPT OF TIME

8.6.1 Introduction

Time and timing are most important factors in digital transmission. For this discussion consider a binary end instrument sending out in series a continuous run of marks and spaces. Those readers who have some familiarity with Morse will recall that the spaces between dots and dashes told the operator where letters ended and where words ended. With the sending device or transmitter delivering a continuous series of characters to the line, each consisting of 5, 6, 7, 8, or 9 elements (bits) to the character, let the receiving device start its print cycle when the transmitter starts sending. If the receiver is perfectly in step with the transmitter, ordinarily one could expect good printed copy and few, if any, errors at the receiving end.

It is obvious that when signals are generated by one machine and received by another, the speed of the receiving machine must be the same or very close to that of the transmitting machine. When the receiver is a motor-driven device, timing stability and accuracy are dependent on the accuracy and stability of the speed of rotation of the motors used.

Most simple data/telegraph receivers sample at the presumed center of the signal element. It follows, therefore, that whenever a receiving device accumulates timing error of more than 50% of the period of one bit, it will print in error.

The need for some sort of synchronization is shown in Figure 8.9a. A 5-unit code is employed and three characters transmitted sequentially

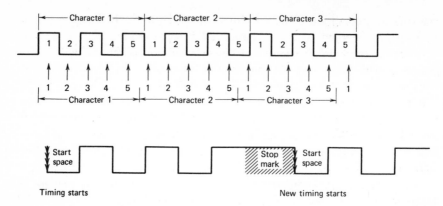

Figure 8.9 (*a*) 5-unit synchronous bit stream with timing error; (*b*) 5-unit start and stop stream of bits with a 1.5-unit stop element.

are shown. Sampling points are shown in the figure as vertical arrows. Receiver timing begins when the first pulse is received. If there is a 5% timing difference between the transmitter and receiver, the first sampling at the receiver will be 5% away from the center of the transmitted pulse. At the end of the tenth pulse or signal element the receiver may sample in error. The eleventh signal element will indeed be sampled in error and all subsequent elements are errors. If the timing error between transmitting machine and receiving machine is 2%, the cumulative error in timing would cause the receiving device to print all characters in error after the twenty-fifth bit.

8.6.2 Asynchronous and Synchronous Transmission

In the earlier days of printing telegraphy, start-stop transmission or asynchronous operation was developed to overcome the problem of synchronism. Here timing starts at the beginning of a character and stops at the end. Two signal elements are added to each character to signal the receiving device that a character has begun and ended.

As an example consider a 5-element code such as CCITT No. 2 (see Figure 8.1). In the front of a character an element is added called a start space and at the end of each character a stop mark is inserted. Send the letter A in Figure 8.1. The receiving device starts its timing sequence on receiving element No. 1, a space or 0, then a 11000 is received; the A is selected, then the stop mark is received, and the timing sequence stops.

On such an operation accumulation of timing errors can take place only inside each character.

Suppose the receiving device is again 5% slower or faster than its transmitting counterpart; now the fifth information element will be no more than 30% displaced in time from the transmitted pulse and well inside the 50% or halfway point for correct sampling to take place.

In start-stop transmission information signal elements are each of the same duration, which is the duration or pulse width of the start element. The stop element has an indefinite length or pulse width beyond a certain minimum.

If a steady series of characters are sent, then the stop element is always the same width or has the same number of unit intervals. Consider the transmission of two A's, 0110001011000111111 ——————→ 11111. The start space (0) starts the timing sequence for 6 additional elements, which are the 5 code elements in the letter A and the stop mark. Timing starts again on the mark-to-space transition between the stop mark of the first A and the start space of the second. Sampling is carried out at pulse center for most asynchronous systems. One will note that at the end of the second A a continuous series of marks is sent. Thus the signal is a continuation of the stop element or just a continuous mark. It is the mark-to-space transition of the start element that tells the receiving device to start timing a character.

Minimum lengths of stop elements vary. The example discussed above shows a stop element of 1-unit interval duration (1 bit). Some are 1.42-unit intervals, others are of 1.5- and 2-unit intervals duration. The proper semantics of data/telegraph transmission would describe the code of the previous paragraph as a 5-unit start-stop code with a 1-unit stop element.

A primary objective in the design of telegraph and data systems is to minimize errors received or to minimize the error rate. There are two prime causes of errors. These are noise and improper timing relationships. With start-stop systems a character begins with a mark-to-space transition at the beginning of the start space. 1.5-unit intervals later the timing causes the receiving device to sample the first information element which simply is a mark or space decision. The receiver continues to sample at 1-bit intervals until the stop mark is received. In start-stop systems the last information bit is most susceptible to cumulative timing errors. Figure 8.9*b* is an example of a 5-unit start-stop bit stream with a 1.5-unit stop element.

Another problem in start-stop systems is that of mutilation of the start element. Once this happens the receiver starts a timing sequence on the next mark-to-space transition it sees and thence continues to print in error until, by chance, it cycles back properly on a proper start element.

Synchronous data/telegraph systems do not have start and stop elements but consist of a continuous stream of information elements or bits such as in figure 8.9a. The cumulative timing problems eliminated in asynchronous (start-stop) systems are present in synchronous systems. Codes used on synchronous systems are often 7-unit codes with an extra unit added for parity, such as the ASCII or CCITT No. 5 codes. Timing errors tend to be eliminated by virtue of knowing the exact rate at which the bits of information are transmitted.

If timing error of 1% were to exist between transmitter and receiver, not more than 100 bits could be transmitted until the synchronous receiving device was 100% apart in timing from the transmitter and up to all bits received were in error. Even if timing accuracy were improved to 0.05%, the correct timing relationship between transmitter and receiver would exist for only the first 2000 bits transmitted. It follows, therefore, that no timing error at all can be permitted to accumulate since anything but absolute accuracy in timing would cause eventual malfunctioning. In practice the receiver is provided with an accurate clock which is corrected by small adjustments as explained below.

8.6.3 Timing

All currently used data transmission systems are synchronized in some manner. Start-stop synchronization has been discussed. Fully synchronous transmission systems all have timing generators or clocks to maintain stability. The transmitting device and its companion receiver at the far end of the data circuit must be a timing system. In normal practice the transmitter is the master clock of the system. The receiver also has a clock which in every case is corrected by one means or another to its transmitter equivalent at the far end.

Another important timing factor which must also be considered is the time it takes a signal to travel from the transmitter to the receiver. This is called propagation time. With velocities of propagation as low as 20,000 mi/s, consider a circuit 200 mi in length. The propagation time would then be 200/20,000 s or 10 ms. Ten milliseconds is the time duration of one bit at a data rate of 100 bps. Thus the receiver in this case must delay its clock by 10 ms to be in step with its incoming signal.

Temperature and other variations in the medium may affect this delay. One can also expect variations in the transmitter master clock as well as other time distortions due to the medium.

There are basically three methods of overcoming these problems. One is to provide a separate synchronizing circuit to slave the receiver to the transmitter's master clock. This wastes bandwidth by expending a voice channel or subcarrier just for timing. A second method, which was used fairly widely up to several years ago, was to add a special synchronizing pulse for groupings of information pulses, usually for each character. This method was similar to start-stop synchronization and lost its appeal largely owing to the wasted information capacity for synchronizing. The most prevalent system in use today is one that uses transition timing. With this type of timing the receiving device is automatically adjusted to the signaling rate of the transmitter and adjustment is made at the receiver by sampling the transitions of the incoming pulses. This offers many advantages, most important of which is that it automatically compensates for variations in propagation time. With this type of synchronization the receiver determines the average repetition rate and phase of the incoming signal transition and adjusts its own clock accordingly.

In digital transmission the concept of a transition is very important. The transition is what really carries the information. In binary systems the space-to-mark and mark-to-space transitions (or lack of transitions) placed in a time reference contain the information. Decision circuits regenerate and retime in sophisticated systems and care only *if* a transition has taken place. Timing cares *when* it takes place. Timing circuits must have memory in case a long series of marks or spaces are received. These will be periods of no transition but they carry meaningful information. Likewise, the memory must maintain timing for reasonable periods in case of circuit outage. Keep in mind that synchronism pertains to both frequency and phase and that the usual error in high stability sytems is a phase error (i.e., the leading edges of the received pulses are slightly advanced or retarded from the equivalent clock pulses of the receiving device).

High stability systems once synchronized need only a small amount of correction in timing (phase). Modem internal timing systems may be as stable as 1×10^8 or greater at both the transmitter and the receiver. Before a significant error condition can build up owing to a time rate difference at 2400 bps, the accumulated time difference between transmitter and receiver must exceed approximately 2×10^{-4} s. This figure neglects phase. Once the transmitter and receiver are synchronized and the circuit is shut down, then the clock on each end must drift apart by at least 2×10^{-4} s before significant errors take place. Again this means that the leading edge of the receiver clock equivalent timing pulse is 2×10^{-4} in advance or retarded from the leading edge of the received pulse from the distant end. Often an idling signal is sent on synchronous data circuits during conditions of no traffic to maintain timing. Other high stability systems need to resynchronize only once a day.

Bear in mind that we are considering dedicated circuits only, not switched synchronous data. The problems of synchronization of switched data immediately come to light. Two such problems are the following:

- No two master clocks are in perfect phase sync.
- The propagation time on any two paths may not be the same.

Thus such circuits will need a time interval for synchronization at each switching event before traffic can be passed.

To sum up, synchronous data systems use high stability clocks, and the clock at the receiving device is undergoing constant but minuscule corrections to maintain an in-step condition with the received pulse train from the distant transmitter by looking at mark-space and space-mark transitions.

8.6.4 Distortion

It has been shown that the key factor in data transmission is timing. The signal must be either a mark or space, but that alone is not sufficient. The marks and spaces (or ones and zeros) must be in a meaningful sequence based on a time reference.

In the broadest sense distortion may be defined as any deviation of a signal in any parameter, such as time, amplitude, or wave shape, from that of the ideal signal. For data and telegraph binary transmission, distortion is defined as a displacement in time of a signal transition from the time which the receiver expects to be correct. In other words the receiving device must make a decision whether a received signal element is a mark or a space. It makes the decision during the sampling interval which is usually at the center of where the received pulse or bit should be. Thus it is necessary for the transitions to occur between sampling times and preferably halfway between them. Any displacement of the transition instants is called distortion. The degree of distortion a data signal suffers

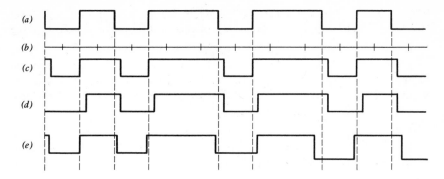

Figure 8.10 Three typical distorted data signals.

as it traverses the transmission medium is a major contributor in determining the error rate that can be realized.

Telegraph and data distortion is broken down into two basic types, systematic and fortuitous. Systematic distortion is repetitious and is broken down into bias distortion, cyclic distortion, and end distortion (more common in start-stop systems). Fortuitous distortion is random in nature and may be defined as a distortion type in which the displacement of the transition from the time interval in which it should occur is not the same for every element. Distortion caused by noise spikes in the medium or other transients may be included in this category. Characteristic distortion is caused by transients in the modulation process which appear in the demodulated signal.

Figure 8.10 shows some examples of distortion. Figure 8.10a is an example of a binary signal without distortion and Figure 8.10b shows the sampling instants, which occur ideally in the center of the pulse to be sampled. From this we can see that the displacement tolerance is nearly 50%. This means that the point of sample could be displaced by up to 50% of a pulse width and still record the mark or space condition present without error. However, the sampling interval does require a finite amount of time so that in actual practice the displacement permissible is somewhat less than 50%. Figures 8.10c and 8.10d show bias distortion. An example of spacing bias is shown in Figure 8.10c, where all the spacing impulses are lengthened at the expense of the marking impulses. Figure 8.10d shows the reverse of this; the marking impulses are elongated at the expense of the spaces. This latter is called marking bias. Figure 8.10e shows fortuitous distortion, which is a random type of distortion. In this

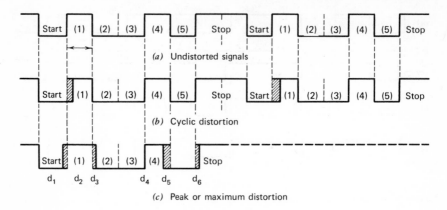

Figure 8.11 Distorted telegraph signals illustrating cyclic and peak distortion. The peak distortion in the character shown appears a ttransition d_5.

case the displacement of the signal element is not the same as the time interval in which it should occur for every element.

Figure 8.11 shows distortion that is more typical of start and stop transmission. Figure 8.11a is an undistorted start and stop signal. Figure 8.11b shows cyclic or repetitive distortion typical of mechanical transmitters. In this type of distortion the marking elements may increase in length for a period of time and then the spacing elements will increase in length. Figure 8.11c shows peak distortion. Identifying the type of distortion present on a signal often gives a clue to the source or cause of distortion. Distortion measurement equipment measures the displacement of the mark-to-space transition from the ideal of a digital signal. If a transition occurs too near to the sampling point, the signal element is liable to be in error.

8.6.5 Bits, Bauds, and Words per Minute

There is much confusion among transmission engineers in handling some of the semantics and simple arithmetic used in data and telegraph transmission.

The bit has been defined previously. Now the term "words per minute" will be introduced. A word in our telegraph and data language consists of six characters, usually five letters, numbers, or graphics and a space. All signal elements transmitted must be counted such as "carriage return" and "line feed."

Let us look at some arithmetic, for instance:

1. A channel is transmitting at 75 bps using a 5-unit start and stop code with a 1.5-unit stop element. Thus for each character there are 7.5 unit intervals (7.5 bits). Therefore the channel is transmitting at 100 wpm.

$(75 \times 60)/(6 \times 7.5) = 100$ wpm.

2. A system transmits in CCITT No. 5 code at 1500 wpm with parity. How many bits per second are being transmitted?

$1500 \times 8 \times 6/60 = 1200$ bps.

The baud is the unit of modulation rate. In binary transmission systems the baud and bit per second are synonymous. Thus a modem in a binary system transmitting to the line 110 bps has a modulation rate of 110 bauds. In multilevel or M-ary systems the number of bauds is indicative of the number of transitions per second. The baud is more meaningful to the transmission engineers concerned with the line side of a modem. This concept will be discussed more at length further on.

8.7 DATA INTERFACE

Up to this point data have been considered as a serial stream of bits or a continuous run of marks and spaces. The actual data end intrument transmits and receives data in a different form. Instead of bits in sequence or in series, high speed data end instruments usually receive data from the line as parallel characters. Thus a data interface is necessary to provide for serial-to-parallel conversion as well as other control functions.

Such an interface is carried out in intermediate equipment which provides line buffering. The equipment is usualy mounted with the data end instrument such as a card punch or card reader, etc. The intermediate equipment receives the serial data stream from the data communication equipment (modem) as well as timing and other control signals for reception. It stores the data, converts it to parallel data, and provides the data to the end instrument on command in parallel form. The reverse holds true for transmission.

The intermediate equipment interfaces with the line or data set, modem, or data communications equipment. This interface should follow CCITT Rec. V.24 (White Books) or RS-232C (EIA). The data signals at either point are polar, usually $+$ and -6 volts dc. The electrical characteristics are defined in Rec. V.24. Figures 8.12a and 8.12b show typical intermediate equipment and end instrument interconnection for card punches and card readers, respectively.

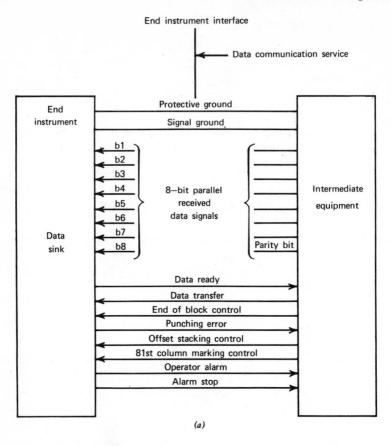

Figure 8.12 End instrument interface interchange circuits for (*a*) card punches and (*b*) card readers.

8.8 DATA INPUT/OUTPUT DEVICES

The following paragraphs are meant to give the reader a broad-brush familiarity with data subscriber equipment, which is more often referred to as input/output devices. Such equipments convert user information (data or messages) into electrical signals or vice versa. Many refer to the input/output devices as the human interface. Electrically a data subscriber terminal consists of the end instrument, the intermediate equipment (buffering equipment), and a device called a modem or communication set. The intermediate equipment has been discussed in Section 8.7. Modems are

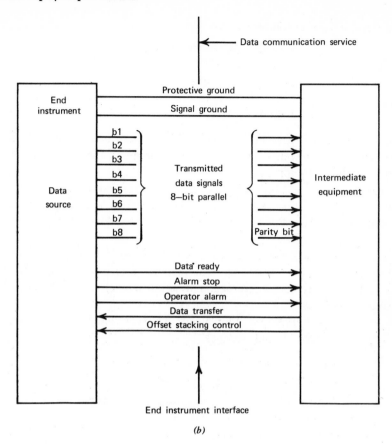

(b)

Figure 8.12 (*continued*)

described in Section 8.13. The data source is the input device and the data sink is the output device.

Data input/output devices handle paper tape, punched cards, magnetic tape, and printed page copy. Input devices* may be broken down into the following categories:

> Keyboard sending units
> Card readers
> Paper tape readers
> Magnetic tape readers

* Compatible with the data rates under discussion.

Output devices* are as follows:

> Printers (paper, hard copy)
> Card punches
> Paper tape punches
> Magnetic tape recorders
> Visual displays

Further, these devices may be used as on-line devices or off-line. Off-line devices are not connected directly to the communication system, but serve as auxiliary equipment.

Off-line devices are used for tape or card preparation for eventual transmission. In this case a keyboard is connected to a card punch for card preparation. Also the keyboard may be connected to a paper tape punch.

Once tape or cards are prepared, they then are handled by on-line equipment, either tape readers or card readers. Intermediate equipment or line buffers supply timing, storage, and serial-parallel conversion to the input/output devices.

The following table provides equivalence between telegraph and data terminology of input/output devices.

Data	Telegraph
Keyboard	Keyboard
Tape reader	Transmitter-distributor†
Printer	Teleprinter
Tape punch	Perforator, reperforator
Visual displays	—

8.9 DIGITAL TRANSMISSION ON AN ANALOG CHANNEL

8.9.1 Introduction

There are two fundamental approaches to the practical problem of data transmission. The first approach is often to design and construct a complete new network expressly for the purpose of data transmission. The second approach is to adapt the many existing telephone facilities for data transmission. The paragraphs that follow deal with the second approach.

* Compatible with the data rates under discussion.
† The distributor performs the parallel-to-serial equivalent conversion of data transmission.

Transmission facilities designed to handle voice traffic have characteristics which make it difficult to transmit dc binary digits or bit streams. To permit the transmission of data over voice facilities (i.e., the telephone network), it is necessary to convert the dc data into a signal within the voice frequency range. The equipment which performs the necessary conversion to the signal is generally termed a data modem. Modem is an acronym for MOdulator-DEModulator.

8.9.2 Modulation-Demodulation Schemes

A modem modulates and demodulates. The types of modulation used by present-day modems may be one or combinations of the following:

Amplitude modulation, double sideband (DSB)
Amplitude modulation, vestigial sideband (VSB)
Frequency shift modulation (FSK)
Phase shift modulation (PSK)

AMPLITUDE MODULATION-DOUBLE SIDEBAND

With this modulation technique binary states are represented by the presence or absence of an audio tone or carrier. More often it is referred to as on-off telegraphy. For data rates up to 1200 bps, one such system uses a carrier frequency centered at 1600 Hz. For binary transmission amplitude modulation has significant disadvantages which include (1) susceptibility to sudden gain change, and (2) inefficiency in modulation and spectrum utlization, particularly at higher modulation rates.

AMPLITUDE MODULATION, VESTIGIAL SIDEBAND VSB

An improvement in the amplitude modulation, double sideband technique results from the removal of one of the information-carrying sidebands.

Since the essential information is present in each of the sidebands, there is no loss of content in the process. The carrier frequency must be preserved to recover the dc component of the information envelope. Therefore, digital systems of this type use vestigial sideband modulation in which one sideband, a portion of the carrier and a "vestige" of the other sideband, is retained. This is accomplished by producing a double sideband signal and filtering out the unwanted sideband components. As a result, the signal takes only about three-fourths of the bandwidth required for a double sideband system. Typical vestigial sideband data modems are operable up to 2400 bps in a telephone channel. Data rates up to 4800 bps are achieved using multilevel (*M*-ary) techniques. The carrier frequency is usually located between 2200 and 2700 cycles.

FREQUENCY SHIFT MODULATION (FSK)

A large number of data transmission systems utilize frequency shift modulation. The two binary states are represented by two different frequencies and are detected by using two frequency tuned sections, one tuned to each of the two bit frequencies. The demodulated signals are then integrated over the duration of a bit and upon the result a binary decision is based.

Digital transmission using frequency shift modulation has the following advantages. (1) The implementation is not much more complex than an AM system. (2) Since the received signals can be amplified and limited at the receiver, a simple limiting amplifier can be used whereas the AM system requires sophisticated automatic gain control in order to operate over a wide level range. Another advantage is that FSK can show a 3- or 4-dB improvement over AM in most types of noise environment, particularly at distortion threshold (i.e., at the point where the distortion is such that good printing is about to cease). As the frequency shift becomes greater, the advantage over AM improves in a noisy environment.

Another advantage of FSK is its·immunity from the effects of non-selective level variations even when extremely rapid. Thus it is used almost exclusively on worldwide high frequency radio transmission where rapid fades are a common occurrence. In the United States it has nearly universal application for the transmission of data at the lower data rates (i.e., 1200 bps and below).

PHASE MODULATION

For systems using higher data rates, phase modulation becomes more attractive. Various forms are used such as two-phase, relative phase, and quadrature phase.

A two-phase system uses one phase of the carrier frequency for one binary state and the other phase for the other binary state. The two phases are 180° apart and are detected by a synchronous detector using a reference signal at the receiver which is of known phase with respect to the incoming signal. This known signal is at the same frequency as the incoming signal carrier and is arranged to be in phase with one of the binary signals.

In the relative phase system, a binary "one" is represented by sending a signal burst of the same phase as that of the previous signal burst sent. A binary "zero" is represented by a signal burst of a phase opposite to that of the previous signal transmitted. The signals are demodulated at the receiver by integrating and storing each signal burst of one bit period for comparison in phase with the next signal burst.

In the quadrature phase system, two binary channels (2 bits) are phase multiplexed onto one tone by placing them in phase quadrature as shown in the sketch below. An extension of this technique places two binary channels on each of several tones spaced across the voice channel of a typical telephone circuit.

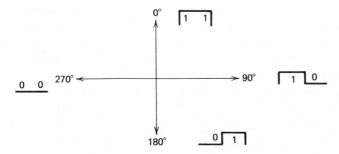

Some of the advantages of phase modulation are as follows:

1. All available power is utilized for intelligence conveyance.
2. The demodulation scheme has good noise rejection capability.
3. The system yields a smaller noise bandwidth.

A disadvantage of such a system is the complexity of equipment required.

8.9.3 Critical Parameters

The effect of the various telephone circuit parameters on the capability of a circuit to transmit data is a most important consideration. The following discussion is to familiarize the reader with the problems most likely to be encountered in the transmission of data over analog circuits (e.g., the telephone network) and to make some generalizations in some cases, which can be used to help in planning the implementation of data systems.

ENVELOPE DELAY DISTORTION (GROUP DELAY)

Delay distortion on a telephone circuit is caused by the inequality of propagation times at different frequencies within the passband, whether the voice channel or the group or supergroup passbands. The principal source of delay distortion is the cumulative effect of the many filters used in carrier systems, and hence the magnitude of delay distortion is generally dependent on the number of carrier modulation stages in a transmission path rather than the length of the path. Figure 8.13 is a typical response curve of frequency versus differential delay in milliseconds of a voice channel due to frequency multiplex equipment only. Maximum delays are noted at band edge, and the delay at the center portion is minimum.

Delay distortion is more often defined as envelope delay distortion by data transmission engineers. Envelope delay is defined as the derivative of the phase characteristics with respect to frequency. It is a measure of the time required to propagate a change in the envelope of a wave (the actual information-bearing part of the signal) through the system. Another term used to describe delay distortion is group delay.

Delay distortion is a major limitation to modulation rate. The shorter the pulse width (the width of one bit of data), the more critical will be the parameters of envelope delay distortion. As will be discussed later, it is desirable to keep the delay distortion in the band of interest below the period of 1 bit.

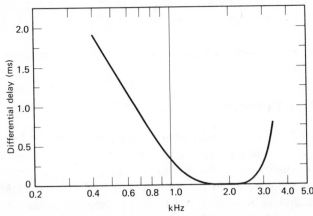

Figure 8.13 Typical differential delay across a voice channel, FDM equipment back-to-back.

AMPLITUDE RESPONSE (ATTENUATION DISTORTION)

Another parameter that seriously affects the transmission of data and which can place very definite limits on the modulation rate is that of amplitude response. Ideally all frequencies across the passband of the channel of interest should suffer the same attenuation. Place a -10-dBm signal at any frequency between 300 and 3400 Hz and the output at the receiving end of the channel may be -23 dBm, for example, at any and all frequencies in the band; we would then describe a fully flat channel. Such a channel has the same loss or gain at any frequency within the band. This type of channel is ideal but would be unachievable in a real, working system. In Rec. G.132, CCITT recommends no more than 9 dB of amplitude distortion relative to 800 Hz between 400 and 3000 Hz. This figure, 9 dB, describes the maximum variation that may be expected from the reference level at 800 Hz. This variation of amplitude response often is called attenuation distortion. A conditioned channel, such as a Bell System C-4 channel, will maintain a response of -2 to$+$ 3 dB from 500 to 3000 Hz and $-$ 2 to $+$ 6 dB from 300 to 3200 Hz. Channel conditioning is discussed in Section 8.12.

Considering tandem operation, the deterioration of amplitude response is arithmetically summed when sections are added. This is particularly true at band edge in view of channel unit transformers and filters which account for the upper and lower cutoff characteristics.

Amplitude response is also discussed in Section 1.9.3. Figure 8.14 illustrates a typical example of amplitude response across FDM carrier equipment (see Chapter 3) connected back-to-back at the voice channel input/output.

NOISE

Another important consideration in the transmission of data is that of noise. All extraneous elements appearing at the voice channel output that were not due to the input signal are considered to be noise. For convenience noise is broken down into four categories:

- Thermal
- Crosstalk
- Intermodulation
- Impulse

Thermal noise, often called resistance noise, white noise, or Johnson noise, is of a Gaussian nature or fully random. Any system or circuit

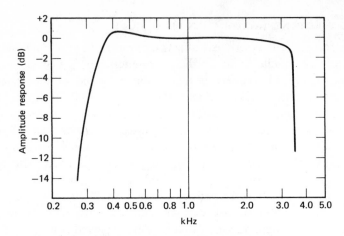

Figure 8.14 Typical amplitude versus frequency response across a voice channel, channel modulator, demodulator back-to-back, FDM equipment.

operating at a temperature above absolute zero inherently will display thermal noise. It is caused by the random motions of discrete electrons in the conduction path.

Crosstalk is a form of noise caused by unwanted coupling from one signal path into another. It may be due to direct inductive or capacitive coupling between conductors or between radio antennas. (Refer to Section 1.9.6.)

Intermodulation noise is another form of unwanted coupling usually caused by signals mixing in non-linear elements of a system. Carrier and radio systems are highly susceptible to intermodulation noise particularly when overloaded. (Refer to Section 1.9.6.)

Impulse noise is a primary source of errors in the transmission of data over telephone networks. It is sporadic and may occur in bursts or discrete impulses called "hits." Some types of impulse noise are natural, such as that from lightning. However, man-made impulse noise is ever-increasing, such as from automobile ignition systems, power lines, etc. Impulse noise may be a high level in telephone switching centers due to dialing, super-vision, and switching impulses which may be induced or otherwise coupled into the data transmission channel.

For our discussion of data transmission, only two forms of noise will be considered: random or Gaussian noise and impulse noise. Random noise measured with a typical transmission measuring set appears to have a relatively constant value. However, the instantaneous value of the noise fluctuates over a wide range of amplitude levels. If the instantaneous noise voltage is of the same magnitude as the received signal, the receiving detection equipment may yield an improper interpretation of the received signal and an error or errors will occur. For a proper analytical approach to the data transmission problem, it is necessary to assume a type of noise which has an amplitude distribution which follows some predictable pattern. White noise or random noise has a Gaussian distribution and is considered representative of the noise encountered on the analog telephone channel (i.e., the voice channel). From the probability distribution curve shown in Figure 8.15 of Gaussian noise we can make some accurate predictions. It may be noted from this curve that the probability of occurrence of noise peaks which have amplitudes 12.5 dB above the rms level is 1 in 10^5. Hence if we wish to ensure an error rate of 10^{-5} in a particular system using binary polar modulation, the rms noise should be at least 12.5 dB below the signal level (Ref. 4, p. 114; Ref. 5, p. 6). This simple analysis is valid for the type of modulation used, assuming that no other factors are degrading the operation of the system and that a cosine shaped receiving filter is used. If we were to interject envelope delay distortion, for example, into the system, we could translate the degradation into an equivalent signal-to-noise ratio improvement necessary to restore the desired error rate. For example, if the delay distortion were the equivalent of one pulse width, the signal-to-noise ratio improvement required for the same error rate would be about 5 dB or the required signal-to-noise ratio would now be 17.5 dB.

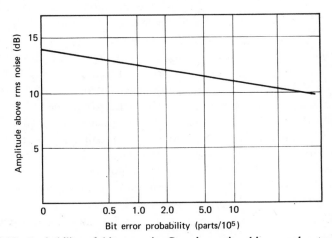

Figure 8.15 Probability of bit error in Gaussian noise, binary polar transmission.

For reasons that will be discussed later, let us assume that the signal level is -10 dBm at the zero transmission level point of the system; then the rms noise measured at the same point would be -27.5 dBm to retain the error rate of 1 in 10^5.

In order for the above figure to have any significance, it must be related to the actual noise found in a channel. CCITT recommends no more than 50,000 pW of noise psophometrically weighted on an international connection made up of six circuits in a chain. However, CCITT states (Rec. G. 142A and 142D) that for data transmission at as high a modulation rate as possible without significant error rate, a reasonable circuit objective for maximum random noise would be -40 dBm0p for leased circuits (impulse noise not included) and -36 dBm0p for switched circuits without compandors. This figure obviously appears quite favorable when compared to the -27 dBm0 (-29.5 dBm0p) required in the example above. However, other factors which will be developed later will consume much of the noise margin that appears available.

Whereas random noise has an rms value when we measure level, impulse noise is another matter entirely. It is measured as the number of "hits" or "spikes' per interval of time over a certain threshold. In other words it is a measurement of the recurrence rate of noise peaks over a specified level. The word "rate" should not mislead the reader. The recurrence is not uniform per unit time, as the word "rate" may indicate, but we can consider a sampling and convert it to an average.

The Bell System circuit objective on leased lines is 15 counts in 15 min at -69 dBm with equivalent C-message weighting.

CCITT states than "In any four-wire international exchange the busy hour impulsive noise counts should not exceed 5 counts in 5 minutes at a threshold level of -35 dBm0" (Rec. Q. 45).

Remember that random noise has a Gaussian distribution and will produce peaks at 12.5 dB over the rms value (unweighted) 0.001% of the time on a data bit stream for an equivalent error rate of 1×10^{-5}. It should be noted that some references use 12 dB, some 12.5 dB, and others 13 dB. The 12.5 dB above the rms random noise floor should establish the impulse noise threshold for measurement purposes. We should assume in a well designed data transmission system traversing the telephone network, that the signal-to-noise ratio of the data signal will be well in excess of 12.5 dB. Thus impulse noise may well be the major contributor to degrade the error rate.

Care must be taken when measuring impulse noise. A transient such as an impulse noise spike on a band-limited system (which our telephone network most certainly is) tends to cause "ringing." Here the initial impulse noise spike causes what we might call a main bang or principal spike followed by damped subsidiary spikes. If we are not careful, these subsidiary spikes, that ringing effect, may also be counted as individual hits in our impulse noise count total. To avoid this false counting, impulse noise meters have a built-in dead time after each count. It is a kind of damping. Bell System, for example, specifies a 150-ms dead time after each count. This limits the counting capabilitity of the meter to no more than 6 or 7 cpm (counts per minute).

In this damping or dead time period, missed (real) impulse noise hits may seem to be a problem. For example, the Bell System offers that the average improved (increased) sensitivity to measure "all" hits is only 0.9 dB with a standard deviation of 0.76 dB (Ref 24).

The period of measurement is also important. How long should the impulse noise measurement set remain connected to a line under test to give an accurate count? It appears empirically that 30 min is sufficient. However, a good estimate of error can be made and corrected for if that period of time is reduced to 5 min. This is done by reducing the threshold (on paper) of the measuring set. From Ref. 24, the standard deviation for a 5-min period is about 2.2 dB. Thus 95% of all 5-min measurements will be within ± 3.6 dB of a 30-min measurement period.

To clarify this, remember that impulse noise distributions are log normal and impulse noise level distributions are normal. With this in mind we can relate count distributions, which can be measured readily, to level distributions. The mean of the level distribution is the threshold value that the impulse noise level meter was set to record the count distribution (in dBm, dBmp, dBrnC, or whatever unit). The set has a count associated with that threshold which is simply the median of the observed count distribution. The sigma, σ_1, standard deviation of the impulse noise level distribution is estimated by the expression: $\sigma_1 = m\sigma_D$, where m is the inverse slope of the peak amplitude distribution in decibels per decade of counts and average 7.0 dB. σ_D is the standard deviation of the log normal count distribution, which is the square root of the $\log_{10}$ of the ratio of the average number of counts to the median count, or

$$\sigma_1 = \sqrt{\log_{10} \frac{\text{average count}}{\text{median count}}}$$

where the median is not equal to zero. For instance, if we measured 10 cpm at a given threshold, the 1-cpm threshold would be 7 dB above the 10-cpm threshold.

When an unduly high error rate has been traced to impulse noise, there are some methods for improving conditions. Noisy areas may be bypassed, repeaters may be added near the noise source to improve signal-to-impulse noise ratio, or in special cases pulse smearing techniques may be used. This latter approach uses two delay distortion networks which complement each other such that the net delay distortion is zero. By installing the networks at opposite ends of the circuit, impulse noise passes through only one network and is therefore smeared because of the delay distortion. The signal is unaffected because it passes through both networks.

MULTIPATH DISTORTION

This is a phenomenon of radio transmission which results from the difference in time of arrival of signals which have traveled along different paths in going from a transmitter to a receiver. The strongest signal usually is the first to arrive at the receiver, having taken the shortest path. This is followed by signals of lesser amplitude which arrive later, with the result that the end of a transmitted pulse appears to be elongated or stretched. In the case of longer delay times and shorter pulse widths, it results in unwanted second pulse or noise spikes. Pulse elongation is most common and the modulation rate, therefore, must be limited to something less than an equivalent, which is one-half of the elongation or "smear."

For tropospheric scatter circuits, representative figures are from 0.1 to 1 μs, and where multiple reflections are involved, up to 10 μs. On HF radio circuits delays of 1–3 ms are observed frequently. However, maximums of 6, 12, and even 40 ms also have been observed.

An HF circuit with a 3-ms multipath delay must limit its modulation rate to an equivalent of 6-ms pulse width on any one carrier or subcarrier; or $1/6 \times 10^{-3} = 1000/6 = 166$ bauds.

Multipath delay on LOS (line of sight) microwave paths is not considered to be serious enough to limit the modulation rates covered in this discussion.

LEVELS AND LEVEL VARIATIONS

The design signal levels of telephone networks traversing FDM carrier systems are determined by average talker levels, average channel occupancy, permissible overloads during busy hours, etc. Applying constant amplitude digital data tone(s) over such an equipment at 0 dBm0 on each channel would result in severe overload and intermodulation within the system.

Loading does not affect hard wire systems except by increasing crosstalk. However, once the data signal enters carrier multiplex (voice) equipment, levels must be considered carefully and the resulting levels most probably have more impact on the final signal-to-noise ratio at the far end than anything else. CCITT* recommends -10 dBm0 in some cases, and -13 dBm0 when the proportion of nonspeech circuits on an international carrier circuit exceeds 10 or 20%. For multichannel telegraphy -8.7 dBm0 for the composite level, or for 24 channels, each channel would be adjusted for -22.5 dBm0. Even this loading may be too heavy if a high proportion of the voice channels are loaded with data. Depending on the design of the carrier equipment, cutbacks to -13 dBm0 or less may be advisable.

In a properly designed transmission system the standard deviation of the variation in level should not exceed 1.0 dB/circuit. However, data communication equipment should be able to withstand level variations in excess of 4 dB.

FREQUENCY TRANSLATION ERRORS

Total end-to-end frequency translation errors on a voice channel being used for data or telegraph transmission must be limited to 2 Hz (CCITT Rec. G. 135). This is an end-to-end requirement. Frequency translations occur mostly owing to carrier equipment modulation and demodulation steps. Frequency division multiplex carrier equipment widely uses single sideband, suppressed carrier techniques. Nearly every case of error can be traced to errors in frequency translation (we refer here to deriving the group, supergroup, mastergroup, and its reverse process; see also Chapter 3) and carrier reinsertion frequency offset, the frequency error being exactly equal to the error in translation and offset or the sum of several such errors. Frequency locked (e.g., synchronized) or high stability master carrier generators (1×10^{-7} or 1×10^{-8}, depending on the system), with all derived frequency sources slaved to the master source, usually are employed to maintain the required stability.

Although 2 Hz seems to be a very rigid specification, when added to the possible back-to-back error of the modems themselves, the error becomes more appreciable. Much of the trouble arises with modems that employ sharply tuned filters. This is particularly true of telegraph equipment. But for the more general case, high speed data modems can be designed to withstand greater carrier shifts than those that will be encountered over good telephone circuits.

* CCITT Recs. G.151C, H51, H23, and V2.

PHASE JITTER

The unwanted change in phase or frequency of a transmitted signal due to modulation by another signal during transmission is defined as phase *jitter*. If a simple sinusoid is frequency or phase modulated during transmission, the received signal will have sidebands. The amplitude of these sidebands compared to the received signal is a measure of the phase jitter imparted to it during transmission.

Phase jitter is measured in degrees of variation peak-to-peak for each hertz of transmitted signal. Phase jitter shows up as unwanted variations in zero crossings of a received signal. It is the zero crossings that data modems use to distinguish marks and spaces. Thus the higher the data rate, the more jitter can affect error rate on the receive bit stream.

The greatest cause of phase jitter in the telephone network is FDM carrier equipment, where it shows up as undesired incidental phase modulation. Modern FDM equipment derives all translation frequencies from one master frequency source by multiplying and dividing its output. To maintain stability, phase lock techniques are used. Thus the low jitter content of the master oscillator may be multiplied many times. It follows, then, that we can expect more phase jitter in the voice channels occupying the higher baseband frequencies.

Jitter most commonly appears on long-haul systems at rates related to the power line frequency (e.g., 60 Hz and its harmonics and submultiples, or 50 Hz and its harmonics and submultiples) or derived from 20-Hz ringing frequency. Modulation components that we define as jitter usually occur close to the carrier from about 0 to about ± 300 Hz maximum.

8.10 CHANNEL CAPACITY

A leased or switched voice channel represents a financial investment. The goal of the system engineer is to derive as much benefit as possible from the money invested. For the case of digital transmission this is done by maximizing the information transfer across the system. This subsection discusses how much information in bits can be transmitted, relating information to bandwidth, signal-to-noise ratio, and error rate. An empirical discussion of these matters is carried out in Section 8.11.

First, looking at very basic information theory, Shannon stated in his bandwidth paper (Ref. 13) that if input information rate to a band-limited channel is less than C (bps), a code exists for which the error rate approaches zero as the message length becomes infinite. Conversely, if the input rate exceeds C, the error rate cannot be reduced below some finite positive number.

The usual voice channel approximates to a Gaussian band-limited channel (GBLC) with additive Gaussian noise. For such a channel consider a signal wave of mean power of S watts applied at the input of an ideal low-pass filter having a bandwidth of W (Hz) and containing an internal source of mean Gaussian noise with a mean power of N watts uniformly distributed over the passband. The capacity in bits per second is given by

$$C = W \log_2 \left(1 + \frac{S}{N} \right)$$

Applying Shannon's "capacity" formula (GBLC) to some everyday voice channel criteria,

$$W = 3000 \text{ Hz}$$
$$\frac{S}{N} = 1023$$

Then $C = 30,000$ bps.

(Remember that bits per second and bauds are interchangeable in binary systems.)

Neither S/N or W is an unreasonable value. Seldom, however, can we achieve a modulation rate greater than 3000 bauds. The big question in advanced design is how to increase the data rate and keep the error rate reasonable.

One important item that Shannon's formula did not take into consideration is intersymbol interference. A major problem of a pulse in a band-limited channel is that the pulse tends not to die out immediately, and a subsequent pulse is interfered with by "tails" from the preceding pulse. This is shown in Figure 8.16.

H. Nyquist provided another approach to the data rate problem, this time using intersymbol interference (the tails in Figure 8.16) as a limit (Ref. 12). This resulted in the definition of the so-called Nyquist rate $= 2W$ elements/s. W is the bandwidth (Hz) of a band-limited channel as shown in Figure 8.16. In binary transmission, we are limited to $2W$ bps. If we let $W = 3000$ Hz the maximum data rate attainable is 6000 bps. Some refer to this as "the Nyquist 2-bit rule."

The key here is that we have restricted ourselves to binary transmission and we are limited to $2W$ bps no matter how much we increase the signal-

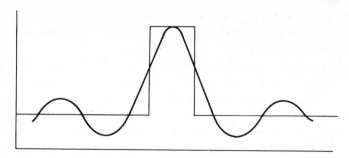

Figure 8.16 Pulse response through a Gaussian band-limited channel (GBLC).

to-noise ratio. The Shannon GBLC equation indicates that we should be able to increase the information rate indefinitely by increasing the signal-to-noise ratio. The way to attain a higher value of C is to replace the binary transmission system with a multilevel system, often termed an M-ary transmission system with $M > 2$. An M-ary channel can pass $2W \log_2 M$ bps with an acceptable error rate. his is done at the expense of signal-to-noise ratio. As M increases (as the number of levels increases), so must S/N increase to maintain a fixed error rate.

8.11   VOICE CHANNEL DATA MODEMS VERSUS CRITICAL DESIGN PARAMETERS

The critical parameters that affect data transmission have been discussed; these are amplitude-frequency response (sometimes called amplitude distortion), envelope delay distortion, and noise. Now we relate these parameters to the design of data modems to establish some general limits or "boundaries" for equipment of this type. The discussion that follows purposely avoids HF radio considerations.

As stated earlier in the coverage of envelope delay distortion, it is desirable to keep the transmitted pulse (bit) length equal to or more than the residual differential envelope delay distortion. Since about 1.0 ms is assumed to be reasonable residual delay after equalization (conditioning), the pulse length should then be no less than approximately 1 ms. This corresponds to a modulation rate of 1000 pulses/s (binary). In the interest of standardization (CCITT Rec. V.22), this figure is modified to 1200 bps.

The next consideration is the usable bandwidth required for the transmission of 1200 bps. This figure is approximately 1800 Hz using modulation methods such as phase shift (PSK), frequency shift (FSK) or double sideband AM (DSB-AM), and somewhat less for vestigial sideband AM (VSB-AM). Since delay distortion of a typical voice channel is at its minimum between 1700 and 1900 Hz, the required band when centered about these points extends from 800 to 2600 or 1000 to 2800 Hz. From the previous discussion we can see from Figure 8.14 that the envelope delay distortion requirement is met easily over the range of 800–2800 Hz.

Bandwidth limits modulation rate (as will be discussed below). However, the modulation rate in bauds and the data rate in bits per second may not necessarily be the same. This is a very important concept.

Suppose a moulator looked at the incoming serial bit stream 2 bits at a time rather than the conventional 1 bit at a time. Now let four amplitudes of a pulse be used to define each of the four possible combinations of two consecutive bits such that

$$A_1 = 00$$
$$A_2 = 01$$
$$A_3 = 11$$
$$A_4 = 10$$

where A_1, A_2, A_3, and A_4 represent the four pulse amplitudes. This form of treating two bits at a time is called di-bit coding (see Section 8.9.2).

Similarly, we could let eight pulse levels cover all the possible combinations of 3 consecutive bits so that with a modulation rate of 1200 bauds it is possible to transmit information at a rate of 3600 bps. Rather than vary amplitude to four or eight levels, phase can be varied. A four-phase system (PSK) could be coded as follows.

$$F_1 = \quad 0° = 00$$
$$F_2 = \quad 90° = 01$$
$$F_3 = \quad 18° = 11$$
$$F_4 = 270° = 10$$

Again, with a four-phase system using di-bit coding, a tone with a modulation rate of 1200 bauds PSK can be transmitting 2400 bps. An eight-phase PSK system at 1200 bauds could produce 3600 bps of information transfer. Obviously this process cannot be extended indefinitely. The limitation comes from channel noise. Each time the number of levels or phases is increased, it is necessary to increase the signal-to-noise ratio to maintain a given error rate. Consider the case of a signal voltage, S, and a noise voltage, N. The maximum number of increments of signal (amplitude) that can be discerned is S/N (since N is the smallest discernible increment).

Add to this the no-signal case and the number of discernible levels now becomes $S/N + 1$ or $S + N/N$; where S and N are expressed as power, the number of levels becomes the square root of this expression. This formula shows that in going from two to four levels or from four to eight levels, approximately 6-dB noise penalty is incurred each time we double the number of levels.

If a similar analysis is carried out for the multiphase case, the penalty in going from two to four phases is 3 dB, and to eight phases 6 dB.

Sufficient background has been developed to appraise the data modem for the voice channel. Now consider a data modem for a data rate of 2400 bps. By using quaternary phase shift keying (QPSK) as described above, 2400 bps is transmitted with a modulation rate of 1200 bauds. Assume the modem uses differential phase detection wherein the detector decisions are based on the change in phase between the last transition and the preceding one.

Assume the bandwidth to be present for the data modem under consideration (for most telephone networks the minimum bandwidth discussed for the sample case is indeed present—1800 Hz). It is now possible to determine if the noise requirements can be satisfied. Figure 8.16 shows that 12.5 dB signal-to-noise ratio (Gaussian noise) is required to maintain an error rate 1×10^{-5} for a binary polar (AM) system. As is well established FSK or PSK systems have about 3-dB improvement. In this case only a 9.5-db signal-to-noise ratio would be needed, all other factors held constant (no other contributing factors).

Assume the input from the line to be -10 dBm0 in order to satisfy loading conditions. To maintain the proper signal-to-noise ratio, the channel noise must be down to -19.5 dBm0.

To improve the modulation rate without expense of increased bandwidth, quaternary phase shift keying (four-phase) is used. This introduces a 3-dB noise degradation factor, bringing the required noise level down to -22.5 dBm0.

Consider now the effects of envelope delay distortion. It has been found that for a four-phase differential system, this degradation will amount to 6 dB if the permissible delay distortion is one pulse length. This impairment brings the noise requirement down to -28.5 dBm0 of average noise power in the voice channel. Allow 1 dB for frequency translation error or other factors, and the noise requirement is now down to 29.5 dBm0.

If the transmit level were -13 dBm0 instead of -10 dBm0, the numbers for noise must be adjusted another 3 dB such that it is now down to -32.5 dBm0 (19.5 dB signal-to-noise ratio). Thus it can be seen that to achieve a certain error rate, for a given modulation rate, several modulation schemes should be considered. It is safe to say in the majority of these schemes the noise requirement will fall somewhere between − 25 and − 40 dBm0. This is safely inside the CCITT figure of − 43 dBm (see Section 8.9.3, noise). More discussion on this matter may be found in Ref. 4.

8.12   CIRCUIT CONDITIONING

Of the critical circuit parameters mentioned above in Section 8.9.3 two that have severe deleterious effects on data transmission can be reduced to tolerable limits by circuit conditioning. These two are amplitude-frequency response (distortion) and envelope delay distortion.

Another name for circuit conditioning is equalization. There are several methods of performing equalization. The most common is to use one or several networks in tandem. Such networks tend to flatten response. In the case of amplitude, they add attenuation increasingly toward channel center and less toward its edges. The overall effect is one of making the amplitude response flatter. The delay equalizer operates fairly much in the same manner. Delay increases toward channel edges parabolically from the center. Delay is added in the center much like an inverted parabola, with less and less delay added as band edge is approached. Thus the delay response is flattened at some small cost to absolute delay which, in most data systems, has no effect. However, care must be taken with the effect of a delay equalizer on an amplitude equalizer and, conversely, the amplitude equalizer on the delay equalizer. Their design and adjustment must be such that the flattening of the channel for one parameter does not entirely distort the channel for the other.

Another type of equalizer is the transversal type of filter. It is useful where it is necessary to select among, or to adjust, several attenuation (amplitude) and phase characteristics. The basis of the filter is a tapped delay line to which the input is presented. The output is taken from a summing network which adds or sums the outputs of the taps. Such a filter is adjusted to the desired response (equalization of both phase and amplitude) by adjusting the tap contributions.

If the characteristics of a line are known, another method of equalization is predistortion of the output signal of the data set. Some devices use a shift register and a summing network. If the equalization needs to be varied, then a feedback circuit from the receiver to the transmitter would be required to control the shift register. Such a type of active predistortion is valid for binary transmission only.

A major drawback of all the equalizers discussed (with the exception of the last with a feedback circuit), is that they are useful only on dedicated or leased circuits where the circuit characteristics are known and remain fixed. Obviously a switched circuit would require a variable automatic equalizer, or conditioning would be required on every circuit in the switched system that would be transmitting data.

Circuits are usually equalized on the receiving end. This is called postequalization. Equalizers must be balanced and must present the proper impedance to the line. Administrations may choose to condition (equalize) trunks and attempt to eliminate the need to equalize station lines; the economy of considerably fewer equalizers is obvious. In addition, each circuit that would possibly carry high speed data in the system would have to be equalized, and the equalization must be good enough that any possible combination will meet the overall requirements. If equalization requirements become greater (i.e., parameters more stringent), then consideration may have to be given to the restriction of the maximum number of circuits (trunks) in tandem.

Conditioning to meet amplitude-frequency response requirements is less exacting on the overall system than envelope delay. Equalization for envelope delay and its associated measurements are time consuming and expensive. Envelope delay in general is arithmetically cumulative. If there is a requirement of overall envelope delay distortion of 1 ms for a circuit between 1000 and 2600 Hz, then in three links in tandem, each link must be better than 333 μs between the same frequency limits. For four links in tandem, each link would have to be 250 μs or better. In practice accumulation of delay distortion is not entirely arithmetical, resulting in a loosening of requirements by about 10%. Delay distortion tends to be inversely proportional to the velocity of propagation. Loaded cables display greater delay distortion than nonloaded cables. Likewise, with sharp filters a greater delay is experienced for frequencies approaching band edge than for filters with a more gradual cutoff.

In carrier multiplex systems channel banks contribute more to the overall envelope delay distortion than any other part of the system. Because Channels 1 and 12 of standard CCITT modulation plan, those nearest the group band edge, suffer additional delay distortion owing to the effects of group and, in some cases, supergroup filters, the systems engineer should allocate channels for data transmission near group and supergroup center. On long-haul critical data systems, the data channels should be allocated to through-groups and through-supergroups, minimizing as much as possible the steps of demodulation back to voice frequencies (channel demodulation).

Automatic equalization for both amplitude and delay shows promise, particularly for switched data systems. Such devices are self-adaptive and require a short adaptation period after switching, on the order of 1–2 s. This can be carried out during synchronization. Not only is the modem clock being "averaged" for the new circuit on transmission of a synchronous idle signal, but the self-adaptive equalizer adjusts for optimum equalization as well. The major drawback of adaptive equalizers is their expense.

8.13 PRACTICAL MODEM APPLICATIONS

8.13.1 Voice Frequency Carrier Telegraph (VFCT)

Narrow shifted FSK transmission of digital data goes under several common names. These are VFTG and VFCT. VFTG stands for voice frequency telegraph and VFCT for voice frequency carrier telegraph.

In practice voice frequency carrier telegraph techniques handle data rates up to 1200 bps by a simple application of FSK modulation. The voice channel is divided into segments or frequency bounded zones or bands. Each segment represents a data or telegraph channel, each with a frequency shifted subcarrier.

For proper end-to-end system interface, it is convenient to use standardized modulation plans, particularly on international circuits. In order for the far-end demodulator to operate with the near-end modulator, it must be tuned to the same center frequency and accept the same shift. Center frequency is that frequency in the center of the passband of the modulator/demodulator. The shift is the number of hertz that the center frequency is shifted up and down in frequency for the mark and space condition. From Table 8.1, by convention the mark condition is the center frequency shifted downward, and the space upward. For modulation rates below 80 bauds bandpasses have either 170-* or 120-Hz bandwidths with frequency shifts of ± 42.5 or ± 30 Hz, respectively. CCITT recommends (R.31) the 120-Hz channels for operating at 50 bauds and below; however, some administrations operate these channels at higher modulation rates. Figure 8.17 shows graphically the partial modulation plan, 120-Hz

* CCITT Rec. R.39.

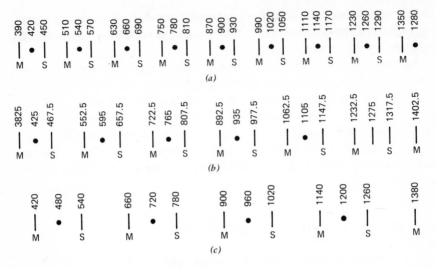

Figure 8.17 CCITT channel frequencies—VFCT (a) 120-Hz spacing, + and − 30-Hz shift; (b) 170-Hz spacing, + and − 42.5-Hz shift; (c) 240-Hz spacing, + and − 60-Hz shift.

spacing in Figure 8.17a; 170-Hz spacing in Figure 8.17b; and 240-Hz spacing in Figure 8.17c. The 240-Hz channel is recommended by the CCITT for 100-baud operation with + and − 60 Hz frequency shift.

The number of tone telegraph or data channels that can be accommodated on a voice channel depends for one thing on the usable voice channel bandwidth. For high frequency radio with a voice channel limit on the order of 3 kHz, 16 channels may be accommodated using 170-Hz spacing (170 Hz between center frequencies). Twenty-four VFCT channels may be accommodated between 390 and 3210 Hz with 120-Hz spacing, or 12 channels with 240-Hz spacing. This can easily meet standard telephone frequency division multiplex carrier channels of 300–3400 Hz.

Some administrations use a combination of voice and telegraph/data simultaneously on a telephone channel. This technique is commonly referred to a "voice plus" or S + D (speed plus derived). There are two approaches to this technique. The first is recommended by CCITT and is used widely by Intelsat orderwires. It places five telegraph channels (Channels 20–24) above a restricted voice band with a roofing filter near 2500 Hz. Speech occupies a band between 300 and 2500 Hz. Above 2500 Hz appear up to five 50-baud telegraph channels.

The second approach removes a slot from the center of the voice channel into which up to two telegraph channels may be inserted. The slot is a 500-Hz band centered on 1275 Hz.

However, some administrations use a slot for telegraphy of frequencies 1680 Hz and 1860 Hz by either amplitude or frequency modulation (FSK). See CCITT Rec. R.43.

The use of speech plus should be avoided on trunks in large networks because it causes degradation to speech and also precludes the use of the channel for higher speed data. In addition the telegraph channels should be removed before going into two-wire telephone service (i.e., at the hybrid or term set); otherwise service drops to half-duplex on telegraph.

8.13.2 Medium Data Rate Modems

In normal practice FSK is used for the transmission of data rates up to 1200 bps. The 120-Hz channel is nominally modified as in Figure 8.17c such that one 240-Hz channel replaces two 120-Hz channels. Administrations use the 240-Hz channel for modulation rates up to 150 bauds. The same process can continue using 480-Hz channels for 300 bauds FSK, and 960-Hz channels for 600 bauds. CCITT Rec. V.23 specifies 600/1200 baud operation in the nominal 4-kHz voice band.

CCITT Recs. V.21, 22, and 23 (White Books, Vol. VIII) recommend the following. V.21, which refers to "200-baud Modem Standardized for Use in the General Switched Telephone Networks," recommends

> Frequency shift $+$ and $-$ 100 Hz
> Center frequency of Channel 1, 1080 Hz
> Center frequency of Channel 2, 1750 Hz
> In each case space (0) is the higher frequency

It also provides for a disabling tone on echo suppressors, a very important consideration on long circuits.

V.22 standardizes modulation rates for synchronous data transmission at 600 and 1200 bauds.

V.23 recommends 600/1200-baud modem standardized for use in the general switched telephone network for application to synchronous or asynchronous systems. Provision is made for an optional backward channel for error control.

For the forward channel the following modulation rates and characteristic frequencies are presented:

	F_0	F_Z	F_A
Mode 1: up to 600 bauds	1500 Hz	1300 Hz	1700 Hz
Mode 2: up to 1200 bauds	1700 Hz	1300 Hz	2100 Hz

The backward channel for error control is capable of modulation rates up to 75 bauds. Its mark and space frequencies are

F_Z	F_A
390 Hz	450 Hz

Refer to Table 8.1 for the mark-space convention ($F_Z =$ mark or binary "1," $F_A =$ space or binary "0").

CCITT has tried to achieve a universality, recommending a modem that can be used nearly anywhere in the world "in the general switched telephone network." It considers worst-case conditions of amplitude-frequency response and envelope delay distortion.

8.13.3   High Data Rate Modems

Section 8.10 put a limit on the data rate in a bandwidth of 3000 Hz. The Nyquist limit was 6000 bits for binary transmission. It was also noted that a practical limit for this bandwidth is about 3000 bauds without automatic equalization. For the binary case this would be equivalent to 1 b/Hz (3000 bauds $\approx$ 3000 bps). For the quaternary case a data rate of 6000 bps may be reached (3000 bauds $\approx$ 6000 bps) or 2 b/Hz. However, most telephone lines do not have 3000 Hz of usable bandwidth available.

Consider the following modems, their required bandwidths, and their modulation schemes, which permit an improved data rate on a telephone channel.

Data Rate (bps)	Modulation Rate (bauds)	Modulation	Bits per Hertz	Bandwidth Required (Hz)
1. 2400 synchronous	1200	Differential four-phase	2	1200
2. 4800 synchronous	1600	Differential four-phase	3	1600

Date Rate (bps)	Modulation Rate (bauds)	Modulation	Bits per Hertz	Bandwidth Required (Hz)
3. 3600 synchronous	1200	Differential four-phase, two-level (combined PSK-AM)	3	1200
4. 2400 synchronous	800	Differential eight-phase	3	800
5. 9600 synchronous	4800	Differential two-phase, two-level	2	2400*

* Uses automatic equalizer.

8.14 SERIAL-TO-PARALLEL CONVERSION FOR TRANSMISSION ON IMPAIRED MEDIA

Often the transmission medium, in most cases the voice channel, cannot support a high data rate even with conditioning. The impairments may be due to poor amplitude-frequency response, envelope delay distortion, or excessive impulse noise.

One step that may be taken in these circumstances is to convert the high speed serial bit stream at the dc level (e.g., demodulated) to a number of lower speed parallel bit streams. One technique widely used on HF radio systems is to divide a 2400-bit serial stream into 16 parallel streams, each carrying 150 bps. If each of the slower streams is di-bit coded (2 bits at a time, discussed in Section 8.11) and applied to a QPSK tone modulator, the modulation rate on each subchannel is reduced in this case to 75 bauds. The equivalent period for a di-bit interval is 1/75 s or 13 ms.

There are two obvious advantages of this technique. First, each subchannel has a comparatively small bandwidth and thus looks at a small and tolerable segment of the total delay across the channel. The impairment of envelope delay distortion is less on slower speed channels. Secondly, there is less of a chance of a noise burst or hit of impulse noise to smear the subchannel signal beyond recognition. If the duration of noise burst is less than half the pulse width, the data pulse can be regenerated, and the pulse will not be in error. The longer one can make the pulse width, the less chance of disturbance from impulse noise. In this case the interval or pulse width has had an equivalent lengthening by a factor of 32.

8.15 PARALLEL-TO-SERIAL CONVERSION FOR IMPROVED ECONOMY OF CIRCUIT USAGE

Long, high quality (conditioned) toll telephone circuits are costly to lease or are a costly investment. The user is often faced with a large number of slow speed circuits (50–300 bps) which originate in one general geographic location, with a general destination to another common geographic location. If we assume these are 75-bps circuits (100 wpm), which are commonly encountered in practice, only 18–24 can be transmitted on a high grade telephone channel by conventional voice frequency telegraph techniques (see Section 8.13.1).

Circuit economy can be affected using a data/telegraph time division multiplexer. A typical application of this type is illustrated in Figure 8.18. It shows one direction of transmission only. Here incoming, slow speed VFCT channels are converted to equivalent dc bit streams. Up to 34 of these bit streams serve as input to a time division multiplexer in the application that is illustrated in the figure. The output of the multiplexer is a 2400-bit synchronous series bit stream. This output is fed to a con-

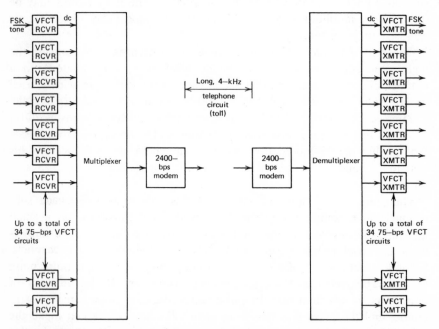

Figure 8.18 Typical application of parallel-to-serial conversion. VFCT, voice frequency carrier telegraph; rec., receiver (converter); xmt, transmitter (keyer); bps, bits per second; FSK, frequency shift keying.

ventional 2400-bit modem. At the far end the 2400 bps serial stream is demodulated to dc and fed to the equivalent demultiplexer. The demultiplexer breaks the serial stream back down to the original 75-bps circuits. Figure 8.18 illustrates the concept. It does not show clocking or other interconnect circuitry. By use of a time division multiplexer, a savings of up to 2 to 1 can be effected. Whereas by conventional VFCT means only about 18 75-bps circuits can be transmitted on a good telephone channel, by means of the multiplexer up to 34 such circuits can now be transmitted on the same channel.

8.16   MODEMS FOR APPLICATION TO CHANNEL BANDWIDTHS IN EXCESS OF 4 kHz

Well equalized (conditioned) telephone circuits are handling data rates in excess of 7500 bps. For data rates in excess of this figure, use of 48-kHz channels becomes attractive. 48 kHz is the bandwidth of the standard CCITT 12-channel group (e.g., 60–108 kHz). Group equalization in a 36-kHz portion of the band can bring envelope delay distortion (group delay) down to 50 μs. From previous reasoning, it can be assumed that the channel will support a 20-kilobaud modulation rate. It should also be noted that 50 μs is considerably greater than the multipath delay encountered on troposcatter circuits (10 μs); hence operation will be possible over most wideband media—coaxial cable, troposcatter, satellite, and radiolink.

The industry is now standardizing on data rates of 75×2^n, where n is any whole number. Thus a 48-kHz channel can support 19.2 kbs of two-level FSK or PSK or 38.4 kbs of four-level AM or PSK.

Taking the above discussion into account as well as some of the other design rules set forth in the chapter, the following specifications are suggested for a 48-kHz data modem:

Modulation rate	19.2 kilobauds
Data rate	38.4 kb/s
Input/output to the line	60–108 kHz
Envelope delay distortion	
1. within 200 μs	64–104 kHz
2. within 50 μs	66–102 kHz
Transmit level to the line	$+1$ dBm0

(derived from -10 dBm0 $+ 10 \log n$, where $n = $ number of channels $= 12$)

Noise (equivalent white noise)	-20 dBm0 with flat weighting

Quaternary differential phase shift modulation
    (with data input di-bit coded)

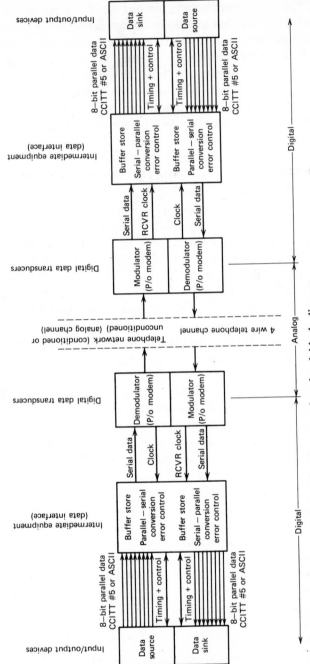

Figure 8.19 Standard data transmission system—functional block diagram.

396

8.17 DATA TRANSMISSION SYSTEM— FUNCTIONAL BLOCK DIAGRAM

Figure 8.19 is a simplified functional block diagram showing a typical data transmission system end-to-end. The diagram is meant to be representative and conceptual. For instance, card transmission would require the substitution of Figures 8.11 and 8.12 at each end. The system shown in the figure uses an eight-level code operating in a synchronous mode. Sections 8.3.2 and 8.6.2 cover eight-level codes and synchronous transmission, respectively, as well as clocking requirements. The receive clock in this case is corrected by averaging transitions on the incoming bit stream. Master clocks are contained in the modems. Timing problems are covered in Section 8.6.3. The data interface is discussed in Section 8.7 and input/output devices in Section 8.8. Analog (telephone) channel criteria are covered in Section 8.9 and subsequently. The diagram shows the functions of the various building blocks and how the higher speed, more sophisticated data transmission system differs from the more conventional telegraph system.

REFERENCES AND BIBLIOGRAPHY

1. *Reference Data for Radio Engineers*, 5th ed., Howard W. Sams & Co., Indianapolis, 1968.
2. D. H. Hamsher (ed.), *Communication System Engineering Handbook*, McGraw-Hill, New York, 1967.
3. *Transmission Systems for Communications*, 3rd ed., Bell Telephone Laboratories, 1964.
4. W. R. Bennett and J. R. Davey, *Data Transmission*, McGraw-Hill, New York, 1965.
5. *Data Transmission, Parameters and Capabilities*, International Telephone and Telegraph Co., Federal Laboratories, October 1961.
6. *Understanding Telegraph Distortion*, Stelma, Inc., Stamford, Conn., 1962.
7. J. M. Weir, "Digital Data Communication Techniques," *Proc. IRE*, Jan. 1961.
8. CCITT White Books, Mar del Plata, 1968, Vol. III, G. Recommendations.
9. *Ibid.*, Vol. VII, R. Recommendations.
10. *Ibid.*, Vol. VIII, V. Recommendations.
11. R. W. Lucky, J. Salz, and E. J. Weldon, *Principles of Data Communication*, McGraw-Hill, New York, 1968.
12. H. Nyquist, "Certain Topics in Telegraph Transmission Theory," *AIEE Trans. (U.S.)*, **47**, 617–644 (April 1928).
13. C. E. Shannon, "A Mathematical Theory of Communication," *Bell Syst. Tech. J.*, **27**, 379–423 (July 1948); 623–656 (Oct. 1948).
14. J. Martin, *Systems Analysis for Data Transmission*, Prentice-Hall, New York, 1972.

15. W. P. Davenport, *Modern Data Communication*, Hayden Book Co., New York, 1971.

16. S. Goldman, *Information Theory*, Dover Publications, New York, 1968.

17. *DCS Autodin Interface and Control Criteria*, DCAC 370-D-175-1, Defense Communication Agency, Washington, D.C., 1965.

18. M. P. Ristenbatt, "Alternatives in Digital Communications," *Proc. IEEE*, **61**(6), June 1973.

19. *Notes on Transmission Engineering*, 3rd ed., U.S. Independent Telephone Association, New York, 1971.

20. C. L. Cuccia, "Subnanosecond Switching and Ultra-speed Data Communications," *Data and Communications*, Nov. 1971.

21. D. R. Doll, "Controlling Data Transmission Errors," *Data Dynamics*, July 1971.

22. E. N. Gilbert, *Information Theory after 18 Years*, Bell Telephone Monograph, Bell Telephone Laboratories.

23. EIA Standard RS-232C, Electronic Industries Association, Aug. 1969.

24. *Analog Parameters Affecting Voiceband Data Transmission—Description of Parameters*, Bell Syst. Tech. Ref. Publ. 41008, ATT, New York, Oct. 1971.

1975

9.1 INTRODUCTION

A coaxial cable is simply a transmission line consisting of an unbalanced pair made up of an inner conductor surrounded by a grounded outer conductor, which is held in a concentric configuration by a dielectric. The dielectric can be of many different types, such as solid "poly" (polyethylene or polyvinyl chloride), foam, Spirafil, air, or gas. In the case of air/gas dielectric, the center conductor is kept in place by spacers or disks.

Systems have been designed to use coaxial cable as a transmission medium with a capability of transmitting a frequency division multiplex configuration ranging from 120 voice channels to 10,800. Community antenna television (CATV) systems use single cables for transmitted bandwidths on the order of 300 MHz.

Frequency division multiplex was developed originally as a means to increase the voice channel capacity of wire systems. At a later date the same techniques were applied to radio. Then for a time, the 20 years after World War II, radio systems became the primary means for transmitting long-haul, toll telephone traffic. Lately, coaxial cable has been making a strong comeback in this area.

One advantage of coaxial cable systems is reduced noise accumulation when compared to radiolinks. For point-to-point multichannel telephony the FDM line frequency (see Chapter 3) configuration can be applied directly to the cable without further modulation steps as required in radio-links, thus substantially reducing system noise.

In most cases radiolinks will prove more economical than coaxial cable. Nevertheless, owing to the congestion of centimetric radio wave (radiolinks) systems (see Chapter 10), coax is making a new debut. Coaxial cable

399

should be considered in lieu of radiolinks using the following general guidelines:

- In areas of heavy microwave (including radiolink) RFI.
- On high density routes where it may be more economical than radio-links. (Think here of a system that will require 5000 circuits at the end of 10 years.)
- On long national or international backbone routes where the system designer is concerned with noise accumulation.

Coaxial cable systems may be attractive for the transmission of television or other video applications. Some activity has been noted in the joint use of TV and FDM telephone channels on the same conductor. Another advantage in some circumstances is that system maintenance costs may prove to be less than for equal capacity radiolinks.

One deterrent to the implementation of coaxial cable systems, as with any cable installation, is the problem of getting right-of-way for installation, and its subsequent maintenance (gaining access), particularly in urban areas. Another consideration is the possibility of damage to the cable once it is installed. Construction crews may unintentionally dig up or cut the cable. For more details on the choice, coaxial cables or radiolinks, refer to Section 9.12.

9.2 BASIC CONSTRUCTION DESIGN

Each coaxial line is called a "tube." A pair of these tubes is required for full-duplex, long-haul application. One exception is the CCITT small bore coaxial cable system where 120 voice channels, both "go" and "return," are accommodated in one tube. For long-haul systems more than one tube is included in a sheath. In the same sheath filler pairs or quads are included, sometimes placed in the interstices, depending on the size and lay-up of the cable. The pairs and quads are used for orderwire and control purposes as well as for local communication. Some typical cable lay-ups are shown in Figure 9.1. Coaxial cable is usually placed at a depth of 90–120 cm, depending on frost penetration, along the right-of-way. Tractor-drawn trenchers or plows normally are used to open the ditch where the cable is placed, using fully automated procedures.

Cable repeaters are spaced uniformly along the route. Secondary or "dependent" repeaters are often buried. Primary power feeding or "main" repeaters are installed in surface housing. Cable lengths are factory-cut so that the splice occurs right at repeater locations.

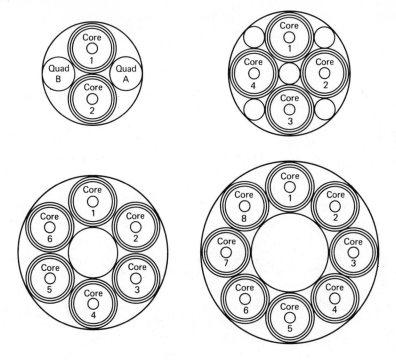

Figure 9.1 Some basic coaxial lay-ups.

9.3 CABLE CHARACTERISTICS

For long-haul transmission, standard cable sizes are as follows:

in.	mm
0.047/0.174	1.2/4.4 (small diameter)
0.104/0.375	2.6/9.5

The fractions express the outside diameter of the inner conductor over the inside diameter of the outer conductor. For instance, for the large bore cable, the outside diameter of the inner conductor is 0.104 in. and the inside diameter of the outer conductor is 0.375 in. This is shown in Figure 9.2. As can be seen from the equation in Figure 9.2, the ratio of the diameters of the inner and outer conductors has an important bearing on attenuation. If we can achieve a ratio of $b/a = 3.6$, a minimum attenuation per unit length will result.

For air dielectric cable pair, $\epsilon = 1.0$
outside diameter of inner conductor = $2a$
inside diameter of outer conductor = $2b$
Attenuation constant (dB)/mi

$$\alpha = 2.12 \times 10^{-5} \frac{\sqrt{f}\left(\frac{1}{a} + \frac{1}{b}\right)}{\log b/a} \qquad (9.1)$$

where a = radius of inner conductor and b = radius of outer conductor.

Characteristic impedances (Ω)

$$Z = \left(\frac{138}{\sqrt{\epsilon}}\right) \log \frac{b}{a} = 138 \log \frac{b}{a} \text{ in air} \qquad (9.2)$$

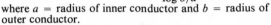

Figure 9.2 Basic electrical characteristics of coaxial cable.

The characteristic impedance of coaxial cable is $Z_0 = 138 \log (b/a)$ for an air dielectric. If $b/a = 3.6$, then $Z_0 = 77 \ \Omega$. Using dielectric other than air reduces the characteristic impedance. If we use the disks mentioned above to support the center conductor, the impedance lowers to 75 Ω.

Figure 9.3 is a curve giving attenuation per unit length in decibels versus frequency for the two most common types of coaxial cable discussed in this chapter. Attenuation increases rapidly as a function of frequency and is a function of the square root of frequency as shown in Figure 9.2. The transmission system engineer is basically interested in how much bandwidth is available to transmit an FDM line frequency configuration (Chapter 3). For instance, the 0.375-in. cable has an attenuation of about 5.8 dB/mi at 2.5 MHz and the 0.174-in. cable, 12.8 dB/mi. At 5 MHz the 0.174-in. cable has about 19 dB/mi and the 0.375-in. cable, 10 dB/mi. Attenuation is specified for the highest frequency of interest.

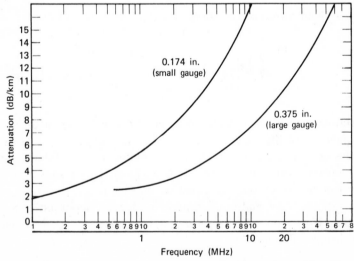

Figure 9.3 Attenuation-frequency response per kilometer of coaxial cable.

Coaxial cable can transmit signals down to dc, but in practice, frequencies below 60 kHz are not used because of difficulties of equalization and shielding. Some engineers lift the lower limit to 312 kHz. The high frequency limit of the system is a function of the type and spacing of repeaters as well as cable dimensions and the dielectric constant of the insulating material. It will be appreciated from Figure 9.3 that the gain frequency characteristics of the cable follows a root frequency law, and equalization and "preemphasis" should be designed accordingly.

9.4 SYSTEM DESIGN

Figure 9.4 is a simplified application diagram of a coaxial cable system in long-haul, point-to-point multichannel telephone service. To summarize system operation, an FDM line frequency (Chapter 3) is applied to the coaxial cable system via a line terminal unit. Dependent repeaters are spaced uniformly along the length of the cable system. These repeaters are fed power from the cable itself. In the ITT design (Ref. 1), the dependent repeater has a plug-in automatic level control unit. In temperate zones where cable laying is sufficient and where diurnal and seasonal temperature variations are within the "normal' (a seasonal swing of $\pm 10°C$), a plug-in level control (regulating) unit is incorporated in every fourth dependent amplifier (see Figure 9.5). We use the word "dependent" for the dependent repeater (DA in Figure 9.5) for two reasons. It depends on a terminal or main repeater for power and it provides to the terminal or main repeater fault information.

Let us examine Figures 9.4 and 9.5 at length. Assume we are dealing with a nominal 12-MHz system on a 0.375-in. (9.5-mm) cable. Up to 2700 voice channels can be transmitted. To accomplish this, two tubes are required, one in each direction. Most lay-ups, as shown in Figure 9.1, have more than two tubes. Consider Figure 9.4 from left to right. Voice channels in a four-wire configuration connect with the multiplex equipment in both the "go" and "return" directions. The output of the multiplex equipment is the line frequency (baseband) to be fed to the cable. Various line frequency configurations are shown in Figures 3.11, 3.18, and 3.19. The line signal is fed to the terminal repeater, which performs the following functions:

- Combines the line control pilots with the multiplex line frequency.
- Provides "preemphasis" to the transmitted signal, distorting the output signal such that the higher frequencies get more gain than the lower frequencies, such as shown in Figure 9.3.
- Equalizes the incoming wide band signal.
- Feeds power to dependent repeaters.

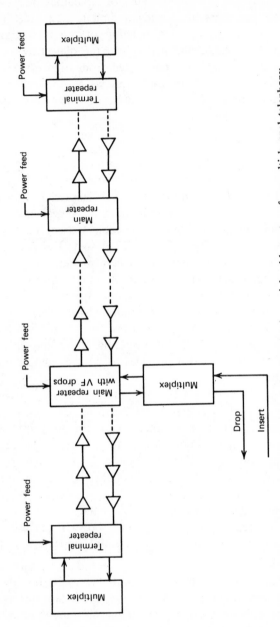

Figure 9.4 Simplified application diagram of a long-haul coaxial cable system for multichannel telephony.

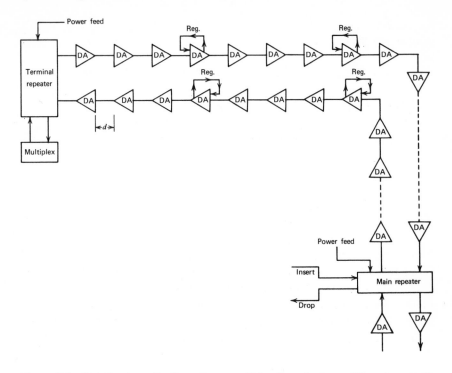

Figure 9.5 Detail of application diagram. DA, dependent amplifier (repeater); Reg., regulation circuitry; d, distance between repeaters.

The output of the terminal repeater is a preemphasized signal with required pilots along with power feed. In the ITT design this is a dc voltage up to 650 V with a stabilized current of 110 mA. A main (terminal) repeater feeds, in this design, up to 15 dependent repeaters in each direction. Thus a maximum of 30 dependent repeaters appear in a chain for every main or terminal repeater. Other functions of a main repeater are to equalize the wide band signal and to provide access for drop and insert of telephone channels by means of through-group filters.

Figure 9.5 is a blowup of a section of Figure 9.4 showing each fourth repeater with its automatic level regulation circuitry. Distance "d" between "DA" repeaters (dependent amplifier) is 4.5 km or 2.8 mi for a nominal 12-MHz system (0.375-in. cable). Amplifiers have gain adjustments of ± 6 dB, equivalent to varying repeater spacing ± 570 m (1870 ft).

As can be seen from the above, the design of coaxial cable systems for both long-haul multichannel telephone service as well as CATV systems has become, to a degree, a "cookbook" design. Basically, system design involves the following:

- Repeater spacing as a function of cable type and bandwidth.
- Regulation of signal level.
- Temperature effects on regulation.
- Equalization.
- Cable impedance irregularities.
- Fault location or the so-called supervision.
- Power feed.

Other factors are, of course, right-of-way for the cable route with access for maintenance and the laying of the cable. With these factors in mind, consult Tables 9.1 and 9.2, which review the basic parameters of the Bell System approach (9.1) and the CCITT approach (9.2).

For the 0.375-in. coaxial cable systems practical noise accumulation is less than 1 pWp/km, whereas radiolinks allocate 3 pWp/km. These are good guideline numbers to remember for gross system considerations. Noise in coaxial cable systems derives from the active devices in the line (e.g., the repeaters) as well as the terminal equipment, both line condition-

Table 9.1   Characteristics of "L" Coaxial Cable Systems

	"L" System Identifier			
Item	L1	L3	L4	L5
Max. design line length	4000 mi	4000 mi	4000 mi	4000 mi
No. of 4-kHz FDM VF channels	600	1860	3600	10,800
TV NTSC	Yes	Yes plus 600 VF	No	Not stated
Line frequency	60–2788 kHz	312–8284 kHz	564–17,548 kHz	1590–68,780 kHz
Nominal repeater spacing	8 mi	4 mi	2 mi	1 mi
Power feed points	160 mi or every 20 rptrs	160 mi or 42 rptrs	160 mi or every 80 rptrs	75 mi or every 75 rptrs

Notes.   1. Cable type of all "L" systems, 0.375 in.
2. Number of VF channels expressed per pair of tubes, one tube "go" and one tube "return."

Table 9.2   Characteristics of CCITT Specified Coaxial Cable Systems (Large Diameter Cable)

Item	Nominal Top Modulation Frequency				
	2.6 MHz	4 MHz	6 MHz	12 MHz	60 MHz
CCITT Rec.	G.337A	G.338	G.337B	G.332	G.333
Repeater type	Tube	Tube	Tube	Transistor	Transistor
Video capability	No	Yes	Yes	Yes	Not stated
Video + FDM capability	No	No	No	Yes	Not stated
Nominal re-peater spacing	6 mi/ 9 km	6 mi/ 9 km	6 mi/ 9 km	3 mi/ 4.5 km	1 mi/ 1.55 km
Main line reg. pilot	2604 kHz	4092 kHz	See CCITT Rec. J.72	12,435 kHz	12,435/4287 kHz
Auxiliary reg. pilot(s)		308, 60 kHz	See Rec. J.72	4287, 308 kHz	61,160, 40,920 and 22,372 kHz

Note.   Cable type for all systems, 0.104/0.375 in. = 2.6/9.5 mm.

ing and multiplex. Noise design of these devices is a trade-off between thermal and intermodulation (IM) noise. IM noise is the principal limiting parameter forcing the designer to install more repeaters per unit length with less gain per repeater.

Refer to Chapter 3 for CCITT recommended FDM line frequency configurations, in particular Figures 3.18 and 3.19 valid for 12-MHz systems. CCITT pilot frequencies and system levels are covered in Section 9.7.

9.5   REPEATER DESIGN—AN ECONOMIC TRADE-OFF FROM OPTIMUM

9.5.1   General

Consider a coaxial cable system 100 km long using 0.375-in. cable capable of transmitting up to 2700 VF channels in an FDM/SSB configuration (12 MHz). At 12 MHz cable attenuation per kilometer is approximately 8.3 dB (from Figure 9.3). The total loss at 12 MHz for the 100-km cable section is $8.3 \times 100 = 830$ dB. Thus one approach the system design engineer might take would be to install a 830-dB amplifier at the front end of the 100-km section. This approach is rejected out of hand. Another approach would be to install a 415-dB amplifier at the front end and another at the 50-km point. Suppose the signal level was -15 dBm composite at the originating end. Thus -15 dBm $+$ 415 dB $= +400$ dBm or $+370$ dBW. Remember that $+60$ dBW is equivalent to a megawatt; otherwise we would have an amplifier with an output of 10^{37} W or 10^{31} Mw. Still another approach is to have 10 amplifiers with 83-dB gain, each spaced at 10-km intervals. Another would be 20 amplifiers or $830/20 = 41.5$ dB each; or 30 amplifiers at $830/30 = 27.67$ dB, each spaced at 3.33-km intervals. As we shall see later, the latter approach begins to reach an optimum from a noise standpoint keeping in mind that the upper limit for noise accumulation is 3 pWp/km. The gain most usually encountered in coaxial cable amplifiers is 30–35 dB.

If we remain with the 3 pWp/km criterion, in nearly all cases radiolinks (Chapter 5) will be installed because of their economic advantage. Assuming 10 full-duplex RF channels per radio system at 1800 VF channels per RF channel, the radiolink can transmit 18,000 full-duplex channels, and do it probably more cheaply on an installed cost basis. On the other hand, if we can show noise accumulation less on coaxial cable systems, these systems will prove in at some number of channels less than 18,000 if the reduced cumulative noise is included as an economic factor. There are other considerations, such as maintenance and reliability, but let us discuss noise further.

Suppose we design our coaxial cable systems for no more than 1 pWp/km. Most long-haul coaxial cable systems being installed today meet this figure. However, we will use the CCITT figure of 3 pWp/km in some of the examples that follow.

In Chapter 1, noise for this discussion consists of two major components, namely,

Thermal noise (white noise)
Intermodulation (IM) noise

Coaxial cable amplifier design, to reach a goal of 1 pWp/km of noise accumulation, must walk a "tightrope" between thermal and intermodulation noise. It is also very sensitive to overload, with its consequent impact on intermodulation noise.

The purpose of the abbreviated and highly simplified discussion in this section is to give the transmission system engineer some appreciation of coaxial repeater design. For a deeper analysis, the reader should refer to Ref. 2, Chaps. 12–16, and Ref. 3, Chaps. 3–7.

9.5.2 Thermal Noise

From Section 1.11, thermal noise threshold, P_n, may be calculated for an active, two-port, device as follows:

$$P_n = -174 \text{ dBm/Hz} + NF + 10 \log B_w \qquad (9.3)$$

where B_w = bandwidth (Hz)
NF_{dB} = noise figure of the amplifier

Restating the equation above for a voice channel with a nominal B_w of 3000 Hz, we then have

$$P_n = -139 \text{ dBm} + NF \quad (\text{dBm/3 kHz}) \qquad (9.4)$$

Assume a coaxial cable system with identical repeaters, each with gain G_r, spaced at equal intervals along a uniform cable section. Here G_r exactly equals the loss of the intervening cable between repeaters. The noise output of the first repeater is $P_n + G_r$ (in dBm). For N repeaters in cascade, the total noise (thermal) output of the Nth repeater is

$$P_n + G_r + 10 \log N \quad (\text{dBm}) \qquad (9.5)$$

An important assumption all along is that the input/output impedance of the repeaters just equals the cable impedance, Z_0.

9.5.3 Overload and Margin

The exercise of this section is to develop an expression for system noise and discuss methods of reducing it. In Section 9.5.2 we developed a term for thermal noise for a string of cascaded amplifiers ($P_n + G_r + 10 \log N$). The next step is to establish 0 dBm as a reference, or more realistically -2.5 dBmp, because we are dealing with a voice channel nominally 3 kHz wide and we want it weighted psophometrically (see Section 1.9.6). Now we can establish a formula for a total thermal noise level as measured at the end of a coaxial cable system with N amplifiers in cascade:

$$P_t = P_n + G_r + 10 \log N - 2.5 \text{ dBmp} \qquad (9.6)$$

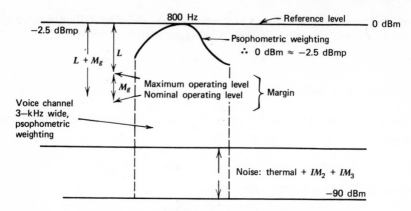

Figure 9.6 Graphic representation of reference level, signal levels, and noise levels in a coaxial cable system. (*Note.* Levels are not drawn to scale.)

As before we assume that all the amplifiers are identical and spaced at equal intervals and that the gain of each is G_r, which is exactly equal to the loss of the intervening cable between each amplifier.

Examining the equation above, we see that the operating level is high. The next step is to establish an operating level which should never be exceeded and call it L. A margin to that level must also be established to take into account instability of the amplifiers caused by aging effects, poor maintenance, temperature variations, misalignment, and so forth. The margin to the maximum operating level point is M_g. All units are in decibels. A more realistic equation can now be written for total thermal noise including a suitable margin:

$$P_t = P_n + G_r + 10 \log N - 2.5 \text{ dBmp} + L + M_g \qquad (9.7)$$

These levels are shown graphically in Figure 9.6.

A number of interesting relationships can be developed if we consider a hypothetical example. CCITT permits 3-pWp/km noise accumulation (CCITT Rec. G.222). Allow 2 pWp of that figure to be attributed to thermal noise. If we were to build a system 100 km long, we could then accumulate 200 pWp of thermal noise. Now set 200 pWp equal to P_t in the previous equation. First convert 200 pWp to dBmp (-67 dBmp). Thus

$$-67 \text{ dBmp} = P_n + G_r + 10 \log N - 2.5 \text{ dBmp} + L + M_g$$
$$P_r + G_r + 10 \log N + L + M_g = -64.5 \text{ dBmp}$$

Let us assign some numbers to the equation that are somewhat reasonable. To the 100-km system install 20 repeaters at equal intervals. Cable loss is 5 dB/km, or 500 dB total loss at the highest operating frequency. Thus repeater gain, G_r is 25 dB, with $N = 20$. Let

$$L + M = 15 \text{ dB}$$

From Section 9.5.2

$$P_n = -139 + NF$$

Thus

$$-139 + NF + 25 + 10 \log 20 + 15 = -64.5 \text{ dBmp}$$
$$NF = 21 \text{ dB or less}$$

This is a noise figure that is fairly easy to meet.

Let us examine this exercise a little more closely and see if we cannot derive some important relationships that can offer the system and amplifier design engineer some useful guidance.

1. By doubling the length of the system, system noise increases 3 dB, or by doubling the number of amplifiers, G_r being held constant, system noise doubles (i.e., $10 \log 2N$).

2. By making the terms L and M_g smaller, or in other words, increasing the maximum operating level, reducing the margin, system thermal noise improves on a decibel for decibel basis.

3. Of course by reducing NF, system noise may also be reduced. But suppose NF turned out to be very small in the calculations, a figure that could not be met or would imply excessive expense. Then we would have to turn to other terms in the equation, such as reducing terms G_r, L, and M_g. However, there is little room to maneuver with the latter two, 15 dB in the example. That leaves us with G_r. Of course, reducing G_r is at the cost of increasing the number of amplifiers (or increasing the size of the cable to reduce attenuation, etc.). As we reduce G_r, the term $10 \log N$ increases because we are increasing the number of amplifiers, N. The trade-off between the term $10 \log N$ and G_r occurs where G_r is between 8 and 9 dB.

Another interesting relationship is that of the attenuation of the cable. It will be noted that the loss in the cable is approximately inversely proportional to cable diameter. As an example, let us assume that the loss of a cable section between repeaters is 40 dB. By increasing the cable diameter

25%, the loss of the cable section becomes $40/(1 + 0.25) = 32$ dB. In our example above, by increasing cable diameter, repeater gain may be decreased with the consequent improvement in system noise (thermal). *(Note.* The examples given above are given as exercises and may not necessarily be practicable owing to economic constraints.)

9.5.4 Intermodulation Noise

The second type of noise to be considered in coaxial cable system and repeater design is intermodulation (IM) noise. IM noise on a multichannel FDM system may be approximated by a Gaussian distribution (see Section 3.4.4) and consists of second, third, and higher order intermodulation products. Included in these products, in the wide band systems we cover here, are second and third harmonics. IM products (e.g., intermodulation noise) are a function of the nonlinearity of active devices* (see Section 1.9.6).

To follow our argument on intermodulation noise in coaxial cable repeater design, the reader is asked to accept the following (Ref. 1). If a simple sinusoid wave is introduced at the input port of a cable amplifier, the output of the amplifier could be expressed by an equation with three terms, the first of which is linear representing the desired amplification. The second and third terms are quadratic and cubic representing the non-linear behavior of the amplifier (i.e., second and third order products). On the basis of this power series, for each 1-dB change of fundamental input to the amplifier, the second harmonic changes 2 dB, and the third harmonic 3 dB. Furthermore, for two waves A and B, a second order sum $(A + B)$ or difference $(A - B)$ is equivalent to the second harmonic of A at the output plus 6 dB. Likewise, the sum of $A + B + C$ would be equivalent to the level of the third harmonic level at the output plus 15.6 dB of one of the waves. We consider that all inputs are of equal level. The situation for $2A + B$ would be equivalent to $3A + 9.6$ dB, and so forth. These last three power series may be more clearly expressed when set down as follows, where P_H is the harmonic/intermodulation power:

$$P_{H(A \pm B)} = P_{H(2A)} + 6 \text{ dB}$$
$$P_{H(A \pm B \pm C)} = P_{H(3A)} + 15.6 \text{ dB}$$
$$P_{H(2A \pm B)} = P_{H(3A)} + 9.6 \text{ dB}$$

---

* IM products may also be produced in passive devices, but to simplify the argument in this chapter, we have chosen to define IM products as those derived in active devices.

For this discussion let IM_2 and IM_3 express the nonlinearity of a repeater; they are, respectively, the power of the second and third harmonics corresponding to a 0 dBm fundamental (-2.5 dBmp). Adjusted to the maximum operating level L (see Figure 9.4), the second harmonic power, P_{2A}, is

$$P_{2A} = (IM_2 + 10 \log N + L) \quad (\text{dBmp}) \tag{9.9}$$

L is assumed to be positive as in our argument in the preceding Section.

Now decrease the applied signal level to a repeater by L dB, assuming the power of the fundamental of a wave at the 0 TLP (test level point) was L dBmp. It follows that, by decreasing the applied power L dB so that the fundamental of a wave is now 0 dBmp at the 0 test level point (TLP), the magnitude of the fundamental is decreased L dB at the input of the first amplifier.

For every decrease of 1 dB in the fundamental, the second harmonic decreases by 2 dB. Therefore the new power of the second harmonic amplitude will be decreased by $2L$ dB, or

$$P_{2A} = (IM_2 + 10 \log N + L - 2L) \quad \text{dBmp} \quad \text{(for the system)}$$

or the second harmonic noise power level.

$$P_{2A} = IM_2 + 10 \log N - L \text{ dBmp} \tag{9.10}$$

Let us consider the $A + B$ product. For fundamentals of equal magnitudes, such a product in a single repeater will be 6 dB higher than a $2A$ product. It also varies by 2 dB per 1 dB variation in both fundamentals and adds in power addition as a function of the number of repeaters. Thus

$$P_{(A+B)} = IM_3 + 10 \log N - L + 6 \text{ dBmp} \tag{9.11}$$

Similarly for the $3A$ condition (e.g., third harmonic) of a wave fundamental, A,

$$P_{3A} = IM_3 + 10 \log N - 2L \quad (\text{dBmp}) \tag{9.12}$$

For the $A + B - C$ condition,

$$P_{(A+B-C)} = IM_3 + 20 \log N - 2L + 15.6 \text{ dBmp} \tag{9.13}$$

Again we find "$2L$" because third order products vary 3 dB for every 1-dB change in the fundamental. We use 20 log rather than 10 log assuming that the products add in phase (i.e., voltage-wise) versus the number of repeaters. The 10 log term represented power addition.

For the $2A - B$ condition,

$$P_{2A-B} = IM_3 - 2L + 20 \log N + 9.6 \text{ dBmp} \tag{9.14}$$

9.5.5   Total Noise and its Allocation

In summary there are three noise components to be considered:

- Thermal noise.
- Second order IM noise.
- Third order IM noise.

Let us consider them two at a time. If a system is thermal and second order IM noise limited, minimum noise is achieved allowing an equal contribution. For the 3 pWp/km case, we would assign 1.5 pW to each component.

For thermal and third order IM noise limited systems, twice the contribution is assigned to thermal noise as to third order IM noise. Again for the 3-pWp case, 2 pWp is assigned to thermal noise and 1 pWp to third order IM noise.

Expressed in decibels with P_t equal to total noise, the following table expresses these relationships in another manner:

System	Noise assigned to thermal	to IM
Thermal and second order limited	$P_t - 3$	$P_t - 3$
Thermal and third order limited	$P_t - 1.8$	$P_t - 4.8$
Overload limited	P_t	—

The parameter L is established such that these apportionments can be met by adjusting repeater spacing and repeater design. As an example in practice Figure 9.7 shows noise allocation of the North American L4 system (Ref. 5).

9.6   EQUALIZATION

9.6.1   Introduction

Consider the result of transmitting a signal down a 12-MHz coaxial cable system with the amplitude-frequency response shown in Figure 9.3. The noise per voice channel would vary from an extremely low level for the channels assigned to the very lowest frequency segments of the line frequency (baseband) to extremely high levels for those channels that were assigned to the spectrum near 12 MHz. For long systems there would be

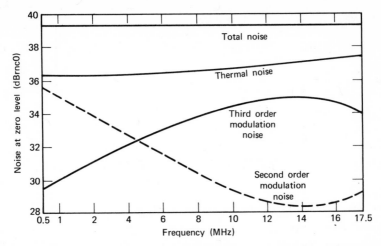

Figure 9.7 Allocation of noise in a practical system. Accumulated noise over 4000 mi of the North American L4 system. Copyright © 1969, by American Telephone and Telegraph Co.; from *Bell Syst. Tech. J.*, 830 (April, 1969).

every reason to believe that these higher frequencies would be unusable if nothing was done to correct the cable to make the amplitude response more uniform as a function of frequency. Ideally we would wish it to be linear.

Equalization of a cable deals with the means used to assure that the signal-to-noise ratio in each FDM telephone channel is essentially the same no matter what its assignment in the spectrum (see Chapter 3). In the following discussion we consider both fixed and adjustable equalizers.

9.6.2 Fixed Equalizers

The coaxial cable transmission system design engineer has three types of cable equalization available which fall into the category of fixed equalizers. These are as follows:

- Basic equalizers
- Line build-out (LBO) networks
- Design deviation equalizers.

The basic equalizer is incorporated in every cable repeater. It is designed to compensate for the variation of loss frequency characteristic of uniform

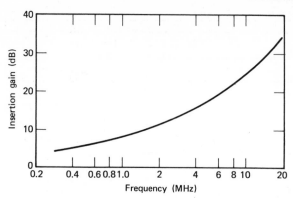

Figure 9.8 Repeater gain characteristic for North American L4 system.

cable sections. This is done by simply making the fixed gain proportional to the square root of frequency, matching loss for the nominal length. For the North American L4 system the gain characteristic is shown in Figure 9.8. For the case of 12-MHz cable, the section nominal length would be 4.5 km (CCITT Rec. G.332).

The key word in the preceding paragraph is *uniform*. Unfortunately some cable sections are not uniform in length. It is not economically feasible to build tailor-made repeaters for each nonuniform section. This is what line build-out networks are used for. Such devices are another class of fixed equalizer for specific variations of nominal repeater spacing. One way of handling such variations is to have available LBO equalizers for 5%, 10%, 15%, etc., of the nominal distance.

The third type of equalizer compensates for design deviation of the nominal characteristics standard for dependent repeaters and actual loss characteristic of the cable system in which the repeaters are to be installed. Such variation is systematic such that the third level of equalization, the design deviation equalizer, is installed one for each 10, 15, or 20 repeaters to compensate for gross design deviation over that group of repeaters.

9.6.3 Variable Equalizers

Figure 9.9a shows the change of loss of cable as a function of temperature variation and 9.9b shows approximate earth temperature variations with time. Adjustable equalizers are basically concerned with gain frequency variations with time. Besides temperature, variations due to aging of components may also be a problem; however, this is much less true with

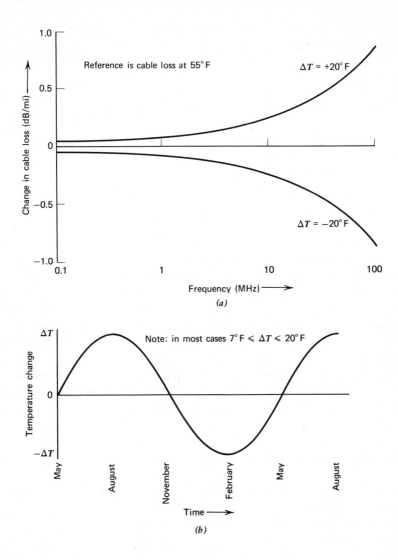

Figure 9.9 (*a*) Change in loss of 1 mi of $\frac{3}{8}$ -in. coaxial cable for $\pm 20°F$ change in temperature. (*b*) Approximate earth temperature variation with time at 4-ft depth. Copyright © 1970, by Bell Telephone Laboratories.

transistorized equipment. Cable loss per kilometer shown in Figure 9.3 is loss at a mean temperature. The $\frac{3}{8}$-in. cable used in "L" system applications has a variation of $\pm 20°\text{F/mi}$ at 20 MHz or about $\pm 0.6\%$ dB/mile at 60 MHz (Ref. 2). This loss can be estimated as $0.11\%/°\text{F}$.

The primary purpose of automatic regulation is to compensate for the gain variation due to temperature changes. Such automatic regulation usually is controlled by a pilot tone at the highest cable frequency. For instance, in the ITT cable design for 12 MHz (Ref. 1), "The pilot controlled system will always apply exact compensation. . . . at the pilot frequency of 12,435 kHz, an error may occur at other frequencies. On a single amplifier this error is very small but will add systematically along the route." Such error is usually corrected by manually adjustable equalizers.

In the North American L4 system (Ref. 5) the regulation is controlled by a 11,648 kHz line pilot. The gain frequency characteristic is varied to compensate for temperature associated changes in loss of a regulated section. Another regulator is controlled by a thermistor buried in the ground near the repeater to monitor ground temperature. This latter regulator provides about half the temperature compensation necessary.

The L4 system also uses an additional repeater called an equalizing repeater spaced up to 54 mi apart. It includes six networks for adjusting the gain frequency characteristics to mop up collective random deviations in the 54-mi section. The equalization of the repeater is done remotely from manned stations while the system is operational.

9.7 LEVEL AND PILOT TONES

Intrasystem levels are fixed by cable and repeater system design. These are the L and M_g established in Section 9.5.

Modern 12-MHz systems display an overload point of $+ 24$ dBm or more.* Remember that the

> Overload point = equivalent peak power level
> $\qquad\qquad\qquad$ + relative sending level + margin

The margin is M_g as in Section 9.5, and L may be related to relative sending level.

M_g can be reduced, depending on how well system regulation is maintained. System pilots, among other functions (covered in Section 9.8),

---

* CCITT Rec. G.223 calls for at least a $+20$-dBm overload point.

provide a means for automatic gain control of some or all cable repeaters so as to compensate (partially or entirely; see Section 9.6) for transmission loss deviations due to temperature effects on the system and aging of active components (e.g., repeaters).

Typically pilot levels are -10 dBm0. The level is a compromise, bearing in mind system loading, to minimize the pilots' contribution to intermodulation noise in a system that is multichannel in the frequency domain (FDM). Another factor tending to force the system designer to increase pilot level is the signal-to-noise ratio of the pilot tone required to effectively actuate level regulating circuitry. Pilot level adjustment as the injection point usually requires a settability better than 0.1 dB. Internal pilot stability should display a stability improved over desired cable system level stability. If system level stability is to be ± 1 dB, then internal pilot stability should be better than ± 0.1 dB.

The number of system pilots assigned and their frequencies depend on bandwidth and the specific system design. Commonly 12,435 kHz is used for regulation and 13.5 MHz for supervisory. In the same system an auxiliary pilot is offered at 308 kHz and, as an option, a frequency comparison pilot at 300 kHz.

The only continuous in-band pilot in the L4 system is located at 11,648 kHz. Supervisory pilot tones are transmitted in the band 18.50–18.56 MHz. An L multiplex synchronizing pilot is located at 512 kHz.

9.8 SUPERVISORY

The term supervisory in coaxial cable system terminology refers to a method of remotely monitoring repeater condition at some manned location. As mentioned above, the L4 system uses 16 pilot tones brought up on command giving status of 16 separate, buried repeaters.

The ITT method uses a common oscillator frequency (13.5 MHz) and relies on time separation to establish identity of each repeater being monitored. An interrogation signal is injected at the terminal repeater or other manned station. At the first dependent repeater the signal is filtered off and, after a delay, regenerated and passed on to the next repeater. Simultaneously on receipt of the regenerated pulse, a switch is closed connecting a local oscillator signal to the output of the repeater for a short time interval. This local oscillator pulse is transmitted back to the terminal or other manned station. The delay added at each repeater is added to the natural delay of the intervening cable. This added delay allows for a longer return pulse from each repeater, thereby simplifying circuitry. This same interrogating pulse, delayed, regenerated, and then passed on to the next repeater, carries on down the line of dependent repeaters causing returning "tone bursts" originating from successive amplifiers along the cable route.

The tone burst response pulses are rectified and fed to a counter at the manned station. The resulting count is compared with the expected count and an alarm is indicated if there is a discrepancy. The faulty amplifier is identified automatically. Table 9.3 gives basic operating parameters of the system. Such a system can be used for coaxial cable system segments up to 280 km in length.

Table 9.3   Supervisory System Parameters (ITT)

Response oscillator frequency	13.5 MHz
Response oscillator level	-20 dBm0
Response pulse duration	10 μs
Response pulse delay	250 μs
Interrogation pulse repetition rate	6/s
Interrogation pulse amplitude	$\sim$0.5 V
Interrogation pulse rise time	$\sim$10 μs

9.9   POWERING THE SYSTEM

Power feeding of buried repeaters in the ITT system permits the operation of 15 dependent repeaters from each end of a feed point (12-MHz cable). Thus up to 30 dependent repeaters can be supplied power between power feed points. A power feed unit at the power feed point (see Figure 9.2) provides up to 650 V dc voltage between center conductor and ground using 110-mA stabilized direct current. Power feed points may be as far apart as 140 km (87 mi) on large diameter cable.

9.10   60-MHz COAXIAL CABLE SYSTEMS

Wide band coaxial cable systems are presently being implemented due to the ever-increasing demand for long-haul, toll quality telephone channels. Such systems are designed to carry 10,800 FDM nominal 4-kHz channels. The line frequency configuration for such a system, as recommended in CCITT Rec. G.333, is shown in Figure 9.10. To meet long-haul

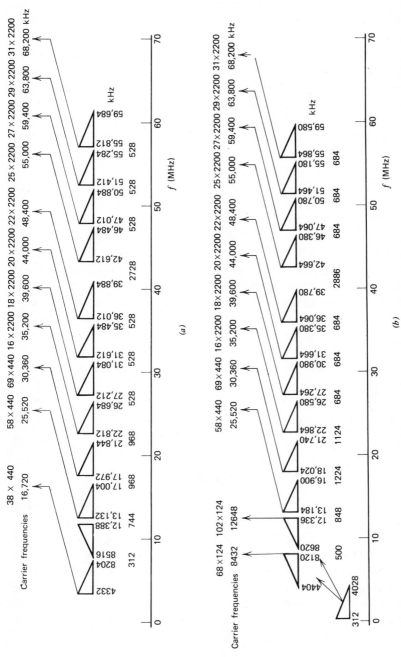

Figure 9.10 Line frequency allocation recommended for 40-MHz and 60-MHz systems on 2.6/9.5-mm coaxial cable pairs using (a) Plan 1 and (b) Plan 2 (from CCITT Rec. G.333).

421

noise objectives the large diameter cable is recommended (e.g., 2.6/9.5 mm).

When expanding a coaxial cable system, a desirable objective is to use the same repeater locations as with the old cable and add additional repeaters at intervening locations. For instance if we have 4.5-km spacing for a 12-MHz system and our design shows that we need three times the number of repeaters for an equal length 60-MHz system, then repeater spacing should be at 1.5-km intervals.

The ITT 12-MHz system uses 4.65-km spacing. Thus its 60-MHz system will use 1.55-km (0.95-mi) spacing with a mean cable temperature of 10°C. The attenuation characteristic of the large gauge cable is shown in Figure 9.11. This is an extension of Figure 9.3.

Repeater gain for the ITT system is nominally 28.5 dB at 60 MHz and can be varied ± 1.5 dB. Line build-out networks allow still greater tolerance. The overload point, following CCITT Rec. 223, is taken at + 20 dBm with a transmit level of − 18 dBm.

System pilot frequency is 61.160 MHz for regulation. A second pilot frequency of 4.287 MHz corrects the level of the lower frequency range. Pilot regulation repeaters are installed at from 7 to 10 nonregulated repeaters, with deviation equalization at every twenty-fourth repeater. All repeaters have temperature control (controlled by the buried ambient).

Power feeding is planned at every 100 km (63 mi). Thus 64 repeaters will be fed remotely using constant direct current feed over the conductors. Each repeater will tap off about 15 V requiring 2 w. Thirty-two repeaters at 2×15 V each will require 960 V. An additional 120 V dc is required for pilot regulated repeaters plus one repeater with deviation equalization. Added to this is the 50-V IR drop on the cable. The total feed voltage adds to 1226 V dc. Fault location is similar to that for the 12-MHz ITT system. (For the ITT 60-MHz system, see Ref. 14.)

9.11 THE L5 COAXIAL CABLE TRANSMISSION SYSTEM

A good example of a 60-MHz coaxial cable transmission system that is presently operational, carrying traffic, is the L5 system operating on a transcontinental route in North America (see Table 9.1). In its present lay-up, it consists of 22 tubes, of which 20 are on-line and 2 are spare. Each tube has the capacity to transmit 10,800 VF channels in one direction. For full-duplex operation, two tubes are required for 10,800 VF channels, or the total system capacity is $[(22 − 2)/2] \times 10,800$ or 108,000 VF channels.

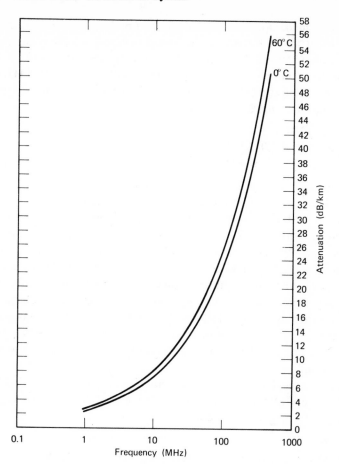

Figure 9.11   Attenuation of large diameter coaxial cable (0.375 in.).

The system is designed for a 40-dBrnC0 (8000-pWp) noise objective in the worst VF channel at the end of a 4000-mi (6400-km) system. This is a noise accumulation of 8000/6400 or 1.25 pWp/km. The system design is such that it is second order modulation and thermal noise limited. Repeater overload is at the 24-dBm point.

Compared to the L4 system, L5 provides three times the voice channel capacity at three times the L4 spectrum, twice the repeaters with a small deterioration in noise accumulation.

The modulation plan is an extension of that shown in Chapter 3, Figure 3.21. The basis of the plan is the development of the "jumbo group" (JG) made up of six master groups (Bell System FDM hierarchy). Keep in mind that the basic mastergroup consists of 600 voice channels (in this case) or 10 standard supergroups and occupies the band 564–3084 kHz. The basic jumbo group occupies the band 564–17,548 kHz with a level control pilot at 5888 kHz. The three jumbo groups are assigned the following line frequencies:

JG 1   3,124–20,108 kHz
JG 2   22,068–39,056 kHz
JG 3   43,572–60,556 kHz

Equalizing pilots are at 2976, 20,992, and 66,048 kHz as transmitted to the line. There is a temperature pilot at 42,880 kHz.

The basic jumbo group frequency generator is built around an oscillator which has an output of 5.12 MHz. This oscillator has a drift rate of less than 1 part in 10^{10}/day after aging and a short-term stability of better than 1 part in 10^8/ms. Excessive frequency offset is indicated by an alarm.

Automatic protection of the 10 operating systems is afforded by the LPSS (line protection switching system) on a 1:10 basis. A maximum length of switching span is 150 mi. Power feeds are at 150-mi intervals, feeding power in both directions. Thus a power span is 75 mi long, or 75 repeaters. Power is 910 mA on each cable, + and − 1150 V operating against ground.

The basic repeater is a fixed gain amplifier, spaced at 1-mi intervals. Typically, every fifth repeater is a regulating repeater and this regulation is primarily for temperature compensation.

9.12   COAXIAL CABLE OR RADIOLINK—THE DECISION

9.12.1   General

One major decision that the transmission system engineer often faces is whether to install on a particular point-to-point circuit a radiolink or coaxial cable.

What are the factors that will determine the choice? Most obviously they fall into two categories, technical and economic. Table 9.4 compares the two media from a technical viewpoint. These comparisons can serve as a fundamental guide for making a technical recommendation in the selection of facility. System mixes may also be of interest. (Refer to Chapter 5 for a discussion of radiolink engineering.)

The discussion that follows is an expansion of some of the points covered in Section 9.1. Table 9.4 summarizes the factors listed below.

Table 9.4   Comparison of Coaxial Cable Versus Radio Link

Item	Cable	Radio
Land acquisition	Requires land easements or right-of-way along entire route and recurring maintenance access later	Repeater site acquisition every 30–50 km with building, tower, access road at each site
Insert and drop	Insert and drop at any repeater. Should be kept to minimum. Land buys, building required at each insert location	Insert and drop at more widely spaced repeater sites
Fading	None aside from temperature variations	Important engineering parameter
Noise accumulation	Less, 1 pWp/km	More, 3 pWp/km
Radio frequency interference (RFI)	None	A major consideration
Limitation on number of carriers or basebands transmitted	None	Strict, band-limited plus RFI ambient limitations
Repeater spacing	1.5, 4.5, 9 km	30–50 km
Comparative cost of repeaters	Considerably lower	Considerably higher
Power considerations	High voltage dc in milliampere range	48 V dc static no-break at each site in ampere range
Cost versus traffic load	Full load proves more economical than radiolink	Less load proves more economical than cable
Multiplex	FDM-CCITT	FDM-CCITT
Maintenance and engineering	Lower level, lower cost	Higher level, higher cost
Terrain	Important consideration in cable laying	Can jump over, even take advantage of difficult terrain

9.12.2   Land Acquisition as a Limitation to Coaxial Cable Systems

Acquisition of land detracts more from the attractiveness of the use of coaxial cable than any other consideration; it adds equally to the attractiveness of selecting radiolinks (LOS microwave). With a radiolink system large land areas are jumped and the system engineer is not concerned with what goes on between. One danger that many engineers tend to overlook is that of the chance building of a structure in the path of the radio beam after installation on the routes has been completed.

Cable, on the other hand, must physically traverse the land area that intervenes. Access is necessary after the cable is laid, particularly at repeater locations. This may not be as difficult as it first appears. One method is to follow parallel to public highways, keeping the cable lay on public land. Otherwise, with a good public relations campaign, easement or rights-of-way often are not hard to get.

This leads to another point. The radiolink relay sites are fenced. Cable lays are marked, but the chances of damage by the farmer's plow or construction activity are fairly high.

9.12.3 Fading

Radiolinks are susceptible to fading. Fades of 40 dB on long hops are not unknown. Overbuilding a radiolink system tends to keep the effect of fades on system noise within specified limits.

On coaxial cable systems signal level variation is mainly a function of temperature variation. Level variations are well controlled by regulators controlled by pilot tones and, in some cases, auxiliary regulators controlled by ground ambient.

9.12.4 Noise Accumulation

Noise accumulation has been discussed in Section 9.1. Either system will serve for long-haul backbone routes and meet the minimum specific noise criteria established by CCITT/CCIR. However, in practice, the enginering and installation of a radio system may require more thought and care to meet those noise requirements. Modern coaxial cable systems have a design target of 1 pWp/km of noise accumulation. With care, using IF repeaters, radio systems can meet the 3-pWp criterion. Besides fading, radiolinks, by definition, have more modulation steps and thus are noisier.

9.12.5 Group Delay—Attenuation Distortion

Group delay is less of a problem with radiolinks. Figure 9.3 shows amplitude response of a cable section before amplitude equalization. The cable plus amplifiers plus amplitude equalizers add to the group delay problem.

It should be noted that for video transmission on cable an additional modulation step is required to translate the video to the higher frequencies and invert the band (see Section 12.8.2). While on radiolinks, video can be transmitted directly without additional translation or inversion besides RF modulation.

9.12.6 Radio Frequency Interference (RFI)

There is no question as to the attractiveness of broadband buried coaxial cable systems over equivalent radiolink systems when the area to be traversed by the transmission medium is one of dense RFI. Usually these areas are built-up metropolitan areas with high industrial/commercial activity. Unfortunately, as a transmission route enters a dense RFI area, land values increase disproportionately, as do construction costs for cable laying. Yet the trade-off is there.

9.12.7 Maximum VF Channel Capacity

In heavily populated areas of highly developed nations frequency assignments are becoming severely limited or unavailable. Although some of the burden on assignment will be removed as the tendency toward usage of the millimeter region of the spectrum is increased, coaxial cable remains the most attractive of the two for high density FDM configurations.

If it is assumed that there are no RFI or frequency assignment problems, a radiolink can accommodate up to eight carriers in each direction (CCIR Rec. 384-1) with 2700 VF channels per carrier. Thus the maximum capacity of such a system is $8 \times 2700 = 21,600$ VF channels.

Assume a 12-MHz coaxial cable system with 22 tubes, 20 operative or 10 "go" and 10 "return." Each coaxial tube has a capacity of 2700 channels. Thus the maximum capacity is $10 \times 2700 = 27,000$ VF channels.

It should be noted that the radio system with a full 2700 VF channels may suffer from some multipath problems. Coaxial cable systems have no similar interference problems. However, cable impedance must be controlled carefully when splicing cable sections. Such splices usually are carried out at repeater locations.

Consider now 60-MHz cable systems with 20 active tubes, 10 "go" and 10 "return." Assume 10,000 channels capacity per tube; thus $10,000 \times 10 = 100,000$ VF channels or equivalent to five full radiolink systems.

9.12.8   Repeater Spacing

As discussed in Chapter 5, a high average for radiolink repeater spacing is 50 km (30 mi), depending on drop and insert requirements as well as an economic trade-off between tower height and hop distance. For coaxial cable systems, repeater separation depends on the highest frequency to be transmitted, ranging from 9.0 km for 4-MHz systems to 1.5 km for 60-MHz systems. A radiolink repeater is much more complex than a cable repeater.

Coaxial cable repeaters are cheaper than radiolink repeaters, considering tower, land, and access roads for radiolinks. However, much of this advantage for coaxial cables is offset because radiolinks require many fewer repeaters. It also should be kept in mind that a radiolink system is more adaptable to difficult terrain.

9.12.9   Power Considerations

The 12-MHz ITT coaxial cable system can have power feed points separated by as much as 140 km (87 mi) using 650 V dc at 150 mA. In a 140-km section of a radiolink route at least four power feed points would be required, one at each repeater site. About 2 A is required for each transmitter receiver combination using standard 48 V dc battery, usually with static no-break power. Power also will be required for tower lights and perhaps for climatizing equipment enclosures.

9.12.10   Engineering and Maintenance

Cable systems are of "cook book" design. Radiolink systems require a greater engineering effort prior to and during installation. Likewise, the level of maintenance of radio systems is higher than cable.

9.12.11   Multiplex Modulation Plans

Interworking, tandem working of radio and coaxial cable systems is made easier because both broadband media use the same standard CCITT or L system modulation plans; see Chapter 3.

REFERENCES AND BIBLIOGRAPHY

1. P. J. Howard, M. F. Alarcon, and S. Tronsli, "12-Megahertz Line Equipment," *Electrical Communication* (ITT), **48** (Nos. 1 and 2 in one issue), 1973.

2. *Transmission Systems for Communications*, 4th ed., Bell Telephone Laboratories, American Telephone and Telegraph Co., New York, 1971.

3. William A. Rheinfelder, *CATV System Engineering*, TAB Books, Blue Ridge Summit, Pa., 1970.

4. P. Norman and P. J. Howard, "Coaxial Cable System for 2700 Circuits," *Electrical Communication* (ITT), **42**(4), 1967.

5. "The L-4 Coaxial System," *Bell Syst. Tech. J.*, **48**(4), entire issue (April 1969).

6. CCITT White Books, Mar del Plata, 1968, Vol. III (one star), G. Recommendations, particularly the G.200 and G.300 series (see Appendix A).

7. J. A. Lawlor, *Coaxial Cable Communication Systems, Management Overview*, ITT, New York, Feb. 1972 (A Technical Memorandum).

8. *Lenkurt Demodulator*, Lenkurt Electric Co., San Carlos, Calif.: June 1967, May and June 1970, and May 1971.

9. *Data Handbook for Radio Engineers*, 5th ed., Howard W. Sams & Co., Indianapolis, 1968.

10. F. J. Herr, "The L5 Coaxial System Transmission System Analysis," *IEEE Trans. Commun.*, Feb. 1974.

11. F. C. Kelcourse and T. A. Tarbox, "Design of Repeatered Lines for Long-Haul Co-axial Systems," *IEEE Trans. Commun.*, Feb. 1974.

12. E. H. Angell and M. M. Luniewicz, "Low Noise Ultralinear Line Repeaters for the L5 Coaxial System," *IEEE Trans. Commun.*, Feb. 1974.

13. Y.-S. Cho et al., "Static and Dynamic Equalization of the L5 Repeatered Line," *IEEE Trans. Commun.*, Feb. 1974.

14. L. Becker, "60-Megahertz Line Equipment," *Electrical Communication* (ITT), **48** (Nos. 1 and 2 in one issue), 1973.

10 | MILLIMETER WAVE TRANSMISSION

10.1 GENERAL

Denominating frequency bands for radio communication is arbitrary. The microwave line-of-sight or centimeter band discussed in Chapter 5 covered the spectrum from 150 MHz (200 cm) to 13 GHz (2.3 cm). For the sake of discussion in this chapter, millimeter waves will encompass that region of the electromagnetic spectrum from 13 GHz (23 mm) to over 100 GHz (3 mm).*

One major concern of the radio transmission engineer is propagation, and this is the primary topic throughout this chapter. Millimeter wave transmission through the atmosphere is more adversely affected by certain propagation properties than its centimeter counterpart. These properties are the absorption and scattering of a wave as it is transmitted through the atmosphere. It is the result of this phenomenon that is the principal reason why the transmission engineer is reluctant to use these frequencies for point-to-point communication. Much emphasis will be given to rainfall attenuation; it should be noted that there was some discussion of fading due to rainfall in Chapter 5.

With all its apparent shortcomings, millimeter wave transmission has been given a second look in the last 5 years for various reasons. Probably the primary reason is the increasing congestion in the centimetric wave bands in certain areas of the world. Another consideration is the need for much greater bandwidths (Chapter 11) to accommodate digital transmission. The third factor is that of equipment development; again much of this development is a result of military research and technology. Development of millimeter wave transmission has reached a point equivalent to

---

* The ITU designates 3–30 GHz as centimetric waves and 30–300 GHz as millimetric waves (Ref. 1).

430

about where centimetric wave development was in the late 1940s, when
that region of the spectrum was opened for use in multichannel, point-
to-point transmission.

10.2 PROPAGATION

Consider that we are dealing with propagation of millimetric waves through
the atmosphere. Therefore, in addition to free-space attenuation, several
other factors affecting path loss may have to be considered. As expressed
in Ref. 2, these are as follows:

" *a*) The gaseous contribution of the homogeneous atmosphere due to resonant
and non-resonant polarization mechanisms,
   b) The contribution of inhomogeneities in the atomosphere,
   c) The particulate contributions due to rain, fog, mist and haze (dust,
smoke and salt particles in the air)."

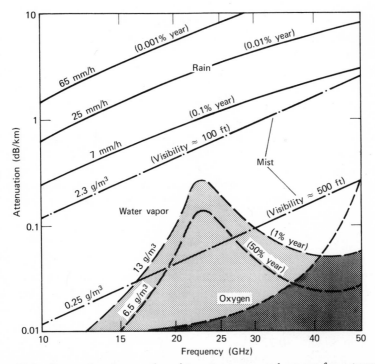

Figure 10.1 Attenuation due to rain, mist, water vapor, and oxygen for a temperate
maritime climate at ground level (Ref. 2). Courtesy Institute of Telecommunication
Sciences, Office of Telecommunications, U.S. Department of Commerce.

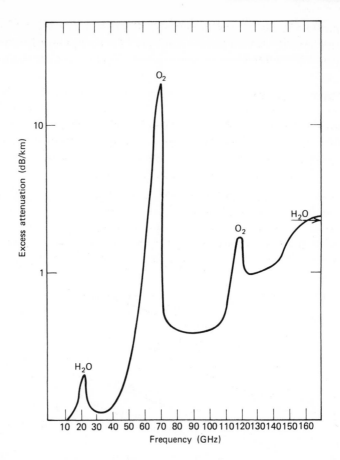

Figure 10.2 Areas of the millimetric spectrum of high attenuation due to oxygen and water vapor. Values are excess attenuation and are estimates (after Hogg, Ref. 3).

Figure 10.1 illustrates these properties diagrammatically. Under (a) above we are dealing with the propagation of a wave through the atmosphere under the influence of several molecular resonances, such as water vapor at 22 and 183 GHz, oxygen with lines around 60 GHz, and a single oxygen line at 119 GHz. These points with their relative attenuation are shown in Figure 10.2.

Other gases display resonant lines as well, such as N_2O, SO_2, O_3, NO_2, and NH_3, but because of their low density in the atmosphere, they have negligible effect on propagation.

The major offender is precipitation [under (b) and (c) above] attenuation. It exceeds that of all other sources of attenuation in the atmosphere above 18 GHz. Rainfall and its effect on propagation is covered in Section 10.3.

The immediate concern in this section is the total loss from absorption and scattering. It will be remembered that when an incident electromagnetic wave passes over an object which has dielectric properties different from the surrounding medium, some energy is absorbed and some scattered. That which is absorbed heats the absorbing material; that scattered is quasi-isotropic and relates to the wavelength of the incident wave. The smaller the scatterer, the more isotropic in direction with respect to the wavelength of the incident energy.

The ideal for the transmission system engineer would be to establish a formula valid anywhere in the world that would provide path loss in decibels. In free space such a formula is available (5.1). This classic formula is

$$\text{Attn}_{dB} = 32.5 + 20 \log F + 20 \log D$$

where D = hop or path length (km)
F = operating frequency (MHz)

From the strictly simplified engineering point of view it can be seen that millimetric wave propagation is ideal in free space for communication between satellites, from satellite to deep space probe, etc. Of course factors F and D must be considered, but no other variables enter the calculation.

For millimetric transmission through the atmosphere the problem takes on considerably greater complexity. To the free-space loss formula (with F in GHz) several terms are added:

$$\text{Attn}_{dB} = 92.45 + 20 \log F_{\text{GHz}} + 20 \log D_{\text{km}} + a + b + c + d + e$$

where a = loss (dB) due to water vapor
b = loss (dB) due to mist and fog
c = loss (dB) due to oxygen (O_2)
d = sum of the absorption losses due to other gases
e = losses (dB) due to rainfall

Notes. a varies with relative humidity, temperature, pressure, and altitude. The transmission engineer would like to think that water vapor content is linear with these parameters and that the atmosphere is homogeneous.

c and d are assumed to vary linearly with atmospheric density thus directly with atmospheric pressure, and are also a function of altitude (e.g., it is assumed that the atmosphere is homogeneous).

b and e vary with the density of rainfall cell or cloud, and the size of raindrops or water particles such as in fog or mist. In this case the atmosphere is most certainly not homogeneous. (Droplets less than 0.01 cm in diameter are considered mist/fog, more than 0.01 cm, rain.)

With reference to Figure 10.2, terms a, b, c, and d can be disregarded except in regions indicated in the figure. However, it is just these "forbidden" regions of the electromagnetic spectrum that, in themselves, may offer an advantage under special circumstances. Suppose you were designing a ship-to-ship secure communication system. Such attenuation, like that encountered in the region of 60 GHz, the oxygen line area, adds an excess attenuation, on the order of 15 dB/km to the free-space loss. That is 15 dB of additional security. Such additional attenuation would help ensure that overshoot of a concentrated beam would not be successfully intercepted by an enemy. This shows that good advantage can be taken of even the so-called "forbidden" regions in millimeter transmission, in this case, the factor c in the equation above.

Figure 10.2 gives total gaseous absorption in decibels per kilometer as a function of frequency. The comparatively high absorption bands of oxygen and water vapor are patently evident. As can be seen from the figure, certain frequency bands are relatively "open." These openings are often called windows. Three such windows are suggested for point-to-point service as follows:

Band (GHz)	Excess Attenuation by Atmospheric Absorption (dB/km)
28– 42	<0.13
75– 95	<0.4
125–140	<1.8

10.3 RAINFALL LOSS

Of all the factors a–e in equation 10.1, factor e is the principal one affecting path loss provided the so-called forbidden regions are avoided. However, even at 22 GHz (1.35 cm), the oxygen vapor line, excess loss accumulates at only 0.15 dB/km, and for a 10-km path only 1.5 dB must be added to the free-space loss to compensate for this loss. This is negligible when compared to free-space loss itself, for example, 128.4 dB for the first kilometer, accumulating thence 6 dB approximately each time path length is doubled (i.e., add 6 dB for 2 km, 12 dB for 4 km, etc). Thus a 10-km path would have a free-space loss of 148.4 dB plus 1.5 dB added for water vapor loss (22 GHz), or a total of 149.9 dB.

Rain is another matter. It has been common practice to express path loss due to rain as a function of precipitation rate. Such a rate depends on liquid water content and the fall velocity of the drops. The velocity, in turn, depends on raindrop size. Thus our interest in rainfall boils down to drop size and drop size distribution for specific rainfall rates. All this information is designed to lead the transmission engineer to fix an excess attenuation due to rainfall on a particular path as a function of time and time distribution. This is a familiar approach that was used in Chapter 5 for fading.

One source (Ref. 3) suggests limiting path length for millimetric transmission to 1 km. Further on we will consider this aspect for system design. However, such a measure may be overly severe and very costly. Another approach would be to get more rainfall information and its effect on higher frequency transmission systems as a function of wavelength. These should then be applied to good statistics available for the area in which a millimeter wave transmission system is to be installed.

Research carried out so far usually has dealt with rain on a basis of rainfall in millimeters per hour. Often this has been done with rain gauges using collected rain averaging over a day or even periods of days. For millimeter path design such statistics are not suitable for paths where we desire a propagation reliability well over 99.9% and do not wish to resort to overconservative design procedures (i.e., assign excessive fade margins).

As pointed out in Chapter 5 several weeks of light drizzle will affect the overall long-term propagation reliability much less than several good downpours that are short-lived (20 min duration, for instance). It is simply this downpour activity for which we need statistics. Such downpours are cellular in nature. How big are the cells? What is the rainfall rate in the cell? What are the size of drops and their distribution?

Hogg (Ref. 3) suggests the use of new, high speed rain gauges with outputs readily available for computer analysis. These gauges can provide minute-by-minute analysis of the rate of fall, something lacking with the older type of gauges. Of course it would be desirable to have several years of statistics for a specific path to provide the necessary information on fading which will govern system parameters such as repeater spacing, antenna size, diversity, and so forth.

Some such information of the type we are looking for is now available and is indicative of a great variation of short-term rainfall rate from one geographical location to another. For instance in one period of measurement it was found that Miami, Florida has maximum rain rates about 20 times greater than those of heavy showers occurring in Oregon, the region of heaviest rainfall in the United States. In Miami, a point rainfall rate may even exceed 700 mm/h. The effect of 700 mm/h on two millimetric frequencies can be extrapolated from Figure 10.3. In this figure the rainfall rate in millimeters per hour extends to 100, which at 100 mm/h provides an excess attenuation of from 25–30 dB/km.

When identical systems were compared (Ref. 3) at 30 GHz with repeater spacing of 1 km with equal desired signal levels (e.g., a 30-dB signal-to-noise ratio), 140 min of total time below the desired level was obtained at Miami, Fla., 13 min at Coweeta, N.C., 4 min at Island Beach, N.J., 0.5 min at Bedford, England, and less than 0.5 min at Corvallis, Oregon, As pointed out, such outages may be improved by increasing transmitter output power, improving receiver noise figure, increasing antenna size, use of diversity, etc.

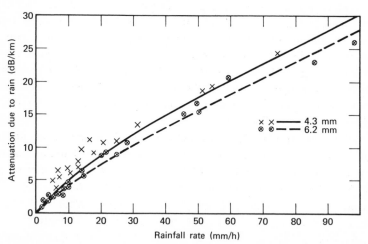

Figure 10.3 Measurements made by Bell Telephone Laboratories of attenuation due to rainfall at wavelengths of 6.2 and 4.3 mm compared with corresponding calculated values (solid and dashed lines (Ref. 3). Copyright © 1968, by American Telephone and Telegraph Co.

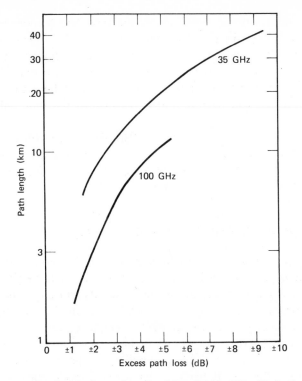

Figure 10.4 Scintillation fading values for 35 and 100 GHz (Ref. 6).

One valid method suggested to lengthen repeater sections (space between repeaters) is by the use of path diversity. This is the most effective form of diversity for downpour rainfall fading. Path diversity is the simultaneous transmission of the same information on paths separated by at least 2 km, the idea being that rain cells affecting one path will have a low probability of affecting the other at the same time. A switch would select the better path of the two. Careful phase equalization between the two paths would be required, particularly for the transmission of high bit-rate data information.

10.4 SCINTILLATION FADING

Scintillation fading is short-term signal fluctuation. Short-term in this context means in the period of 1 min. We might observe from 20 to 50 scintillations or fluctuations in such a time frame. They are brought about by localized departures of temperature, pressure, and homogeneity of the atmosphere causing variations in the refractive index. These inhomogeneities degrade the transmission of broadband information by limiting the useful gain at the receiving station because the coherent phase front will disperse, and short-term refractivity fluctuations limit the useful signal bandwidth that can be transmitted because of the time delays introduced.

Scintillation fading is expressed in decibels about the median signal level. It is a function of frequency and path length. Figure 10.4 gives values of scintillation fading for 35 and 100 GHz. Admittedly for most applications these values are on the conservative side.

10.5 PRACTICAL MILLIMETER WAVE SYSTEMS

For point-to-point multichannel communications, the same radiolink principles will be used as set forth in Chapter 5, namely, heterodyne receivers, some form of angle modulation, low level outputs and parabolic antennas. Wider bandwidths will be used, the 70-MHz standard IF being replaced by a higher frequency. Both 300 and 700 MHz have been used.

The millimeter wave region shows great promise for the use of digital transmission techniques utilizing a number of the advantages that the region provides. Besides its being a region that is wide open for development and that provides greater bandwidths, the greater path attenuation and even the absorption attenuation can be put to advantage. This greater loss can reduce mutual interference problems of overshoot, side lobe radiation, and a general reduction of the effectiveness of RFI. This is another reason why millimeter wave links can be installed today in large metropolitan or built-up industrial areas where there is heavily congested conventional centimeter wave transmission systems. Once away from the built-up areas, the system can interface at IF with conventional centimeter systems, in areas where RFI problems are fewer in the more popular frequency bands.

Millimeter wave antennas will get more gain for less cross-sectional area, beamwidths will be sharper, and with the improved operation of millimeter transmission systems regarding RFI in general, more of these systems will be able to operate "in harmony" in a given geographical area than their centimeter counterparts. Some CATV radiolinks are now operating at 18 GHz.

The obvious problem in long-haul, point-to-point radio systems operating in the conventional analog mode is the accumulation of noise. As proposed in Section 10.3, millimeter wave hops will be much shorter; thus many more repeaters will be required for a given route than for comparable centimeter radiolinks. To avoid excessive noise build-up, one answer is to use digital transmission techniques. PCM is proposed using M-ary (multilevel) modulation such as differential QPSK (quaternary phase shift keying) or eight-level PSK. The regeneration implicit at each repeater obviates much of the noise problem. This transmission mode will be discussed more below.

Bandwidths of 1 and 2 GHz with reasonably flat frequency response and group delay are feasible today as well as low-noise receivers and acceptable transmitter power output all in a solid-state package. For instance millimeter wave power generation has now become feasible using Gunn effect and LSA mode in bulk gallium arsenide (Ref. 4). With IMPATT oscillators RF outputs have been achieved at 50 GHz of 1 W and 120 mW at 140 GHz with a single IMPATT diode. Receiver noise figures of 10 dB are common. A parametric amplifier operating in the 55–65 GHz region gives a 6-dB noise figure across a 670-MHz bandwidth with 14-dB gain (Ref. 4). Circular waveguide using the dominant TE_{11} mode eases the transmission line problem and leads to some modular design concepts.

As mentioned above parabolic antennas will be used as they were in the centimetric wave region. As reviewed in Chapter 5, antenna gain is a function of diameter as well as wavelength. Table 10.1 gives some typical gains that may be expected with conventional feeds (55% efficiency). Dual polarization with polarization discrimination better than 30 dB has also been achieved.

Let us consider a 10-km (6 statute mi) path. Free-space path loss may be taken from Table 10.1. We assume an operating frequency of 40 GHz; given this frequency and path length, the free-space path loss is 144.5 dB. To this figure we must add 10×0.13 dB for gas absorption (see Figure 10.2 and Section 10.2). Assume 100-MHz bandwidth (B_{if}), and we will leave aside for the present the type of modulation used. Allow a 10-dB noise figure for the receiver. With this information, compute the receiver noise threshold.

$$\text{Noise threshold}_{dBW} = -204 + 10 + 80$$
$$= -114 \text{ dBW}$$

$$\text{Path loss} = \text{free-space loss} + \text{excess loss (absoprtion)}$$
$$= 144.5 + 1.3 \text{ dB}$$
$$= 145.8 \text{ dB}$$

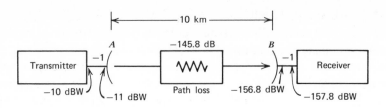

Figure 10.5 Millimeter wave link model.

Assign 100 mw as the output from the transmitter or -10 dBw. Allow 2-dB total line losses for both ends of the link. We can picture link calculations better from the model in Figure 10.5. The imaginary isotropic antennas allow us no gain and we find that we are short 43 dB to reach receiver threshold. Thus at points A and B, parabolic antennas may be installed with 43/2 dB or 21.5 dB each (see Table 10.1).

The transmission design engineer now must assign a margin to the noise threshold. The margin will depend on the type of modulation used. With FM, for example, the advantage gained with the trade-off of thermal noise versus bandwidth is achieved only when the carrier-to-noise (C/N) ratio is 10 dB. Thus, for FM at least, 10 dB must be included to the gain in the model. It is to this figure that we add a margin for propagation conditions or, more properly, for rainfall and scintillation.

Rainfall absorption is the most difficult figure to pin down. Statistics on hourly rainfall are required, given in the form of millimeters of rainfall per hour exceeded by a time percentage. As an example, 20 mm/h of rain is exceeded 0.01% of the time on an annual basis. It would be desirable to have these figures on a 1%, 0.1% and 0.01% basis such that we could select path reliabilities regarding rainfall for 99%, 99.9%, and 99.99% for the year.

Once we have these statistics for rainfall, they can be applied to Figure 10.6. Let us apply 20 mm/h given above to the model. This shows that for 40 GHz 5.5 dB excess attenuation would result 0.01% of the time per kilometer, and 10×5.5 or 55-dB additional margin must be provided for the model link with a 99.99% propagation (rainfall) reliability. From Table 10.1 at 40 GHz using 2-m antennas, the antenna gain would be 2×55.7 dB or 111.4 dB. Remember that we needed 43 dB to reach noise threshold, 10 dB to reach FM improvement threshold, and 55 dB for rainfall margin or 108 dB, leaving 3.4 dB for scintillation fading. This

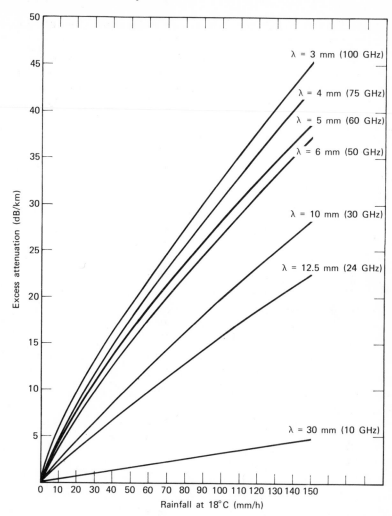

Figure 10.6 Excess path attenuation due to rainfall (after Burrows and Atwood, Ref. 5).

is more than sufficient as shown in Figure 10.4, which is on the conservative side.

Now consider that the rainfall rate was not 20 mm/h, but 25 for 0.01% of the time on an annual basis. From Figure 10.6 we find that we need a 70-dB margin for rainfall fading plus 3 dB for scintillation fading, or 73 dB. This totals to 126 dB (43 + 73 + 10), and it is not feasible to make this

Table 10.1   Parabolic Reflector Gain Versus 10-km Path Loss (Eff. 54%)

Frequency (GHz)	Diameter (m)	Gain (dB)	10-km Free-Space Loss (dB)
20	0.5	37.7	138.5
	1.0	43.7	
	1.5	47.2	
	2.0	49.7	
30	0.5	41	141.9
	1.0	47	
	1.5	50.6	
	2.0	53	
40	0.5	43.7	144.5
	1.0	49.7	
	1.5	53.2	
	2.0	55.7	
50	0.5	45.7	146.5
	1.0	51.7	
	1.5	55.2	
	2.0	57.7	
60	0.5	47.2	148.1
	1.0	53.2	
	1.5	56.7	
	2.0	59.2	
70	0.5	48.6	149.3
	1.0	54.6	
	1.5	58.1	
	2.0	60.6	
80	0.5	49.7	150.5
	1.0	55.7	
	1.5	59.2	
	2.0	61.7	
90	0.5	50.8	151.5
	1.0	56.8	
	1.5	60.3	
	2.0	62.8	
100	0.5	51.7	152.5
	1.0	57.7	
	1.5	61.2	
	2.0	63.7	

up merely by using larger antennas (Table 10.1). However, several alternatives are open to us:

- Reduce propagation reliability.
- Increase power output of the transmitter. (*Note*. If we were to increase it to 1 W, we would only produce 10-dB additional gain.)
- Reduce bandwidth (*Note*. If B_{if} were reduced to 10 MHz, we would only achieve an equivalent gain of 10 dB).
- Use path diversity.
- Possibly use another form of modulation, particularly if we are transmitting digital information, for example, QPSK.
- reduce link length (hop distance). For instance, reducing link length to 8 km would satisfy requirements.

10.6 THE SHORT HOP CONCEPT

As has been suggested earlier in this chapter, one way to ameliorate the excess attenuation problem due to rainfall is to significantly reduce hop length, for instance, to 1 km. An average repeater spacing in the centimetric region discussed in Chapter 5 was 50 km (30 mi), and 50 such repeaters might be an average number required on a 2500-km path. Over the same distance with the short hop concept 2500 repeaters would be required. For analog transmission this number of repeaters in tandem would make accumulated noise intolerable. However, using digital modulation techniques with regeneration at each repeater, the signal delivered at the output of the 2500-km system would be as good as the signal injected at the input.*

This concept fits well with the proposed all-digital network. Obviously, the digital format would be PCM. A 100-Mb/s system which is well within today's technology,* would serve as a vehicle for the transmission of about 1400 nominal 4-kHz analog VF channels.

For centimetric systems 100-m high towers are fairly common and a major cost item. For millimetric wave systems operating with this short hop concept (e.g., 1 km) towers will be proportionately lower (20–30 m), less expensive, and less unsightly. These stubby towers or poles would indeed require very good twist and sway specifications. The entire radio installation would be colocated with the antennas reducing transmission line losses to neglible amount.

---

* Leaving aside jitter, or assuming that the jitter problem has been solved.

Consider a 1-km path using 100-GHz equipment. The free-space path loss is 132.5 dB plus an absorption loss factor to be added of about 0.8 dB (Figure 10.2). This gives a total path loss of 133.3 dB. Assume B_{if} to equal 100 MHz (sufficient bandwidth to permit the transmission of 100 Mb/s allowing 1 b/Hz). Allow a 10-dB noise figure for the receiver, and its noise threshold is calculated to be -118 dBW.

At the transmit end of the path, the transmitter power output is 10 mw $(-20$ dBW). With zero antenna gains, the input to the receiver is -20 dBW $(+)$ -133 dBW $= -153$ dBW with negligible line losses. If the receiver operating point is placed 30 dB above noise threshold (equivalent to -183.3 dBW), the sum of the antenna gains turns out to be 65.3. Therefore, each antenna would require a gain of 65.3/2 or 32.7 dB. From Table 10.1, a 0.5-m antenna provides a 51.7-dB gain. Thus we end up with more than 38 dB over and above requirements, which can be allotted to excess rainfall attenuation and scintillation. From Figure 10.5 we see that this margin is sufficient for about 45 mm/h of rainfall. In regions of less rainfall, the antennas may be smaller. The faded carrier-to-noise ratio is more than sufficient to produce an error rate no greater than 1×10^{-10}, depending on the modulation method used, and still provide a good margin against interference. However, the designer must keep in mind that the receivers must be able to withstand a wide dynamic range of levels, 70 dB or greater.

We have been emphasizing propagation and building in a fade margin sufficient to provide a highly reliable system. One important point has been put aside which only lately is taking on its proper priority. This is equipment reliability (including reliable power). On some centimetric systems the yearly outages due to poor equipment reliability exceed outages due to propagation by as much as 10:1.

In this respect, with approximately 1-km repeater spacing versus 50 km average for centimetric systems, the number of repeaters for a given section utilizing this short hop concept is multiplied 50 times. It certainly follows that equipment and power reliabilities must be multiplied manyfold. Likewise, maintenance must be reduced. This, too, is within our grasp with present-day solid-state and strip-line techniques. Another important feature is low level transmitter power outputs which permit smaller, more compact no-break power sources with sealed batteries floating on-line (Appendix B). Thermoelectric generation may be feasible as a backup during long line voltage outages. Purposely we used 10 mw $(-20$ dBW) as the transmitter output power in our example above.

The approach to reliability for such a system would need to be similar to that used on present undersea multichannel cable systems.

10.7   EARTH-SPACE COMMUNICATION

Another obvious candidate for propagated millimeter wave transmission would be for earth-space communications and, in particular, satellite communications. Nearly all the problems associated with the propagation of millimeter waves come from our atmosphere, such as rain, fog, and gas absorption (refer to Chapter 7).

Consider the free-space loss figures and some parabolic antenna gains related to synchronous satellites given in Table 10.2. The tendency would be to reduce the diameter of the earth station antenna to keep the accuracy of the surface within reasonable bounds, and generally reduce overall cost. By doing so, of course, antenna gain suffers. From Chapter 7, we see that the nominal 30-m "standard" antenna has a gain in excess of 60 dB at 4 GHz, and the free-space loss to a synchronous satellite at that frequency is about 196 dB. Unless we use a 30-m antenna here as well, it would seem that the "standard" earth station criteria may not be met.

Table 10.2   Parabolic Antenna Gains for Synchronous Satellites

Frequency (GHz)	Free-Space Loss (dB)	15-m Diameter Antenna Gain [a]	20-m Diameter [b] Antenna Gain [a]
20	210.5	67.5	70.0
30	214.0	71.0	72.7
40	216.5	73.0	76.0

[a] With 65% efficiency.
[b] *Note.* Surface accuracy requirements severely limit use of large diameter dishes in the high millimeter region.

There are some legislative restrictions and a number of other items in our favor that could reduce these "standard" requirements.

1.   By legislation a millimeter frequency band could be assigned for exclusive use by satellite-earth communication and thereby the + 32-dBW limit of satellite EIRP now in effect for the centimetric range would be increased.

2.   Sky noise decreases with frequency, so we can expect the sky noise component of the earth station noise budget to be well reduced, particularly at high look angles.

3. The "standard" earth station is specified to meet its minimum G/T at 5° elevation angles, where sky noise is the greatest. This angle should be increased to 15 or 20°. The operating area of the satellite would be reduced somewhat as a consequence, but the trade-off in cost and technology should be worthwhile. Likewise, it is just at these low look angles, where the millimeter beam must travel through the most atmosphere, where it will suffer the most outage due to rain and the most gaseous absorption.

Rainfall continues to be the largest deterrent. An earth station in New Jersey operating at 30 GHz with a 99.99% propagation reliability would require 38-dB rainfall margin. Remember that such margins for 'standard" earth stations operating at 4 GHz are on the order of 4–6 dB. Path diversity should reduce this figure significantly if separation between receiving sites is 5 mi (8 km) or greater.

REFERENCES AND BIBLIOGRAPHY

1. *Reference Data for Radio Engineers*, 5th ed., Howard W. Sams & Co., Indianapolis, Ind.
2. H. J. Liebe, *Atmospheric Propagation Properties in the 10 to 75-GHz Region: A Survey and Recommendations*, ESSA Tech. Rep. ERL 130-ITS 91, Boulder, Colo., 1969.
3. D. C. Hogg, "Millimeter-Wave Propagation Through the Atmosphere," *Science*, 1968.
4. Jesse J. Taub, "The Future of Millimeter Waves," *Microwave J.*, Nov. 1973.
5. B. R. Dean and E. J. Dutton, *Radio Meteorology*, Dover, New York, 1968.
6. J. A. Lane, "Scintillation and Absorption Fading on Line-of-Sight Links at 35 and 100 GHz," IEE Conference on Tropospheric Scatter Propagation, London, Oct. 1968.
7. M. C. Thompson et al., *Phase and Fading Characteristics in the 10 to 40-GHz Band*, Institute of Telecommunication Sciences, Boulder, Colo., Oct. 1972.
8. G. E. Weibel and H. O. Dressel, "Propagation Studies in Millimeter-Wave Link Systems," *Proc. IEEE*, April 1967.
9. Frank Dale, "The Reach for Higher Frequencies," *Telecommunications*, July 1970.
10. E. M. Hickin, "18 GHz Propagation," IEE Conference on Tropospheric Propagation, Oct. 1968.
11. R. E. Skerjanec and C. A. Samson, *Rain Attenuation Study for 15 GHz Relay Design*, Report FAA-RD-70-21, Institute of Telecommunication Sciences, Boulder, Colo., May 1970.
12. R. E. Skerjanec and C. A. Samson, *Microwave Link Performance Measurements at 8 and 14 GHz*, Report FAA-RD-72-115, Institute of Telecommunication Sciences, Boulder, Colo., Oct. 1972.
13. F. A. Benson, *Millimetre and Submillimetre Waves*, Iliffe, London, 1969.
14. D. E. Setzer, "Computed Transmission Through Rain at Microwave and Visible Frequencies," *Bell Syst. Tech. J.*, 1973 (Oct. 1970).

15. IEE Conference on Millimeter Wave Propagation (Conference Papers), London, 1972.

16. *Bibliography on Propagation Effects* 10 *GHz to* 1000 *THz*, Telecommunications and Engineering Report 30, Office of Telecommunications, Boulder, Colo., March 1972.

17. L. C. Tillotson, "Use of Frequencies Above 10 GHz for Common Carrier Applications, *Bell Syst. Tech. J.*, 1563 (July–Aug. 1969).

18. C. L. Ruthroff et al., "Short Hop Radio System Experiment," *Bell Syst. Tech. J.*, 1577 (July–Aug. 1969).

19. A. E. Freeny and J. D. Gabbe, "A Statistical Description of Intense Rainfall," *Bell Syst. Tech. J.*, 1789 (July–Aug. 1969).

20. R. G. Medhurst, "Rainfall Attenuation of Cetimeter Waves: Comparison of Theory and Experiment," *IEEE Trans. Antennas Prop.*, **AP-13**(4), 550 (July 1965).

21. D. B. Hodge, "The Characteristics of Millimeter Wavelength Satellite-to-Ground Space Diversity Links," IEE Conference No. 98, London, April 1973.

11 | PCM AND ITS APPLICATIONS

11.1 WHAT IS PCM?

Pulse code modulation (PCM) is a method of modulation in which a continuous analog wave is transmitted in an equivalent digital mode. The cornerstone of an explanation of the functioning of PCM is the sampling theorem, which states:

> If a band-limited signal is sampled at regular intervals of time and at a rate equal to or higher than twice the highest significant signal frequency, then the sample contains all the information of the original signal. The original signal may then be reconstructed by use of a low-pass filter. [Ref. 1, Section 21.]

As an example of the sampling theorem, the nominal 4-kHz channel would be sampled at a rate of 8000 samples/s (i.e., 4000×2).

To develop a PCM signal from one or several analog signals, three processing steps are required: sampling, quantization, and coding. The result is a serial binary signal or bit stream, which may or may not be applied to the line without additional modulation steps. At this point a short review of Chapter 8 may be in order so that terminology such as mark, space, regeneration, and information bandwidth (Shannon and Nyquist) will not be unfamiliar to the reader.

One major advantage of digital transmission is that signals may be regenerated at intermediate points on links involved in transmission. The price for this advantage is the increased bandwidth required for PCM. Practical systems require 16 times the bandwidth of their analog counterpart (e.g., a 4-kHz analog voice channel requires 16×4, or 64 kHz when transmitted by PCM). Regeneration of a digital signal is simplified and particularly effective when the transmitted line signal is binary, whether neutral, polar, or bipolar. An example of bipolar transmission is shown in Figure 11.1.

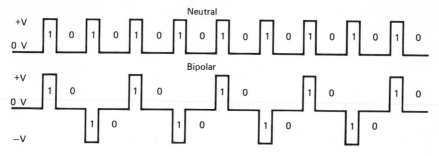

Figure 11.1 Neutral versus bipolar bit streams. The upper drawing illustrates alternate "1"s and "0"s transmitted in a neutral mode; the lower illustrates the equivalent in a bipolar mode.

Binary transmission tolerates considerably higher noise levels (i.e., degraded signal-to-noise ratios) when compared to its analog counterpart (i.e., FDM, Chapter 3). This plus the regeneration capability is a great step forward in transmission engineering. The regeneration that takes place at each repeater by definition recreates a new digital signal; therefore noise, as we know it, does not accumulate. However, there is an equivalent to noise in PCM systems which is generated in the modulation-demodulation processes. This is called quantizing distortion and can be equated in annoyance to the listener with thermal noise. Regarding thermal-intermodulation noise, let us compare a 2500-km conventional analog circuit using FDM multiplex over cable or radio with an equivalent PCM system over either medium.

	FDM/radio/cable	PCM/radio/cable
Multiplex	2,500 pWp	130 pWp equivalent
Radio/cable	7,500 pWp	0 pWp
Total	10,000 pWp	130 pWp equivalent (not dependent on system length, see Sec 11.2.6)

Error rate is another important factor (see Chapter 8). If we can maintain an end-to-end error rate on the digital portion of the system of 1×10^{-5}, intelligibility will not be degraded. A third factor is important in PCM cable applications. This is crosstalk spilling from one PCM system to another or from the send path to the receive path inside the same cable sheath.

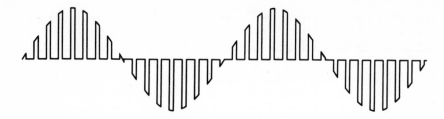

Figure 11.2 A PAM wave as a result of sampling a single sinusoid.

The purpose of this chapter is to provide a background of the problems involved in PCM and its transmission, including the several PCM formats now in use. Practical aspects are stressed later in the chapter, such as the design of interexchange trunks (junctions) and the prove-in distance,* compared with other forms of multiplex or VF cable. Finally, long-distance systems are focused on as well as PCM switching as a method of extending (or shortening) prove-in distance.

* The point at which a PCM system becomes viable economically.

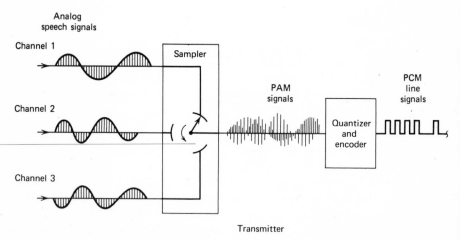

Figure 11.3 A simplified analogy of the formation of a PAM wave. Courtesy

11.2 DEVELOPMENT OF A PCM SIGNAL

11.2.1 Sampling

Consider the sampling theorem given above. If we now sample the standard CCITT voice channel, 300–3400 Hz (a bandwidth of 3100 Hz), at a rate of 8000 samples/s, we will have complied with the theorem and we can expect to recover all the information in the original analog signal. Therefore a sample is taken every 1/8000 s or every 125 μs. These are key parameters for our future argument.

Another example may be a 15-kHz program channel. Here the lowest sampling rate would be 30,000 times/s. Samples would be taken at 1/30,000-s intervals or at 33.3 μs.

11.2.2 The PAM Wave

With at least one exception (i.e., SPADE, Section 7.12) practical PCM systems involve time division multiplexing. Sampling in these cases does not involve just one voice channel, but several. In practice, one system to be discussed samples 24 voice channels in sequence; another samples

GTE Lenkurt Demodulator, San Carlos, Calif.

32 channels. The result of the multiple sampling is a PAM (pulse amplitude modulation) wave. A simplified PAM wave is shown in Figure 11.2, in this case a single sinusoid. A simplified diagram of the processing involved to derive a multiplexed PAM wave is shown in Figure 11.3.

If the nominal 4-kHz voice channel must be sampled 8000 times/s, and a group of 24 such voice channels are to be sampled sequentially to interleave them, forming a PAM multiplexed wave, this could be done by gating. Open the gate for a 3.25-μs period for each voice channel to be sampled successively from Channel 1 through Channel 24. This full sequence must be done in a 125-μs period ($1 \times 10^6/8000$). We call this 125-μs period a *frame*, and inside the frame all 24 channels are successively sampled once.

11.2.3   Quantization

The next step, it would appear, in the process of forming a PCM serial bit stream, would be to assign a binary code to each sample as it is presented to the coder.

Remember from Chapter 8 the discussion of code lengths, or what is more properly called coding "level." For instance, a binary code with four discrete elements (a four-level code) could code 2^4 separate and distinct meanings or 16 characters, not enough for the 26 letters in our alphabet; a five-level code would provide 2^5 or 32 characters or meanings. The ASCII is basically a seven-level code allowing 128 discrete meaning for each code combination ($2^7 = 128$). An eight-level code would yield 256 possibilities.

Another concept that must be kept in mind as the discussion leads into coding is that bandwidth is related to information rate (more exactly to modulation rate) or, for this discussion, to the number of bits per second transmitted. The goal is to keep some control over the amount of bandwidth necessary. It follows, then, that the coding length (number of levels) must be limited.

As it stands, an infinite number of amplitude levels are being presented to the coder on the PAM highway. If the excursion of the PAM wave is between 0 and $+ 1$ V, the reader should ask himself how many discrete values there are between 0 and 1. All values must be considered, even 0.0176487892 V.

The intensity range of voice signals over an analog telephone channel is on the order of 60 dB (see Section 3.13). The 0–1 V range of the PAM highway at the coder input may represent that 60-dB range. Further, it is obvious that the coder cannot provide a code of infinite length (e.g., an infinite number of coded levels) to satisfy every level in the 60-dB range (or a range from -1 to $+1$ V). The key is to assign discrete levels from -1 V through 0 to $+1$ V (60-dB range).

The assignment of discrete values to the PAM samples is called *quantization*. To cite an example, consider Figure 11.4. 16 quantum steps exist between -1 and $+1$ V and are coded as follows:

Step Number	Code	Step Number	Code
0	0000	8	1000
1	0001	9	1001
2	0010	10	1010
3	0011	11	1011
4	0100	12	1100
5	0101	13	1101
6	0110	14	1110
7	0111	15	1111

Examination of Figure 11.4 shows that Step 12 is used twice. Neither time it is used is it the true value of the impinging sinusoid. It is a rounded-off value. These rounded-off values are shown with the dashed line in the figure that follows the general outline of the sinusoid. The horizontal dashed lines show the point where the quantum changes to the next higher or next lower level if the sinusoid curve is above or below that value. Take Step 14 in the curve, for example. The curve, dropping from its maximum, is given two values of 14 consecutively. For the first, the curve is above 14, and for the second, below. That error, in the case of "14," for instance, from the quantum value to the true value, is called *quantizing distortion*. This distortion is the major source of imperfection in PCM systems.

In Figure 11.4, maintaining the $-1-0-+1$ V relationship, let us double the number of quantum steps from 16 to 32. What improvement would we achieve in quantization distortion? First determine the step increment in millivolts in each case. In the first case the total range of 2000 mV would be divided into 32 steps, or 187.5 mV/step. The second case would have 2000/64 or 93.7mV/step.

For the 32-step case, the worst quantizing error (distortion) would occur when an input to be quantized was at the half-step level, or in this

454

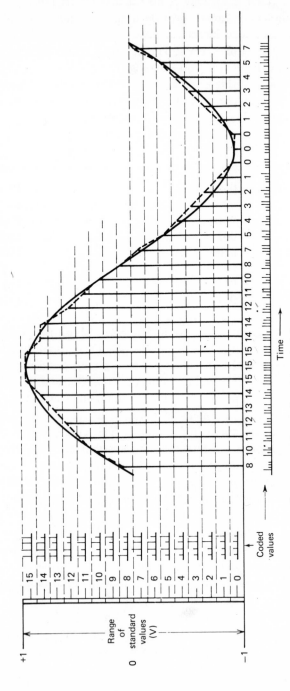

Figure 11.4 Quantization and resulting coding using 16 quantizing steps.

case, 187.5/2 or 93.7 mV above or below the nearest quantizing step. For the 64-step case, the worst quantizing error (distortion) would again be at the half-step level, or 93.7/2 or 46.8 mV. Thus the improvement in decibels for doubling the number of quantizing steps is:

$$20 \log \left(\frac{97.7}{46.8}\right) = 20 \log 2 \text{ or } 6 \text{ dB approximately}$$

This is valid for linear quantization only (see Section 11.2.6). Thus increasing the number of quantizing steps for a fixed range of input values reduces quantizing distortion accordingly. Experiments have shown that if 2048 uniform quantizing steps are provided, sufficient voice signal quality is achieved.

For 2048 quantizing steps, a coder will be required to code the 2048 discrete meanings (steps). Reviewing Chapter 8, we find that a binary code with 2048 separate characters or meanings (one for each quantum step) requires an 11-element code or

$$2^n = 2048; \text{ thus } n = 11$$

With a sampling rate of 8000/s per voice channel, the binary information rate per voice channel will be 88,000 bps. Consider that equivalent bandwidth is a function of information rate; the desirability of reducing this figure is therefore obvious.

11.2.4 Coding

Practical PCM systems use seven- and eight-level binary codes, or

$$2^7 = 128 \text{ quantum steps}$$
$$2^8 = 256 \text{ quantum steps}$$

Two methods are used to reduce the quantum step to 128 or 256 without sacrificing fidelity. These are nonuniform quantizing steps and companding prior to quantizing, followed by uniform quantizing. Keep in mind that the primary concern of digital transmission using PCM techniques is to transmit speech, as distinct from digital transmission covered in Chapter 8, which dealt with the transmission of data and message information. Unlike data transmission, in speech transmission there is a much greater likelihood of encountering signals of small amplitudes than those of large amplitudes.

A secondary, but equally important aspect, is that coded signals are designed to convey maximum information considering that all quantum steps (meanings, characters) will have an equally probable occurrence.

(We obliquely referred to this inefficiency in Chapter 8 because practical data codes assume equiprobability. When dealing with a pure number system with complete random selection, this equiprobability does hold true. Elsewhere, particularly in practical application, it does not. One of the worst offenders is our written language. Compare the probability of occurrence of the letter "e" in written text with "y" or "q.") To get around this problem larger quantum steps are used for the larger amplitude portion of the signal, and finer steps for signals with low amplitudes.

The two methods of reducing the total number of quantum steps can now be labeled more precisely:

- Nonuniform quantizing performed in the coding process.
- Companding (compression) before the signals enter the coder, which now performs uniform quantizing on the resulting signal before coding. At the receive end, expansion is carried out after decoding.

An example of nonuniform quantizing could be derived from Figure 11.4 by changing the step assignment. For instance 20 steps may be assigned between 0.0 and $+ 0.1$ V (another 20 between 0.0 and $- 0.1$, etc.), 15 between 0.1 and 0.2 V, 10 between 0.2 and 0.35 V, 8 between 0.35 and 0.5 V, 7 between 0.5 and 0.75 V, and 4 between 0.75 and 1.0 V.

Most practical PCM systems use companding to give finer granularity (more steps) to the smaller amplitude signals. This is instantaneous companding compared to syllabic companding described in Section 3.13. Compression imparts more gain to lower amplitude signals. The compression and later expansion functions are logarithmic and follow one of two laws, the A law or the "mu (μ) law.

The curve for the A law may be plotted from the formula

$$Y = \frac{AX}{(1 + \log A)} \qquad 0 \le X \le 1/A$$

The curve for the mu law may be plotted from the formula

$$Y = \frac{\log (1 + \mu X)}{\log (1 + \mu)} \qquad -1 \le X \le 1$$

A common expression used in dealing with the "quality" of a PCM signal is the signal-to-distortion ratio (expressed in decibels). Parameters A and μ determine the range over which the signal-to-distortion ratio is comparatively constant. This is the dynamic range. Using a μ of 100 can

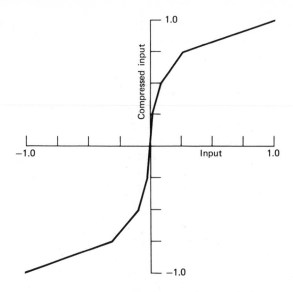

Figure 11.5 Seven-segment linear approximate of the logarithmic curve for the mu law ($\mu = 100$) (Ref. 2). Copyright © 1970, by Bell Telephone Laboratories.

provide a dynamic range of 40 dB of relative linearity in the signal-to-distortion ratio.

In actual PCM systems the companding circuitry does not provide an exact replica of the logarithmic curves shown. The circuitry produces approximate equivalents using a segmented curve, each segment being linear. The more segments the curve has, the more it approaches the true logarithmic curve desired. Such a segmented curve is shown in Figure 11.5.

If the mu law were implemented using a seven (eight)-segment linear approximate equivalent, it would appear as shown in Figure 11.5. Thus upon coding the first three coded digits would indicate the segment number (e.g., $2^3 = 8$). Of the seven-digit code, the remaining four digits would divide each segment in 16 equal parts to further identify the exact quantum step (e.g., $2^4 = 16$).

For small signals the companding improvement is approximately

A law   24 dB

mu law   30 dB

using a seven-level code.

Coding in PCM systems utilizes a straightforward binary coding. Two good examples of this coding are shown in Figures 11.7 and 11.10.

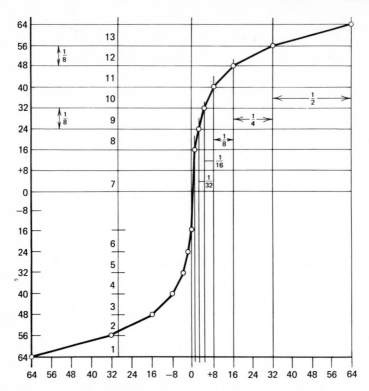

Figure 11.6 A 13-segment approximation of the A-law curve as used on the 24-channel STC system (PSC-24B). The horizontal ordinate represents quantized signal levels. Note that there are many more companding values at the lower signal levels than at the higher signal levels.

The coding process is closely connected to quantizing. In practical systems, whether using A law or mu law, quantizing uses segmented equivalents as discussed above and shown in Figure 11.5. Such segmenting is a handy aid to coding. Consider Figure 11.6, which shows the segmenting used on a 24-channel PCM system (A law) developed by Standard Telephone and Cables (UK). Here there are seven linear approximations (segments) above the origin and seven below, providing a 13-segment equivalent of the A law. It is 13, not 14 (i.e., 7 + 7), because the segments passing through the origin are colinear and are counted as one, not two segments. In this system 6 bits identify the specific quantum level, and a seventh bit identifies whether it is positive (above the origin) or negative (below the origin). The maximum negative step is assigned 0000000, the maximum positive, 1111111. Obviously we are dealing with a seven-level code providing identification of 128 quantum steps, 64 above the origin and 64 below.

The 30 + 2 PCM system also uses a 13-segment approximation of the *A* law, where $A = 87.6$. The 13 segments (14) lead us to an eight-level code. The coding for this system is shown in Figure 11.7. Again, if the first code element (bit) is 1, it indicates a positive value (e.g., the quantum step is located above the origin). The following three elements (bits) identify the segment, there being seven segments above and seven segments below the origin (horizontal axis).

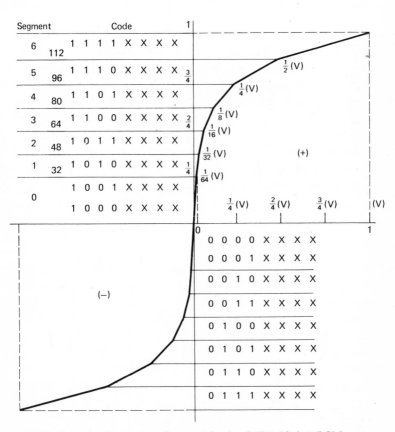

Figure 11.7 Quantization and coding used in the CEPT 30 + 2 PCM system.

The first four elements of the fourth + segment are 1101. The first "1" indicates it is above the horizontal axis (e.g., it is positive). The next three elements indicate the fourth step or

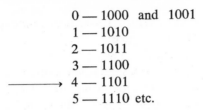

$$0 — 1000 \text{ and } 1001$$
$$1 — 1010$$
$$2 — 1011$$
$$3 — 1100$$
$$\longrightarrow \quad 4 — 1101$$
$$5 — 1110 \text{ etc.}$$

Figure 11.8 shows a "blowup" of the uniform quantizing and subsequent straightforward binary coding of Step 4; the Figure 11.8 shows final segment coding, which is uniform, providing 16 ($2^4 = 16$) coded quantum steps.

The North American D2 PCM system uses a 15-segment approximation of the logarithmic mu law. Again, there are actually 16 segments. The segments cutting the origin are colinear and counted as 1. The quantization in the D2 system is shown in Figure 11.9 for the positive portion of the curve. Segment 5 representing quantizing steps 64–80 is shown blown up in the figure. Figure 11.10 shows the D2 coding. As can be seen in this figure, again the first code element, whether a "1" or whether a "0," indicates if the quantum step is above or below the horizontal axis. The next three elements identify the segment, and the last four elements (bits) identify the actual quantum level inside that segment.

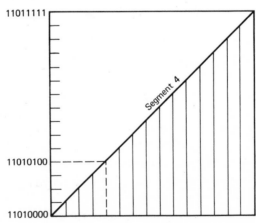

Figure 11.8   CEPT 30 + 2 PCM system, coding of Segment 4 (positive).

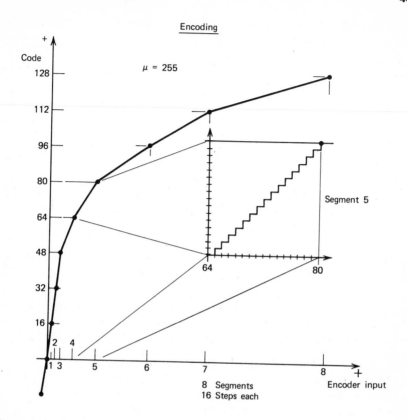

Figure 11.9 Positive portion of the segmented approximation of the mu law quantizing curve used in the North American (ATT) D2 PCM channelizing equipment. Courtesy ITT Telecommunications, Raleigh, N.C.

11.2.5 The Concept of Frame

As shown in Figure 11.3, PCM multiplexing is carried out in the sampling process, sampling several sources sequentially. These sources may be nominal 4-kHz voice channels or other information sources, possibly data or video. The final result of the sampling and subsequent quantization and coding is a series of pulses, a serial bit stream which requires some indication or identification of the beginning of a scanning sequence. This identification tells the far-end receiver when each full sampling sequence starts and ends; it times the receiver. Such identification is called framing. A full sequence or cycle of samples is called a frame in PCM terminology.

Code Level		Digit Number							
		1	2	3	4	5	6	7	8
255	(Peak positive level)	1	0	0	0	0	0	0	0
239		1	0	0	1	0	0	0	0
223		1	0	1	0	0	0	0	0
207		1	0	1	1	0	0	0	0
191		1	1	0	0	0	0	0	0
175		1	1	0	1	0	0	0	0
159		1	1	1	0	0	0	0	0
143		1	1	1	1	0	0	0	0
127	(Center levels)	1	1	1	1	1	1	1	1
126	(Nominal zero)	0	1	1	1	1	1	1	1
111		0	1	1	1	0	0	0	0
95		0	1	1	0	0	0	0	0
79		0	1	0	1	0	0	0	0
63		0	1	0	0	0	0	0	0
47		0	0	1	1	0	0	0	0
31		0	0	1	0	0	0	0	0
15		0	0	0	1	0	0	0	0
2		0	0	0	0	0	0	1	1
1		0	0	0	0	0	0	1	0
0	(Peak negative level)	0	0	0	0	0	0	1*	0

* One digit added to ensure that timing content of transmitted pattern is maintained.

Figure 11.10 Eight-level coding of the North American (ATT) D2 PCM system. Note that actually there are really only 255 quantizing steps because steps "0" and "1" use the same bit sequence, thus avoiding a code sequence with no transitions (i.e., "0's" only) (Ref. 6).

Consider the framing structure of several practical PCM systems: the *ATT D1 System* is a 24-channel PCM system using a seven-level code (e.g., $2^7 = 128$ quantizing steps). To each 7 bits representing a coded quantum step 1 bit is added for signaling. To the full sequence 1 bit is added, called a framing bit. Thus a D1 frame consists of

$$(7 + 1) \times 24 + 1 = 193 \text{ bits}$$

making up a full sequence or frame. By definition 8000 frames are transmitted so the bit rate is

$$193 \times 8000 = 1,544,000 \text{ bps}$$

The *CEPT** 30 + 2 system is a 32-channel system where 30 channels transmit speech derived from incoming telephone trunks and the remaining

* CEPT = Conference European des Postes et Telecommunication.

two channels signaling and synchronization information. Each channel is allotted a time slot and we can speak of time slots (TS) 0–31 as follows:

TS	Type of Information
0	Synchronizing (framing)
1–15	Speech
16	Signaling
17–31	Speech

In TS 0 a synchronizing code or word is transmitted every second frame occupying digits 2–8 as follows

$$0011011$$

In those frames without the synchronizing word, the second bit of TS 0 is frozen at a 1 so that in these frames the synchronizing word cannot be imitated. The remaining bits of time slot 0 can be used for the transmission of supervisory information signals (see Appendix C on signaling).

The North American (ATT) D2 system is a 96-voice channel system made up of four groups of 24 channels each. A multiplexer is required to bring these four groups into a serial bit stream system. The 24-channel basic building block of the D2 system has the following characteristics:

255 quantizing steps, mu-law companding, 15-segment
approximation, with $\mu = 255$, 8-element code

The frame has similar makeup to the D1 system, or

$$8 \times 24 + 1 = 193 \text{ bits per frame}$$

The frame structure is shown in Figure 11.11. Note that signaling is provided by "robbing" Bit 8 from every channel in every sixth frame. For all other frames all bits are used to transmit information coding.

11.2.6   Quantizing Distortion

Quantizing distortion has been defined as the difference between the signal waveform as presented to the PCM multiplex (codec*) and its

---

* The term codec is introduced, meaning *coder-decoder* and is analogous to modem in analog circuits.

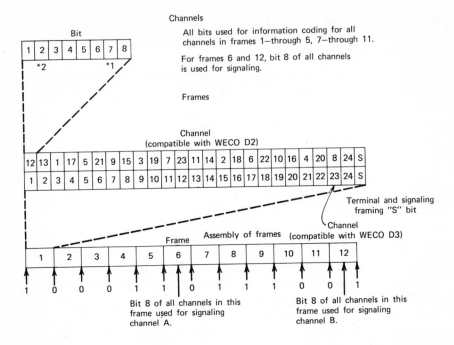

Figure 11.11 The frame structure of the North American (ATT) D2 PCM system for the channel bank. Note the bit "robbing" technique used on each sixth frame to provide signaling information. Courtesy ITT Telecommunications, Raleigh, N.C. *Notes.* (1) If bits 1–6 and 8 are 0, then bit 7 is transmitted as 1. (2) Bit 2 is transmitted as 0 on all channels for transmission if end-to-end alarm. (3) Composite pattern 000110111001, etc.

equivalent quantized value. Quantizing distortion produces a signal-to-distortion ratio (S/D) given by

$$\frac{S}{D} = 6n + 1.8 \text{ dB} \qquad \text{(for uniform quantizing)}$$

where n is the number of bits used to express a quantizing level. This bit grouping is often referred to as a PCM word. For instance, the ATT D1 system uses a 7-bit code word to express a level, and the 30 + 2 and D2 systems use essentially 8 bits.

With a 7-bit code word (uniform quantizing)

$$\frac{S}{D} = 6 \times 7 + 1.8 = 43.8 \text{ dB}$$

Practical S/D values range on the order of 33 dB or better† for average talker levels using A-law logarithmic quantizing. At the lower limit of commercial speech levels where noise is most noticeable, the equivalent contribution of psophometrically weighted noise that the PCM segment contributes in a PCM-analog hybrid system is on the order of 100 pWp. This figure does not increase with PCM system length and assumes only one complete segment. Each time the system decodes to analog voice and recodes to PCM, another 100 pWp must be added.

The 100-pWp figure does relate to the design of the codec and depends on such things as the number of quantizing steps and the type of quantizing employed. It also depends greatly on talker volume. Very loud talkers may suffer clipping with voice peaks outside the quantizing range or in the section of the quantizing curve where quantizing steps are very large. Very low level talkers likewise tend to show a degraded S/D ratio. For instance, if we have an equivalent sinusoid input to a codec of -33.5 dBm, we then could expect an S/D ratio of about 33 dB with a resulting noise floor of -66.5 dBm $(33 + 33.5)$, which equates to about 130 pWp of equivalent noise. However, quantizing noise tends to annoy the telephone user more than conventional noise discussed in preceding chapters.

11.2.7 Idle Channel Noise

An idle PCM channel can be excited by the idle Gaussian noise and crosstalk present on the input analog channel. A decision threshold may be set which would control idle noise if it remains constant. With a constant level input there will be no change in code word output, but any change of amplitude will cause a corresponding change in code word and the effect of such noise may be an annoyance to the telephone listener.

One important overall PCM design decision to control idle channel noise is the selection of either the μ or A values of the logarithmic quantizing curve used. The higher the values of these constants, the more finely granulated are the steps (quantizing steps finer) near the zero signal point. This tends to reduce idle channel noise. Care must also be taken to ensure that hum is minimized at the inputs of voice channels to the PCM equipment (codec).

† Using eight-level coding.

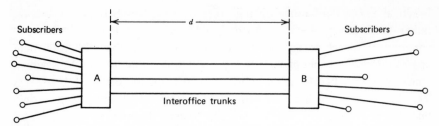

Figure 11.12 Simplified application diagram of PCM as applied to interoffice plant. A and B are switching centers.

11.3 PRACTICAL APPLICATION

11.3.1 General

PCM has found widest application in expanding interoffice trunks (junctions) that have reached exhaust* or will reach exhaust in the near future. An interoffice trunk is one pair of a circuit group connecting two switching points (exchanges). Figure 11.12 sketches the interoffice trunk concept. Depending on the particular application, at some point where distance d is exceeded, it will be more economical to install PCM on existing VF cable plant than to rip up streets and add more VF cable pairs. For the planning engineer, the distance, d, where PCM becomes an economic alternative is called the prove-in distance. d may vary from 8 to 16 km (5 to 10 mi) depending on the location and other circumstances. For distances less than d, additional VF cable pairs should be used for expanding plant.

The general rule for measuring expansion capacity of a given VF cable is as follows:

- For ATT D1/D2 channelizing equipment, 2 VF pairs will carry 24 PCM channels.
- For CEPT 30 + 2 system as configured by ITT, 2 VF pairs plus a phantom pair will carry 30 PCM speech channels.

All pairs in a VF cable may not necessarily be usable for PCM transmission. One restriction is brought about by the possibility of excessive crosstalk between PCM carrying pairs. The effect of high crosstalk levels

---

* Exhaust is an outside plant term meaning that the useful pairs of a cable have been used up (assigned) from a planning point of view.

is to introduce digital errors in the PCM bit stream. Error rate may be related on a statistical basis to crosstalk, which in turn is dependent on the characteristics of the cable and the number of PCM carrying pairs.

One method to reduce crosstalk and thereby increase VF pair usage is to turn to two-cable working, rather than have the "go" and "return" PCM cable pairs in the same cable.

Another item that can limit cable pair usage is the incompatibility of FDM and PCM carrier systems in the same cable. On the cable pairs that will be used for PCM, the following should be taken into consideration:

- All load coils must be removed.
- Build-out networks, bridged taps must also be removed.
- No crosses, grounds, splits, high resistance splices, nor moisture permitted.

The frequency response of the pair should be measured out to 1 MHz and taken into consideration as far out as 2.5 MHz. Insulation should be checked with a megger. A pulse reflection test using a radar test set is also recommended. Such a test will indicate opens, shorts, and high impedance mismatches. A resistance test and balance test using a Wheatstone bridge may also be in order. Some special PCM test sets are available such as the Lenkurt Electric 91100 PCM Cable Test Set using pseudo random PCM test signals and the conventional digital test eye pattern.

11.3.2 Practical System Block Diagram

A block diagram showing the elemental blocks of a PCM transmission link used to expand installed VF cable capacity is shown in Figure 11.13. Most telephone administrations (companies) distinguish between the terminal area of a PCM system and the repeatered line. The term "span" comes into play here. A span line is composed of a number of repeater sections permanently connected in tandem at repeater apparatus cases mounted in manholes or on pole lines along the span. A "span'" is defined as the group of span lines which extend between two office (switching center) repeater points.

A typical span is shown in Figure 11.13. The spacing between regenerative repeaters is important. Section 11.3.1 above mentioned the necessity of removing load coils from those trunk (junction) cable pairs which were to be used for PCM transmission. It is at these load points that the

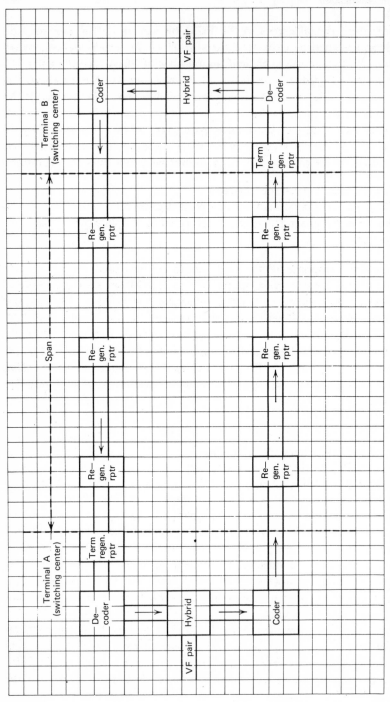

Figure 11.13 Simplified functional block diagram of a PCM link used to expand capacity of an existing VF cable (for simplicity, interface with only one VF pair is shown). Note the spacing between repeaters in the span line.

PCM regenerative repeaters are installed. On a VF line with H-type loading (see Section 2.8.3), spacing between load points is normally about 6000 ft (1830 km). It will be remembered from Chapter 2 that the first load coil out from the exchange on a trunk repair is at half-distance or 3000 ft (915 m). This is provident for a regenerative repeater also must be installed at this point. This spacing is shown in Figure 11.13 (1 space = 1000 ft). The purpose of installing a repeater at this location is to increase the pulse level before entering the environment of an exchange area where the levels of impulse noise may be quite high. High levels of impulse noise induced into the system may cause significant increases in digital error rate of the incoming PCM bit streams, particularly when the bit stream is of a comparatively low level.

Commonly PCM pulse amplitude output of a regenerative repeater is on the order of 3 V. Likewise, 3 V is the voltage on the PCM line cross-connect field at the exchange (terminal area).

A guideline used by Bell Telephone Manufacturing Company (Belgium) is that the maximum distance separating regenerative repeaters is that corresponding to a cable pair attenuation of 36 dB at the maximum expected temperature at 1024 kHz. This frequency is equivalent to the half-bit rate for the CEPT systems (e.g., 2048 kb/s). Actually repeater design permits operation on lines with attenuations anywhere from 4 to 36 dB, allowing considerable leeway in placing repeater points. Table 11.1 gives some other practical repeater spacing parameters for the CEPT-ITT-BTM 30 + 2 system.

The maximum distance is limited by the maximum number of repeaters, which in this case is a function of power feeding and supervisory considerations. For instance, the fault location system can handle up to a maximum of 18 tandem repeaters for the BTM (ITT) configuration.

Table 11.1   Line Parameters for ITT/BTM PCM

Pair Diameter (mm)	Loop Atten. at 1 MHz (dB/km)	Loop Resist-ance (Ω/km)	Voltage Drop (V/km)	Maximum Distance[a] (km)	Total Repeaters	Maximum Distance System (km)
0.9	12	60	1.5	3	18	54
0.6	16	100	2.6	2.25	16	36

[a] Between adjacent repeaters.

Power for the BTM system is via a constant-current feeding arrangement over a phantom pair serving both the "go" and related "return" repeaters, providing up to 150 V dc at the power feed point. The voltage drop per regenerative repeater is 5.1 V. Thus for a "go" and "return" repeater configuration the drop is 10.2 V.

As an example, let us determine the maximum number of regenerative repeaters in tandem that may be fed from one power feed point by this system using 0.8-mm diameter pairs with a 3-V voltage drop in an 1830-m spacing between adjacent repeaters:

$$\frac{150}{(10.2 + 3)} = 11$$

Assuming power fed from both ends and an 1800-m "dead" section in the middle, the maximum distance between power feed points is approximately

$$(2 \times 11 + 1)\, 1.8 \text{ km} = 41.4 \text{ km}$$

Fault tracing for the North American (ATT) T1 system is carried out by means of monitoring the framing signal, the 193rd bit (Section 11.2.6). The framing signal (amplified) normally holds a relay closed when the system is operative. With loss of framing signal, the relay opens actuating alarms. By this means a faulty system is identified, isolated, and dropped from "traffic."

To locate a defective regenerator on the BTM (Belgium)-CEPT system, traffic is removed from the system, and a special pattern generator is connected to the line. The pattern generator transmits a digital pattern with the same bit rate as the 30 + 2 PCM signal but the test pattern can be varied to contain selected low frequency spectral elements. Each regenerator on the repeatered line is equipped with a special audio filter, each with a distinctive passband. Up to 18 different filters may be provided in a system. The filter is bridged across the output of the regenerator, sampling the output pattern. The output of the filter is amplified and transformer-coupled to a fault transmission pair, which is normally common to all PCM systems on the route, span, or section.

To determine which regenerator is faulty, the special test pattern is tuned over the spectrum of interest. As the pattern is tuned through the frequency of the distinct filter of each operative repeater, a return signal will derive from the fault transmission pair at a minimum specified level. Defective repeaters will be identified by absence of return signal or a return level under specification. The distinctive spectral content of the return signal is indicative of the regenerator undergoing test.

11.3.3 The Line Code

PCM signals as transmitted to the cable are in the bipolar mode (biternary), as shown in Figure 11.1. The marks or "1"s have only a 50% duty cycle. There are several advantages to this mode of transmission:

- No dc return is required; thus transformer coupling can be used on the line.
- The power spectrum of the transmitted signal is centered at a frequency equivalent to half the bit rate.

It will be noted in bipolar transmission that the "0"s are coded as absence of pulses and the "1"s are alternately coded as positive and negative pulses with the alternation taking place at every occurrence of a "1." This mode of transmission is also called alternate mark inversion (AMI).

One drawback to straightforward AMI transmission is that when a long string of "0"s is transmitted (e.g., no transitions), a timing problem may come about because repeaters and decoders have no way of extracting timing without transitions. The problem can be alleviated by forbidding long strings of "0"s. Codes have been developed which are *b*ipolar but with *N* *z*eros *s*ubstitution; they are called BNZS codes. For instance, a B6ZS code substitutes a particular signal for a string of 6 "0"s.

Another such code is the HDB3 code (*h*igh *d*ensity *b*inary *3*), where the 3 indicates that it substitutes for binary formations with more than 3 consecutive "0"s. With HDB3, the second and third zeros of the string are transmitted unchanged. The fourth "0" is transmitted to the line with the same polarity as the previous mark sent, which is a "violation" of the AMI concept. The first "0" may or may not be modified to a "1" to assure that the successive violations are of opposite polarity.

11.3.4 Signal-to-Gaussian-Noise Ratio on PCM Repeated Lines

As we mentioned earlier, noise accumulation on PCM systems is not an important consideration. This does not mean that Gaussian noise (nor crosstalk, impulse noise) is not important. Indeed, it may affect error performance expressed as error rate (see Chapter 8). Error rate, from one point of view, is cumulative. A decision in error, whether "1" or "0," made anywhere in the digital system, is not recoverable. Thus such an incorrect decision made by one regenerative repeater adds to the existing error rate on the line, and errors taking place in subsequent repeaters further down the line add in a cumulative manner tending to deteriorate the received signal.

In a purely binary transmission system, if a 20-dB signal-to-noise ratio is maintained, the system operates nearly error free. In this respect, consider Table 11.2.

Table 11.2   Error Rate of a Binary Transmission System Versus Signal-to-Rms Noise Ratio

Error Rate	S/N(dB)	Error Rate	S/N(dB)
10^{-2}	13.5	10^{-7}	20.3
10^{-3}	16	10^{-8}	21
10^{-4}	17.5	10^{-9}	21.6
10^{-5}	18.7	10^{-10}	22
10^{-6}	19.6	10^{-11}	22.2

As discussed in Section 11.3.3, PCM, in practice, is transmitted on-line with alternate mark inversion. The marks have a 50% duty cycle permitting energy concentration at a frequency of half the transmitted bit rate. Thus it is advisable to add 1 or 2 dB to the values shown in Table 11.1 to achieve a desired error rate on a practical system.

11.3.5   The Eye Pattern

The "eye" pattern provides a convenient method of checking the quality of a digital transmission line. A sketch of a typical eye pattern is shown in Figure 11.14. Any oscilloscope can produce a suitable eye pattern provided it has the proper rise time. Most quality oscilloscopes now available on the market do have. The scope should either terminate or bridge the repeatered line or output of a terminal repeater. The display on the oscilloscope contains all the incoming bipolar pulses superimposed one on the other.

Eye patterns are indicative of decision levels. The greater the eye opens, the better defined is the decision (whether "1" or "0" in the case of PCM). The opening is often referred to as the decision area (lightly crosshatched in the figure). Degradations reduce the area. Eye patterns are often measured off in the vertical, giving a relative measure of margin of decision.

Amplitude degradations shrink the eye in the vertical. Among amplitude degradations can be included echos, intersymbol interference, and decision threshold uncertainties.

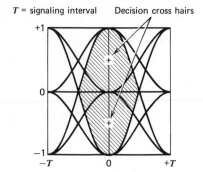

T = signaling interval   Decision cross hairs

Figure 11.14 Sketch of an "eye" pattern.

Horizontal shrinkage of the eye pattern is indicative of timing degradations (i.e., jitter and decision time misalignment).

Noise is the other degradation to be considered. Usually noise may be expressed in terms of some improvement in signal-to-noise ratio to bring the operating system into the bounds of some desired objective; see Table 11.2, for example. This ratio may be expressed as 20 × log* of the ideal eye opening (in the vertical as read on the oscilloscope's vertical scale) to the degraded reading.

11.4 HIGHER ORDER PCM MULTIPLEX SYSTEMS

Using the 24-channel D2 channel bank as a basic building block, higher order PCM systems are being developed in North America. For instance, four D-2 channel banks are multiplexed by a M1-2 multiplexer placing 6.312 Mb/s on a single wire pair (T2 digital line). Figure 11.15a is a simplified block diagram in the first step in the development of a higher order PCM system.

Two major elements make up a higher order system: multiplexers and repeatered line. ATT identifies its repeatered lines as follows:

T1 (deriving from D1 or D2 channel banks)	1.544 Mb/s
T2 (output of multiplexer M1-2)	6.312 Mb/s
T3 (output of multiplexer M2-3)	46.304 Mb/s

Figure 11.15b shows this hierarchy diagrammatically.

The higher order system will also accept data, video (TV), videotelephone, and FDM mastergroups.

---

* Oscilloscopes are commonly used to measure voltage; thus we can measure the degraded opening in voltage units and compare it to the full-scale perfect opening in the same units. A ratio is developed and we take 20 log that ratio to determine S/N.

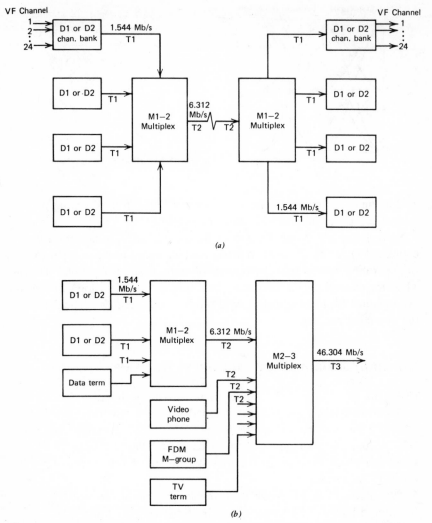

(a)

(b)

Figure 11.15 (a) Development of the 96-channel T2 (ATT) system by multiplexing the 24-channel D1 or D2 channel bank outputs. (b) Development of higher order PCM (ATT plan) (Ref. 2).

To transmit data on the ATT system, sampling is made of the binary serial data transitions, using 3 PCM bits to code each expected transition of the data bit stream. The 3 PCM bits do the coding as follows: for no transition, a series of three "1"s are transmitted. If a transition took place, then

Bit number		Function
1	("0")	Indicates presence of transition
2	("1" or "0")	Indicates in which half of the sampling interval the transition took place
3	("1" or "0")	Indicates the direction of transition, positive or negative going (e.g., "1" to "0" or "0" to "1")

The use of 3 bits for each transition, rather than only 1 bit to indicate just the presence or absence of a transition, prevents ambiguity in case of error and reduces error rate to a degree.

Pulse "stuffing" is an important concept in multiplexers such as the M1-2 which have outputs in the megabit range. With stuffing, the multiplexer output in bits per second is always at a higher bit rate than the sum of the lower speed inputs. This is done by artificially adding bits in the multiplexer (and withdrawing them at the demultiplexer). Stuffing is this artificial addition of bits.

Pulse stuffing simplifies clock design, clocking being the heart of any time division multiplex device. It also simplifies synchronization problems and reduces necessary buffer storage. The problem is not really in the clocking itself but in the variation of propagation time of the medium (see Chapter 8) interconnecting the multiplexer to a common demultiplexer (see Figure 11.15). The variation in propagation time causes the slowing or speedup of the arrival, the arrival bit rate increasing and decreasing with the variation of propagation time. The receiving equipment clocks the incoming bit stream at a constant rate; therefore storage is required.

With stuffing the demultiplexer is informed of the number and location of the stuffed bits. This information is usually passed on a separate data line with the output information. With outputs operating at a speed greater than the inputs, no input information bits can be lost when the storage is exceeded or when storage units are reset.

Future systems will probably be synchronized using master clocks with each terminal area, and each terminal will have the capability of operating independently for periods of time in case of master clock failure. To compensate for slow phase variations due to variation of propagation time, elastic stores will be provided at each terminal.

11.5 LONG-DISTANCE (TOLL) TRANSMISSION BY PCM

11.5.1 General

PCM, with its capability of regeneration, essentially eliminating the accumulation of noise as a signal traverses its transmission media, would appear to be the choice for toll transmission or backbone, long-haul routes. This has not been the case. One must consider the disadvantages of PCM as well. Most important is the competition with FDM systems, the L5 system, for instance (Table 3.1). ATT's L5 provides 10,800 VF channel capacity over long-haul coaxial cable media. The required band-width for this capacity on the cable is 60 MHz. To transmit the same number of channels by PCM would require on the order of 16 times the bandwidth.

Keep in mind the relationship briefly covered in Section 11.1 wherein this 16-multiple concept is shown: a 4-kHz voice channel requires an equivalent PCM bandwidth of 64 kHz, assuming 1-Hz bandwidth per PCM bit transmitted.

Thus a 10,800 VF (4-kHz) channel would require, if transmitted by PCM, about 691.2 MHz. Available bandwidth is still at a premium, whether by cable or radio. Therefore, it is more economical to transmit such large packages of information by FDM techniques and suffer the consequent noise accumulation. These techniques are covered in Chapters 3, 5, and 9. Perhaps when waveguide, millimetric satellite communication, or optical fiber techniques become viable, then PCM will prove useful for high density, long-haul transmission.

11.5.2 Jitter

There is one other important limitation of present-day technology on using PCM as a vehicle for long-haul transmission. This is jitter, more particularly, timing jitter.

A general definition of jitter is "the movement of zero crossings of a signal (digital or analog) from their expected time of occurrence." In Chapter 8 it was called unwanted phase modulation or incidental FM. Such jitter or phase jitter affected the decision process of the zero crossing in a digital data modem. Much of this sort of jitter can be traced to the intervening FDM equipment between one end of a data circuit and the other.

PCM has no intervening FDM equipment, and jitter in PCM systems takes on different characteristics. However, essentially the effect is the same—uncertainty in a decision circuit as to when a zero crossing (transition) took place, or the shifting of a zero crossing from its proper location. In PCM it is more proper to refer to jitter as timing jitter.

The primary source of timing jitter is the regenerative repeater. In the repeatered line jitter may be systematic or nonsystematic. Systematic jitter may be caused by offset pulses (i.e., where the pulse peak does not coincide with regenerator timing peaks, or transitions are offset), intersymbol interference (dependent on specific pulse patterns), and local clock threshold offset. Nonsystematic jitter may be traced to timing variations from repeater to repeater and to crosstalk.

In long chains of regenerative repeaters, systematic jitter is predominant and cumulative, increasing in rms value as $N^{1/2}$, where N is the number of repeaters in the chain. Jitter is also proportional to a repeater's timing filter bandwidth. Increasing the Q of these filters tends to reduce jitter of the regenerated signal, but it also increases error rate due to sampling the incoming signal at nonoptimum times.

The principal effect of jitter on the resulting analog signal after decoding is to distort the signal. The analog signal derives from a PAM pulse train which is then passed through a low-pass filter. Jitter displaces the PAM pulses from their proper location, showing up as undesired pulse position modulation (PPM).

Because jitter varies with the number of repeaters in tandem, it is one of the major restricting parameters of long-haul, high bit rate PCM systems. Jitter can be reduced in future systems by using elastic store at each regenerative repeater (costly) and high-Q phase locked loops.

11.6 PCM TRANSMISSION BY RADIOLINK

11.6.1 General

The transmission of PCM by radiolink is a viable alternative to PCM VF cable pair transmission under the following circumstances:

- Where physical or natural obstructions increase cable cost.
- On long spans, more than 30 mi (50 km) of relatively low VF pair capacity, up to 1200 circuits.
- As alternative routing of a cable system via radio.

Consider a situation where a large number of trunk (junction) routes are presently equipped with PCM. An FDM/FM radiolink is contemplated, and the system engineer is faced with one or several of the above circumstances. To use FM radiolinks with FDM multiplex will prove expensive. The existing PCM will have to be brought to VF (demultiplexed-demodulated) to interface with the new FDM equipment (Chapter 3).

Use of PCM eliminates the additional multiplex equipment cost. Further, PCM channelizing equipment, if we accept groups of 24 or 30 channels at a time, is less expensive on a per-channel basis than FDM equipment.

11.6.2   Modulation, RF Bandwidth and Performance

The most common modulation techniques for PCM over radio are straightforward FSK (frequency shift keying), theoretically requiring about 1 b/Hz of bandwidth, and QPSK (quadrature phase shift keying), requiring 1 Hz of bandwidth for every 2 bits transmitted at the expense of 6 dB improved signal-to-noise ratio when compared to FSK for equal error rates. As in many FM systems, it is usual to modulate the oscillator in the PCM radio transmitter, convert the oscillator output to 70 MHz, the IF, and heterodyne the IF to the output frequency in the 2-, 4-, 6–8-, 11-, and 15-GHz bands. (See Chapter 5.)

Care must be taken with these systems that they do not cause interference to existing FM systems. Special filtering of the RF output of the transmitter is often called for to reduce spurious and to limit sidebands. Increasing the transmission level to 8 (e.g., $m = 8$) or 16 increases bit rate for a given bandwidth resulting in further tightening of signal-to-noise ratio requirements.

Practical systems, however, assign bandwidths of 1 b/Hz in the best of cases. For instance, one system proposed in the United States will carry 21.5 Mb/s in a 30-MHz bandwidth at 6 GHz. Canadian Marconi describes a system as eight-level FM with a bandwidth of 0.44 × bit rate. One way to reduce "necessary" bandwidths is to reduce guard bands and permit some overlapping of power spectra. Otherwise for practical purposes eight-level systems produce 1.55 b/Hz of bandwidth.

One of the major advantages of PCM (digital) radio, as with PCM cable systems, is the ability the system has to periodically regenerate the waveform at each repeater site. Another advantage is that it is little affected by traffic loading, and any mix of voice and data traffic has no effect on system performance.

The disadvantage of PCM radio is that, at present, no more than the equivalent of 120 VF channels load an RF carrier. Again we are up against the fact that a 4-kHz channel is the input for voice information to analog multiplex, and for PCM the same voice channel has an equivalent bandwidth of 64 kHz. One proposed system, on the other hand, will transmit the equivalent of six T2 lines (6 × 96 voice channels) or 596 equivalent voice channels for a line bit rate of 39.6 MB/s requiring a 40-MHz bandwidth of RF. Equivalent FDM/FM systems operate with 1200, 1800, and up to 2700 voice channels on one RF carrier requiring often less bandwidth.

11.7 PCM SWITCHING

The implementation of PCM switching will have a profound impact on PCM transmission in the local area. Local area is discussed in Chapter 2. PCM switching is a form of time division switching where the switching inlets and outlets carry PCM highways in the case of tandem switches. Local switches, depending on the penetration of PCM, whether to the level of subscriber line or not, may be implemented by (1) Subscriber lines to the switch analog with PCM conversion at the switch, or (2) subscriber lines PCM with direct conversion to PCM at the subscriber location. In either case the trunks (junctions) would be PCM.

A PCM switch would operate far differently from an analog switch. Crosspoint usage would be more efficient for they would be operated on a time division mode, rather than the present space division. Call routing would be kept in memory and various crosspoints to the same circuit group would be employed on one call at different time intervals.

Signaling fits well into this arrangement. For analog, space division switches, signaling is digital, whether it is subscriber, line, or interregister (for a more detailed discussion of signaling, see Appendix C). Signaling, like data transmission, must be conditioned to operate over an analog network in most cases, either in the switching equipment or in the transmission equipment. Given a network where both transmission and switching are digital (e.g., PCM), signaling can take on its own characteristic and remain digital. This reduces circuit cost and complexity.

Digital switching, namely, PCM, eliminates switching losses and switching loss variation. Inside the integrated digital network circuit stability may be disregarded as a design limitation. In hybrid networks where both digital and analog systems coexist, stability must be considered at the interface, but only to meet the requirements of the analog side.

Also, by definition, the digital network is four-wire. Reference equivalent can be improved by assigning zero loss to that portion of the network that is digital (subject to overall stability considerations).

Consider a tandem switch that is PCM. The choice is to convert to the PCM mode at the switch or convert in the outlying local exchanges and use PCM as the trunk (junction) transmission media. In most cases the latter would seem desirable. Certainly the prove-in distance mentioned in Section 11.3.1 is reduced.

With a fully digital network, switching and transmission will be much more integrated technologies. Much more equipment will be common to both with concurrent savings of outlay in both disciplines.

REFERENCES AND BIBLIOGRAPHY

1. *Data Handbook for Radio Engineers*, 5th ed., Howard W. Sams & Co., Indianapolis, Section 21.
2. *Transmission Systems for Communications*, 4th ed., Bell Telephone Laboratories.
3. *Lenkurt Demodulator*, Lenkurt Electric Co., San Carlos, Calif.: Nov. 1966, March, April 1968, Sept, 1971, Dec. 1971, Jan. 1972, Oct. 1973.
4. Seminar on Pulse Code Modulation Transmission, Standard Telephone and Cables Limited, Basildon, England, June 1967.
5. *PCM—System Application—* 30 + 2TS, BTM /ITT, Antwerp, Belgium.
6. *Technical Manual–Operations and Maintenance Manual for T324 PCM Cable Carrier System*, ITT Telecommunications, Raleigh, N.C., April 1973.
7. *Principles of Modems* (edited draft), Communication Systems, Inc., Falls Church, Va. (limited circulation).
8. K. W. Catermole, *Principles of Pulse Code Modulation*, Iliffe, London, 1969.
9. W. C. Sain, "Pulse Code Modulation Systems in North America," *Electrical Communication* (ITT), **48**, (1 and 2), 1973.
10. J. V. Marten and E. Brading, 30-channel Pulse Code Modulation System, *Electrical Communication* (ITT), **48**, (1 and 2), 1973.
11. K. E. Fultz and D. B. Penick, "The T1 Carrier System," *Bell Syst. Tech. J.*, **44**, Sept. 1965.
12. R. B. Moore, "T2 Digital Line System," *Proceedings IEEE International Conference on Communications, June* 12, 1973, *Seattle, Wash.*
13. J. R. Davis, "T2 Repeater and Equalization," *Proceedings IEEE International Conference on Communications, June* 12, 1973, *Seattle, Wash.*

14. E. Cookson and C. Volkland, "Taking the Mystery out of Phase Jitter Measurement," *Telephony*, Sept. 25, 1972.

15. "Phase Jitter and its Measurement," CCITT Study Group IV, Question 3/IV, 17–28, Jan. 1972.

16. F. S. Boxal, "Digital Transmission Via Microwave Radio," Parts I and II, *Telecommunications*, April and May 1972.

17. W. H. Smith, "PCM Microwave Links," *Telecommunications*, April 1973.

18. L. Katzschner et al., "An Experimental Local PCM-Switching System," *IEEE Trans. Commun.*, Oct. 1973.

19. A. E. Pinet, "Telecommunication Integrated Network," *IEEE Trans. Commun.*, Aug. 1973.

20. G. C. Hartley et al., *Techniques of Pulse-Code Modulation in Communication Networks*, IEE Monograph Series, Cambridge University Press, 1967.

12 | VIDEO TRANSMISSION

12.1 GENERAL

This chapter provides the basic essentials for designing point-to-point video transmission systems. To understand the video problem the transmission system engineer must first have an appreciation of video and how the standard television video signal is developed. The discussion that follows provides an explanation of the "what and why" of video; there follows a review of critical video transmission parameters, black and white and color transmission standards, video program channel transmission, and the transmission of video over specific media. Finally there is a brief discussion of basic tests of video point-to-point facilities. Television broadcast problems are covered only where they specifically interact with point-to-point transmission.

12.2 AN APPRECIATION OF VIDEO TRANSMISSION

A video transmission system must deal with four factors when transmitting images of moving objects:

- A perception of the distribution of luminance or simply the distribution of light and shade.
- A perception of depth or a three-dimensioned perspective.
- A perception of motion relating to the first two factors above.
- A perception of color (hues and tints).

Monochrome television deals with the first three factors. Color television includes all four factors.

A video transmission system must convert these three (or four) factors into electrical equivalents. The first three factors are integrated to an equivalent electric current or voltage whose amplitude is varied with time. Essentially, at any one moment, it must integrate luminance from a scene in the three dimensions (i.e., width, height, and depth) as a function of time. And time itself is still another variable for the scene is changing in time.

The process of integration of visual intelligence is carried out by "scanning." Horizontal detail of a scene is transmitted continuously and the vertical detail discontinuously. The vertical dimension is assigned discrete values which become the fundamental limiting factor in a video transmission system.

The scanning process consists of taking a horizontal strip across the image on which discrete square elements are scanned from left to right. When the right-hand end is reached another, lower, horizontal strip is explored and so on until the whole image has been scanned. Luminance values are translated on each scanning interval into voltage and current variations and are transmitted over the system. The concept of scanning by this means is shown in Figure 12.1.

NTSC* practice divides an image into 525 horizontal scanning lines. It is the number of scanning lines that determines vertical detail or resolution of a picture.

When discussing picture resolution, aspect ratio is the width-to-height ratio of the video image. The aspect ratio used almost universally is 4:3. In other words a television image 12 in. wide would necessarily be 9 in. high. Thus an image divided into 525 (491) vertical elements would then have 700 (652) horizontal elements to maintain an aspect ratio of 4/3. The numbers in parentheses represent the practical maximum active lines and elements. Therefore the total number of elements approaches something on the order of 250,000. We reach this number because, in practice, vertical detail reproduced is 64–87% of the active scanning lines. A good halftone engraving may have as many as 14,400 elements/in.^{2} compared to approximately 3000/in.^{2} for a 9 by 12 in. television image.

Motion is another variable factor that must be transmitted. The sensation of continuous motion in standard TV video practice is transmitted to the viewer by a successive display of still pictures at a regular rate similar to the method used in motion pictures. The regular rate of display is called the frame rate. A frame rate of 25/s will give the viewer a sense of motion, but on the other hand he will be disturbed by luminance

* National Television Systems Committee (U.S.).

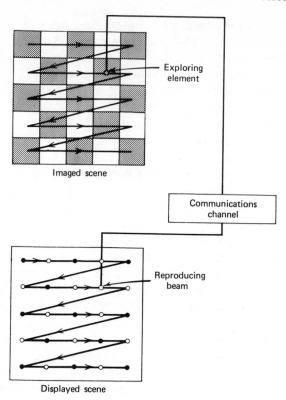

Figure 12.1 The scanning process from television camera to receiver display.

flicker (bloom and decay), or the sensation that still pictures are "flicking" on screen one after the other. To avoid any sort of luminance flicker sensation the image is divided into two closely interwoven (interleaving) parts and each part is presented in succession at a rate of 60/s even though *complete* pictures are still built up at a 30/s rate. It should be noted that interleaving improves resolution as well as improving apparent persistence of the CRT picture tube by tending to reinforce the scanning spots. It has been found convenient to equate flicker frequency to power line frequency. Thus in North American practice where power line frequency is 60 Hz the flicker is 60/s. In Europe it is 50/s to correspond to the 50-Hz line frequency used there.

Following North American practice some other important parameters derive from the previous paragraphs.

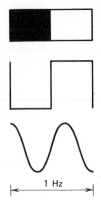

Figure 12.2 Development of a sinusoid wave from the scan of adjacent squares.

1. A field period is 1/60 s. This is the time that is required to scan a full picture on every horizontal line.

2. The second scan covers the lines not scanned on the first period, offset one-half horizontal line.

3. Thus 1/30 s is required to scan all lines on a complete picture.

4. The transit time of exploring and reproducing scanning elements or spots along each scanning line is 1/15750 s (525 lines in 1/30 s) = 63.5 μs.

5. Consider that about 16% of the 63.5 μs is consumed in flyback and synchronization. Therefore only about 53.3 μs are left per line of picture to transmit information.

What will be the bandwidth necessary to transmit images so described? Consider the worst case where each scanning line is made up of alternate black and white squares, each the size of the scanning element. There would be 655 such elements. Scan the picture and a square wave will result with a positive-going square for white and a negative for black. If we let a pair of adjacent square waves be equivalent to a sinusoid (see Figure 12.2), then the baseband required to transmit the image will have an upper cut-off of about 6.36 MHz allowing for no degradation in the intervening transmission system. The lower limit will be a dc or zero frequency.

12.3 THE COMPOSITE SIGNAL

The word composite is confusing in the television industry. On one hand, composite may mean the combination of the full video signal plus the audio subcarrier; the meaning here is narrower. Composite in this

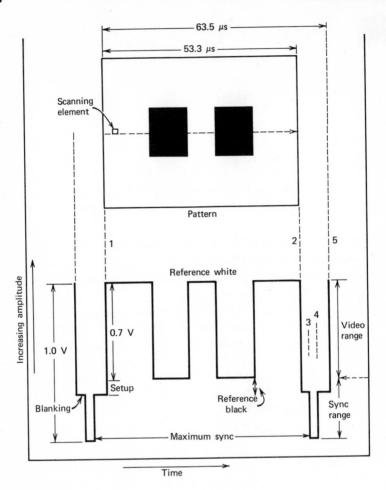

Figure 12.3 Breakdown in time of a scan line.

case deals with the transmission of video information as well as the necessary synchronizing information.

Consider Figure 12.3. An image made up of two black squares is scanned. The total time for the line is 63.5 μs, of which 53.3 μs is available for the transmission of actual video information and 10.2 μs is required for sync and flyback.

During the retrace time or flyback, it is essential that no video information be transmitted. To accomplish this a blanking pulse is superimposed on the video at the camera. The blanking pulse carries the signal voltage into the reference black region. Beyond this region in amplitude is the "blacker than black" region, which is allocated to the synchronizing pulses. The blanking level (pulse) is shown in Figure 12.3.

The maximum signal excursion of a composite video signal is 1.0 V. This 1.0 V is a video/television reference and is always taken as a peak-to-peak measurement. The 1.0 V may be reached at maximum sync voltage and is measured between sync "tips."

Of the 1.0 V peak-to-peak, 0.25 V is allotted for the sync pulses, 0.05 V for the setup, leaving 0.7 V to transmit video information. Thus the video signal varies from 0.7 V for white through gray tonal region to 0 V for black. The best description of the actual video portion of a composite signal is to call it a succession of rapid nonrepeated transients.

The synchronizing portion of a composite signal is exact and well defined. A TV/video receiver has two separate scanning generators to control the position of the reproducing spot. These generators are called the horizontal and vertical scanning generators. The horizontal one moves the spot in the "X" or horizontal direction, and the vertical in the "Y" direction. Both generators control the position of the spot on the receiver and must in turn be controlled from the camera (transmitter) sync generator to keep the receiver in step (synchronization).

The horizontal scanning generator in the video receiver is synchronized with the camera sync generator at the end of each scanning line by means of horizontal sync pulses. These are the sync pulses shown in Figure 12.3 and have the same polarity as the blanking pulses.

When discussing synchronization and blanking, we often refer to certain time intervals. These are discussed as follows:

1.   The time at the horizontal blanking pulse, 2–5 in Figure 12.3, is called the horizontal synchronizing interval.
2.   The interval 2–3 in Figure 12.3 is called the "front porch."
3.   The interval 4–5 is the "back porch."

The "intervals" are important because they provide isolation for over-shoots of video at the end of scanning lines.

Figure 12.4 illustrates the horizontal sync pulses and corresponding porches.

The vertical scanning generator in the video/TV receiver is synchronized with the camera (transmitter) sync generator at the end of each field by means of vertical synchronizing pulses. The time interval between successive fields is called the vertical interval. The vertical synchronizing pulse is

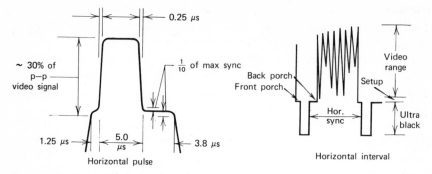

Figure 12.4   Sync pulses and porches.

built up during this interval. The scanning generators are fed by differentiation circuits. Differentiation for the horizontal scan has a relatively short time constant (RC) and that for the vertical a comparatively long time constant. Thus the long-duration vertical sync may be separated from the comparatively short-duration horizontal sync. This method of separation of sync is known as "waveform separation" and is standard in North America.

In the composite video signal (North American standards) the horizontal sync has a repetition rate of 15750/s and the vertical sync has a repetition rate of 60/s.

12.4   CRITICAL VIDEO TRANSMISSION IMPAIRMENTS

The nominal video baseband is divided into two segments: the high frequency segment, that above 15,750 Hz, and the low frequency segment, that below 15,750 Hz. Impairments in the high frequency segment are operative (in general) along the horizontal axis of the received video image. Low frequency impairments are generally operative along the vertical image axis.

Impairment	Cause
High frequency segment	
Undistorted echos	Cyclic gain and phase deviation throughout the passband
Distorted echos	Nonlinear gain and phase deviations, particularly in the higher end of the band

Impairment	Cause
High frequency cutoff effects, ringing	1. Limited bandpass distorts, shows picture transitions, causing overshoot and undershoot
	2. This type of distortion (ringing) may also show up on test pattern from lack of even energy distribution and reduced resolution
Porch distortion (poor reproduction of porches)	Poor attenuation and phase distortion
Porch displacement	Zero wander, dc-restored devices such as clamper circuits
Smearing (blurring of the vertical edges of objects)	1. Coarse variations in attenuation and phase
	2. Quadrature distortion

Low frequency segment

Dc suppression, zero wander, distorted image	Lack of clamping
Low frequency roll-off (gradual shading from top to bottom)	Poor clamping, deterioration of coupling networks
Streaking	Phase and attenuation distortion, usually from transmission medium

Nonlinear distortion

Nonlinearity in the extreme negative region, resulting in horizontal striations or streaking	Compression of synchronizing pulses

Impairments due to noise

Noise, in this case, may be considered an undersirable visual sensation. Noise is considered to consist of three types: single frequency, random, and impulse.

Unwanted pattern on received picture	Single frequency noise
Pattern in alternate fields	Single frequency noise as an integral multiple of field frequency (50 or 60 Hz)
Horizontal or vertical bars	Single frequency noise
It should be noted that the low frequency region is very sensitive to single frequency noise.	

Impairment	Cause
Picture graininess, "snow"	As random noise increases, graininess of picture increases

> *Note.* System design objective of 47 dB signal-to-weighted noise ratio or 44 db signal-to-flat noise ratio, based on 4.2-MHz video bandwidth, meets TASO* rating of excellent picture
> Peak noise should be 37 dB below peak video
> S/N ratios are taken at peak sync tips to rms noise

Noise "hits," momentary loss of sync, momentary rolling, momentary masking of picture	Impulse noise

> *Note.* Bell system limit at receiver input —20 dB reference to 1 V peak-to-peak signal level point, 1 hit/min
> Large amplitudes of impulse noise often masked in black

Weak, extraneous image superimposed on main image	Strong crosstalk
Nonsynchronization of two images causes violent horizontal motion. Effect most noticeable in line and field synchronization intervals	

> *Note.* Limiting loss in crosstalk coupling path should be 58 dB or greater for equal signal levels, design objective 61 dB.
> Crosstalk for video may be defined as the coupling between two television channels

* Television Allocation Study Organization.

The preceding sections present a short explanation of the mechanics of video transmission for a video camera directly connected to a receiving device for display. The primary concern of the chapter, however, is to describe and discuss the problems of point-to-point video transmission. Often the entity responsible for the point-to-point transport of television programs is not the same entity that originated the image transmitted, except in the case of that link directly connecting the studio to a local transmitter (STL links). A great deal of the "why" has now been covered. The remaining parts of the chapter discuss the video transmission problem (the "how") on a point-to-point basis. The medium employed for this purpose may be radiolink, satellite link, coaxial cable, or specially conditioned wire pairs.

12.5 CRITICAL VIDEO PARAMETERS

12.5.1 General

Raw video baseband transmission requires excellent frequency response, particularly from dc to 15 kHz and extending to 4.2 MHz for North American systems and to 5 MHz for European systems. Equalization is extremely important. Few point-to-point circuits are transmitted at baseband because transformers are used for line coupling which deteriorate low frequency response and make phase equalization very difficult.

To avoid low frequency deterioration cable circuits transmitting video have resorted to the use of carrier techniques and frequency inversion using vestigial sideband modulation. However, if raw video baseband is transmitted, care must be taken in preserving its dc component.

12.5.2 Transmission Standard—Level

Standard power levels have developed from what is roughly considered to be the input level to an ordinary television receiver for a noise-free image. This is 1 mV across 75 Ω. With this as a reference TV levels are given in dBmV. For RF and carrier systems carrying video the measurement refers to rms voltage. For raw video it is 0.707 of instantaneous peak voltage, usually taken on sync tips.

Signal-to-noise ratio is normally expressed for video transmission as follows:

$$\frac{S}{N} = \frac{\text{peak signal (dBmV)}}{\text{rms noise (dBmV)}}$$

TASO Picture Ratings (4-MHz bandwidth) are related to S/N ratio (RF) as follows:

1. Excellent (no perceptible snow)   45 dB
2. Fine (snow just perceptible)   35 dB
3. Passable (snow definitively perceptible but not objectionable)   29 dB
4. Marginal (snow somewhat objectionable)   25 dB

12.5.3 Other Parameters

For black and white video systems there are four critical transmission parameters. These are as follows.

- Amplitude-frequency response.
- Envelope delay distortion (group delay).
- Transient response.
- Noise (thermal, intermodulation, crosstalk, and impulse).

Color transmission requires consideration of two additional parameters:

- Differential gain.
- Differential phase.

Descriptions of amplitude-frequency response and envelope delay distortion may be found in Sections 1.9.3 and 1.9.4, respectively. Because video transmission involves such wide bandwidths compared to the voice channel and because of the very nature of video itself, both delay and amplitude requirements are much more stringent.

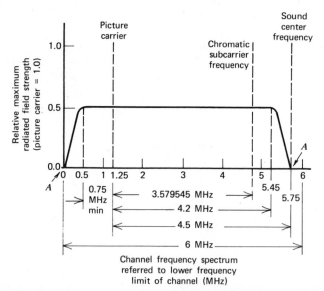

Figure 12.5 Radio frequency amplitude characteristic of television picture transmission. Field strength at points *A* shall not exceed 20 dB below picture carrier. Drawing not to scale.

Transient response is the ability of a system to "follow" sudden, impulsive changes in signal waveform. It usually can be said that if the amplitude-frequency and envelope delay characteristics are kept within design limits, the transient response will be sufficiently good.

Noise is described in Section 1.9.6. Differential gain is the variation in the gain of the transmission system as the video signal level varies (i.e., as it traverses the extremes from black to white). Differential phase is any variation in phase of the color subcarrier as a result of a changing luminance level. Ideally variations in the luminance level should produce no changes in either the amplitude or the phase of the color subcarrier. Table 12.1 summarizes critical transmission parameters for video.

Table 12.1 Critical Parameter Limits

	Input to TV receiver	Point-to-point transmission* Facility-EIA RS-250A Single hop
Amplitude frequency response		±0.1 dB 1–300 kHz ±0.4 dB 0.3–4.3 MHz
Envelope delay distortion		Deferred
Transient response		
$2T \sin^2$ pulse-bar K		1%
Noise (expressed as S/N)		
Hum	45 dB	46 dB
IM (intermodulation)	50 dB	
Crosstalk	50 dB	
Impulse		
Random (thermal + hum + IM)	47 dB	59 dB
Single frequency inteference		60 dB 1000 Hz–4.3 MHz
Differential gain		
10% APL [a]		0.5 dB
50% APL		0.3 dB
90% APL		0.5 dB
Differential phase		
10% APL		±1.0°
50% APL		±0.7°
90% APL		±1.0°
Signal polarity in transmission system		Black negative

[a] APL = average picture level.   *radiolink

12.6 VIDEO TRANSMISSION STANDARDS (Criteria for Broadcasters)

The following outlines video transmission standards from the point of view of broadcasters (i.e., as emitted from TV broadcast transmitters). Figure 12.5 illustrates the components of the emitted wave (North American practice).

12.6.1 Basic Standards

UNITED STATES

(Source: *ITT Reference Data for Radio Engineers 28–13*)

Note. Table 12.2 gives a capsule summary of some national standards as taken from CCIR.

Channel width (transmission)	6 MHz
Video	4.2 MHz
Aural	±25 kHz
(see Figure 12.5)	
Picture carrier location	1.25 MHz above lower boundary of channel
Modulation	AM composite picture and synchronizing signal on visual carrier together with FM audio signal on audio carrier
Scanning lines	525/frame, interlaced 2:1
Scanning sequence	Horizontally from left to right, vertically from top to bottom
Horizontal scanning frequency	15750 Hz for monochrome $2/455 \times$ chrominance subcarrier or 15734.264 ± 0.044 Hz for NTSC color transmission
Vertical scanning frequency	60 Hz for monochrome or $2/525 \times$ horizontal scanning frequency for color or -59.94 Hz
Blanking level	Transmitted at $75 \pm 2.5\%$ of peak carrier level
Reference black level	Black level is separated from the blanking level by $7.5 \pm 2.5\%$ of the video range from blanking level to reference white level
Reference white level	Luminance signal of reference white is $12.5 \pm 2.5\%$ of peak carrier

UNITED STATES—*continued*

Peak-to-peak variation	Total permissible peak-to-peak variation in one frame due to all causes is less than 5%
Polarity of transmission	Negative; a decrease in initial light intensity causes an increase in radiated power
Transmitter brightness response	For monochrome television, RF output varies in an inverse logarithmic relation to the brightness of the scene
Aural transmitter power	Maximum radiated power is 20% (minimum, 10%) of peak visual transmitter power

BASIC EUROPEAN STANDARD

Channel width (transmission)
   Video   5 MHz
   Aural   ±50 kHz
	7 MHz low band, 8 MHz high band

Picture carrier location	1.25 MHz above lower boundary of channel

Note.   Vestigial sideband transmission is used, similar to North American practice

Modulation	AM composite picture and synchronizing signal on visual carrier together with FM audio signal on audio carrier
Scanning lines	625/frame, interlaced 2 : 1
Scanning sequence	Horizontally from left to right, vertically from top to bottom
Horizontal scanning frequency	15625 Hz ± 0.1%
Vertical scanning frequency	50 Hz
Blanking level	Transmitted at 75 ± 2.5% of peak carrier level
Reference black level	Black level is separated from the blanking by 3–6.5% of peak carrier
Peak white level as a percentage of peak carrier	10–12.5%
Polarity of transmission	Negative; a decrease in initial light intensity causes an increase in radiated power
Aural transmitter power	Maximum radiated power is 20% of peak visual power

SOME VARIANCES (see Table 12.2 as well)

United Kingdom

Scanning lines	405/frame interlaced 2:1 on low band
Horizontal scanning frequency	10,125 Hz, low band
Video bandwidth	3 MHz, low band
Nominal RF bandwidth	5 MHz, low band
Aural transmitter power	Maximum radiated power is 25% of peak visual power
Type of sound modulation	AM, low band
Synchronizing level as % of peak carrier	30%
Blanking level as % of peak carrier	30%
Black level	Same as blanking level, low band
Peak white level as % of peak carrier	100%, low band
Sound carrier relative to vision carrier	3.5 MHz low band

France

Scanning lines	819
Nominal video bandwidth	10 MHz
Horizontal scanning frequency	20475 Hz
Nominal RF bandwidth	14 MHz
Sound carrier relative to vision carrier	11.5 MHz
Type of sound modulation	AM
Nominal width of vestigial sideband	2 MHz
Synchronizing level as % of peak carrier	30%
Difference between black level and blanking level as % of peak carrier	5%
Peak white level as % of peak carrier	100%
Percentage of effective radiated power of sound compared to vision	25%

Belgium (same as France except for the following)

Nominal video bandwidth	5 MHz
Nominal RF bandwidth	7 MHz
Sound carrier at	+5.5 MHz
Nominal width of vestigial sideband	Same as rest of Europe
Blanking level as % of peak carrier	22.5–27.5%
Difference between black level and blanking level as % of peak carrier	3–6%

12.6.2 Color Transmission

Three color transmission standards exist. These are

NTSC   National Television System Committee (North America, Japan)
SECAM   Sequential couleur A Memoire (Europe)
PAL   Phase Alternation Line (Europe)

The systems are similar in that they separate the luminance and chrominance information and transmit the chrominance information in the form of two color difference signals which modulate a color subcarrier transmitted within the video band of the luminance signal. The systems vary in the processing of chrominance information.

In the NTSC system, the color difference signals, I and Q, amplitude-modulate subcarriers that are displaced in phase by $n/2$, giving a suppressed carrier output. A burst of the subcarrier frequency is transmitted during the horizontal back porch to synchronize the color demodulator.

In the PAL system, the phase of the subcarrier is changed from line to line which requires the transmission of a switching signal as well a color burst.

In the SECAM system, the color subcarrier is frequency modulated, alternately by the color difference signals. This is accomplished by an electronic line-to-line switch; the switching information is transmitted as a line-switching signal.

Table 12.2 Summary of Some Characteristics of TV Systems

System [a]	No. of Lines	Channel Width (MHz)	Vision Width (MHz)	Separation Vision Sound	Vestigial Sideband	Vision Modulation	Sound Modulation	Sweep Rate	Picture Rate	Field Frame
A	405	5	3	−3.5	.75	Pos	AM	10.125K	25	50
B	625	7	5	+5.5	.75	Neg	FM	15.625K	25	50
C	625	7	5	+5.5	.75	Pos	AM	15.625K	25	50
D	625	8	6	+6.5	.75	Neg	FM	15.625K	25	50
E	819	14	10	±11.15	2.	Pos	AM	20.475K	25	50
F	819	7	5	+5.5	.75	Pos	AM	20.475K	25	50
G	625	8	5	+5.5	.75	Neg	FM	15.625K	25	50
I	625	8	5.5	+6	1.25	Neg	FM	15.625K	25	50
L	625	8	6	+6.5	1.25	Pos	AM	15.625K	25	50
M	525	6	4.2	+4.5	.75	Neg	FM	15.750K	30	60

Source: CCIR Rep. 308, Vol. V, Geneva, 1963; consult report for complete table.

[a] Austria B, Finland B, Luxemborg F, Portugal B, West Germany B, G
Belgium C, F, France E, L, Monaco E, Spain B, U.S., Japan and Canada M
Bulgaria D, Hungary D, Netherlands B, Sweden B
Czechoslovakia, USSR D, Ireland A, Norway A, Switzerland B
Denmark B, Italy B, G, Poland D, U. K. A, I

12.6.3   Standardized Transmission Parameters*
(Point-to-Point Television)

Interconnection at video frequencies:

Impedance	75 Ω unbalanced
Return loss	no less than 24 dB
Input level	1 V P-P
Output level	1 V P-P
Polarity	black-to-white transitions, positive going

Interconnection at IF:

Impedance	75 Ω unbalanced
Input level	0.3 V rms
Output level	0.5 V rms
IF up to 1 GHz	35 MHz
IF above 1 GHz	70 MHz

In a hypothetical reference circuit 2500 km long, signal-to-noise ratios for different systems are as follows:

System (lines)	405	525	625	625	819	819
Video baseband (MHz)	3	4	5	6	5	10
Signal-to-weighted noise (dB)	50	56	52	57	52	50

12.7   METHODS OF PROGRAM CHANNEL TRANSMISSION FOR VIDEO

Composite transmission normally is used on broadcast and CATV distribution. Video and audio carriers are "combined" before being fed to the radiating antenna for broadcast. These audio "subcarriers" are described above in the preceding Section 6.

For point-to-point television transmission the audio program channel generally is transmitted separately on coaxial cable, radiolink, and earth station systems. Separate transmission, usually on a separate subcarrier, provides the following advantages:

- Individual channel level control.
- Greater control over crosstalk.
- Increased guard band between video and audio.
- Saves separation at broadcast transmitter.
- Leaves TV studio as separate channel.
- Permits individual program channel preemphasis.

* Based on CCIR Recs. 421 and 403.

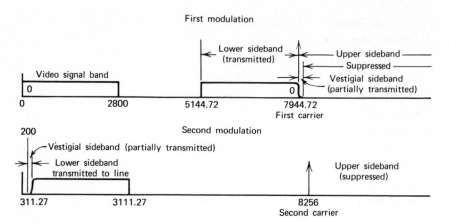

Figure 12.6 Modulation scheme for transmitting television on the North American L1 carrier system (Ref. 14). Copyright © 1961, by American Telephone and Telegraph Co.

12.8 VIDEO TRANSMISSION OVER COAXIAL CABLE

12.8.1 Early System

Early coaxial cable point-to-point video transmission systems reduced usable bandwidth to about 2800 kHz (L1 system). The 2.8 MHz signal was translated, in the case of the L1 system used in North America, to a line frequency of 200 kHz to 3111.27 kHz. Carrier modulation techniques are used. In the case of L1, the line frequency mentioned above is a lower sideband occupying between 311.27 and 3111.27 kHz. The vestigial sideband occupies 200–311.27 kHz. The modulation process is shown in Figure 12.6. The limited bandwidth of the L1 system reduced picture quality and was usable for monochrome video transmission only.

12.8.2 Modern Broadband Coaxial Cable Systems for Video

THE L3 SYSTEM OF NORTH AMERICA

The L3 carrier system was designed for use on 12-MHz coaxial cable. In most applications when L3 is used for video transmission, it shares the same cable with up to 600 FDM telephone channels. The FDM voice channel segment occupies the band from 564 to 3084 kHz. The video signal occupies the region from 3639 to 8500 kHz. The modulation of the video signal, translating it to the indicated segment of the spectrum, is carried out by vestigial sideband methods. The virtual carrier as transmitted on-line is at 4139 kHz and the vestigial sideband occupies the space 3639–4139 kHz. Vestigial sideband is used to avoid some of the problems encountered with the normally used envelope detection regarding video, such as the production of a spurious envelope wherein video signals which exceed a certain value are inverted. In the L3 system, as in most video transmission systems of this type, homodyne detection is used. Here the demodulator is driven by a locally generated carrier, which is synchronous in phase angle and frequency with the carrier component of the transmitted wave. Homodyne detection also makes possible the necessary suppression of the quadrature distortion associated with vestigial sideband transmission. Figure 12.7a and b illustrate the L3 frequency allocation and modulation processes to develop the line frequency.

To ameliorate somewhat the effects of second harmonic distortion a preemphasis network is used in the transmitting terminal to accentuate the amplitude of the high frequency components of the signal before transmission. At the receiving terminal a deemphasis network introduces a complementary frequency characteristic to make the overall transmission characteristic constant with frequency.

Delay and amplitude equalizers are incorporated in the transmit section with an objective of maintaining amplitude in the band of interest to vary no more than $\pm$ 0.02 dB. Phase shift objectives are on the order of $\pm$ 0.1°. Mop-up equalizers also are used on receiver terminals. Repeater spacing for the L3 system is 4 mi (6.4 km).

A 12-MHZ EUROPEAN SYSTEM

12-MHz coaxial cable systems, if used for video transmission, almost always transmit combined FDM telephone channels with the video. One such system modulates a 5.5-MHz video sideband with a 6.799-MHz carrier. An HF signal is produced in the band 6.3–12.3 MHz, using the upper sideband of the modulation process and a vestigial portion of the lower sideband. 1200 FDM telephone channels are transmitted in the lower portion of the band.

The required flatness in envelope delay and amplitude response is maintained in the video transmission band (i.e., 6.3–12.3 MHz) by equalizers built into the coupling and separating filter units for contributions from those units (i.e., each unit is provided with equalizers to flatten response of its own filters). The envelope delay distortion caused by the coaxial line itself and its associated equipment is equalized at the receiving end.

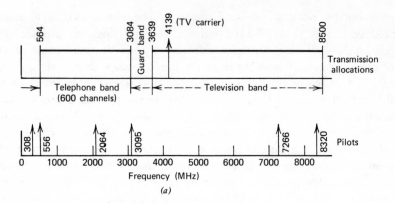

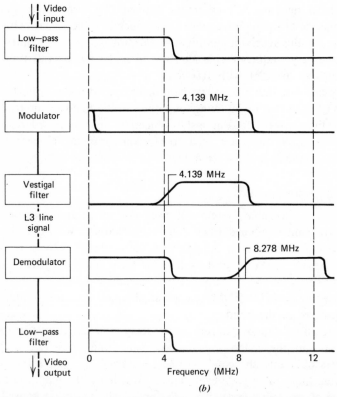

Figure 12.7 (a) Frequency allocations for the North American L3 combined television-telephone transmission system. (b) L3 television terminal modulation processes (Ref. 15). Copyright © 1953, by American Telephone and Telegraph Co.

The television baseband for a modulator/demodulator back-to-back is maintained as follows:

Attenuation distortion	0.2 dB
Envelope delay distortion	25 ns
Differential gain	0.05 dB
Differential phase	0.2°
Signal-to-hum ratio	60 dB
Continuous random noise, unweighted	66 dB (down)

CCITT Rec. G.332 covers this type of transmission under "mixed systems."

12.9 TRANSMISSION OF VIDEO OVER RADIOLINKS

12.9.1 General

Telephone administrations increasingly are expanding their offerings to include other services. One such service is to provide point-to-point broadcast-quality video relay on a lease basis over radiolinks. As covered earlier in this section, video transmission requires special consideration.

The following paragraphs summarize the special considerations a planner must take into account for video transmission over radiolinks.

Raw video baseband modulates the radiolink transmitter. The aural channel is transmitted on a subcarrier well above the video portion. The overall subcarriers are themselves frequency modulated. Recommended subcarrier frequencies may be found in CCIR Rec. 402 and Rep. 289-1. Noise on program (aural) channels is discussed in CCIR Rep. 375.

12.9.2 Bandwidth of the Baseband and Baseband Response

One of the most important specifications in any radiolink system transmitting video is frequency response. A system with cascaded hops should have essentially a flat bandpass in each hop. For example, if a single hop is 3 dB down at 6 MHz in the resulting baseband, a system of five such hops would be 15 dB down. A good single hop should be ± 0.25 dB or greater out to 8 MHz. The most critical area in the baseband for video frequency response is in the low frequency area, 15 kHz and below. Cascaded radiolink systems used in transmitting video must consider response down to 10 Hz.

Modern radiolink equipment used to transport video operate in the 2-GHz band and above. 525-line video requires a baseband in excess of 4.2 MHz plus available baseband above the video for the aural channel. Desirable characteristics for 525-line video then would be a baseband at least 6 MHz wide. 8 MHz would be required for 625-line television assuming that the aural channel would follow the channelization recommended by CCIR Rec. 402.

12.9.3 Preemphasis

Preemphasis, deemphasis characteristics are described in CCIR Rec. 405-1 (also see Figure 5.16).

12.9.4 Differential Gain

Distortion due to differential gain is caused by non-linear elements of the system such as nonlinearity in a receiver discriminator. One result of poor differential gain is cross-modulation of color signals. A specification of ± 0.25 dB of differential gain at full deviation is usually sufficient.

12.9.5 Differential Phase Distortion

The result of this type of distortion is also cross-modulation and often is a modulation of the sound channel by some picture channel element (15-kHz sync, fr example). A satisfactory single hop specification for differential phase is $\pm 0.5°$ per hop at 50% average picture level (APL).

12.9.6 Signal-to-Noise Ratio

The signal-to-noise ratio for 1 hop should be 59 dB (video signal) (EIA RS-250-A). For tandem systems 64 dB S/N ratio is recommended per hop, unweighted. EIA RS-250-A states that system S/N may be degraded to 56 dB for multihop systems. The reader should also consult CCIR Rec. 289-1.

12.9.7 Square Wave Tilt

The departure from horizontal of the top or bottom of a square wave at the field scanning rate shall not exceed $\pm 0.5\%$ (1% total) of the peak-to-peak amplitude. This is specified for one hop. Over a multihop system this figure shall not exceed ± 1.0% (2% total) (EIA RS-250-A).

12.9.8   Radio Link Continuity Pilot

For video transmission the continuity pilot is always above the baseband. CCIR recommends an 8.5-MHz pilot. (Refer to CCIR Rec. 401-2 and Table 5.6.)

12.10   TRANSMISSION OF VIDEO OVER CONDITIONED PAIRS

12.10.1   General

Broadcast-quality video transmission over conditioned pairs has application to interconnect broadcast facilities and the long-distance transmission system. The broadcaster may lease these facilities from a telephone administration to interconnect a master control point and outlying studios or remote program pick-up points. Normally only the video is transmitted on the conditioned pair. The audio is usually transmitted separately on its own program facilities. The cable is designed for installation in ducts and the repeaters in duct-type cabinets or racks.

12.10.2   Cable Description

A conditioned pair video transmission system is composed of a shielded wire pair, terminal equipment, and repeaters. The following description is of a system in wide use in North America. Such a description of a typical system will bring forth the advantages and limitations of video transmission over conditioned pairs.

The line facilities consist of 16-gauge polyethylene-insulated pairs generally referred to as PSV.* At 75°F the loss at 4.5 MHz is 3.52 dB/1000 ft and 18.6 dB/mi. The normal slope variation with temperature is approximately 0.1%/°F. The loss at zero frequency is taken as 0 dB.

Because of the effective shielding of the PSV pairs, there is no limitation as to direction of transmission or number of circuits obtainable within any given size of cable. Noise considerations cause the requirement that

---

* Pair shield video.

the PSV pairs be separated from the remainder of the cable conductors at building entrances. In this case the shielded video pair is brought to the video equipment under a separate sheath.

The charactristic impedance of the video cable is nearly pure resistive, 124 Ω, at frequencies above 500 kHz. The resistive component increases to 1000 Ω at 60 Hz. The reactive component is about the value of the resistive component at 60 Hz, and drops to nearly zero at the higher frequencies.

12.10.3   Terminal Equipment and Repeaters

A transmitting terminal is provided to match the video output of the line, secure the proper level, and predistort the signal as a first step in conditioning the PSV cable. Equalization is basically one of amplitude. The impedance is from 75 Ω unbalanced to 124 Ω balanced.

Amplitude equalization for this type of transmission facility is such that levels are described as fractions. The level unit is dBV (decibels relative to 1 V). For instance, a voltage level may be expressed as − 10/+ 5 dBV. Such a "fraction" describes attenuation-frequency characteristic or "slope." The numerator refers to a level of zero frequency and the denominator of the fraction refers to the reference high frequency, in this case 4.5 MHz. Voltages here are peak-to-peak. Zero frequency may be taken to mean a very low frequency, 30 Hz.

This method of designating level is a useful tool when amplitude-frequency response and its equalization are of primary concern. At the transmitting end of a PSV link we would expect a zero slope and the transmitter output would be 1.0 V p-p (0 dBV) or 0/0 dBV. To equalize the line the transmitter must predistort the signal. After equalization the level may be described assuming 15-dB equalization, as − 10/+ 5 dBV. After traversing 33 dB of cable, the level may then be described as − 10/− 28 dBV. Repeaters provide both amplification and equalization. It would appear that most systems would require custom design. To simplify engineering and standardize components, PSV systems often are built in blocks. The system is configured with fixed-length repeater sections, plug-in equalizers, and standardized receivers. Repeater spacing is on the order of 4.5 mi (7.2 km).

The receiver provides both gain and equalization. It also includes a clamper to correct for low frequency distortion. Equalization is usually variable when the block approach is used for residual gain deviations.

12.11 BASIC TESTS FOR VIDEO QUALITY

12.11.1 Window Signal

The window signal when viewed on a picture monitor is a large square or rectangular white area with a black background. The signal is actually a sine-squared pulse. As such it has two normal levels, reference black and reference white. The signal usually is adjusted so that the white area covers $\frac{1}{4}$ to $\frac{1}{2}$ the total picture width and $\frac{1}{4}$ to $\frac{1}{2}$ the total picture height. This is done in order to locate the maximum energy content of the signal in the lower portion of the frequency band.

A number of useful checks derive from the use of a window signal and a picture monitor. These include the following:

1. Continuity or level check: with a window signal of known white level, the peak-to-peak voltage of the signal may be read on a calibrated oscilloscope using a standard roll-off characteristic (i.e., EIA, etc.).

2. Sync compression or expansion measurements: comparison of locally received window signals with that transmitted from the distant end with respect to white level and horizontal sync on calibrated oscilloscopes using standard roll-off permits evaluation of linearity characteristics.

3. Test and adjustment can be made at clamper amplifiers and low frequency equalizers to minimize streaking by observing the test signal on scopes using the standard roll-off characteristics at both the vertical and horizontal rates.

4. Indication of ringing: with a window signal the presence of ringing may be detected by using properly calibrated wide band oscilloscopes and adjusting the horizontal scales to convenient size. Both amplitude and frequency of ringing may be measured by this method.

12.11.2 Sine-Squared Test Signal

The sine-squared test signal is a pulse type of test signal which permits an evaluation of amplitude-frequency response, transient response, envelope delay, and phase. An indication of the high frequency amplitude characteristic can be determined by the pulse width and height, and the phase characteristic by the relative symmetry about the pulse axis. However, this test signal finds principal application in checking transient response and phase delay. The sine-squared signal is far more practical than a square wave test signal to detect overshoot and ringing. The pulse used for checking video systems should have a repetition rate equal to the line frequency, and a duration, at half amplitude, equal to one-half the period of the nominal upper cutoff frequency of the system.

12.11.3 Multiburst

This test signal is used for a quick check of gain at a few determined frequencies. A common form of multiburst consists of a burst of peak white (called white flag) which is followed by bursts of six sine wave frequencies from 0.5 to 4.0 MHz (for NTSC systems) plus a horizontal sync pulse. All these signals are transmitted during one line intervals. The peak white or white flag serves as a reference. For system checks a multiburst signal is applied to the transmit end of a system.

At the receiving point the signal is checked on an oscilloscope. Measurements of peak-to-peak amplitudes of individual bursts are indicative of gain. A multiburst image on a scope gives a quick check of amplitude-frequency response and changes in setup. Figure 12.8 illustrates a typical NTSC type multiburst signal.

12.11.4 Stair Steps

For the measurement of differential phase and gain a 10-step stair-step signal is often used. Common practice (in the United States) is to super-impose 3.6 MHz on the 10-steps that extend progressively from black to white level. The largest amplitude sine wave block is adjusted on the

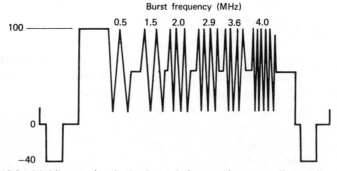

Figure 12.8 Multiburst signal (horizontal frequencies normally used).

oscilloscope to 100 standard (i.e., IRE-IEEE) divisions and is made a reference block. Then the same 3.6 MHz sine waves from the other steps are measured in relation to the reference block. Any difference in amplitudes of the other blocks represents differential gain. By the use of a color analyzer in conjunction with the above, differential phase may also be measured.

The stair-step signal may be used as a linearity check without the sine wave signal added. The relative height between steps is in direct relation to signal compression or nonlinearity.

12.11.5 Vertical Interval Test Signals (VITS)

The VITS makes use of the vertical retrace interval for the transmission of test signals. In the United States, the FCC specifies the interval for United States use as the last 12 μs of line 17 through line 20 of the vertical blanking interval of each field. For whichever interval boundaries specified, test signals transmitted in the interval may include reference modulation levels, signals designed to check performance of the overall transmission system or its individual components, and cue and control signals related to the operation of television broadcast stations. These signals are used by broadcasters because, by necessity, they are inserted at the point of origin. Standard test signals are used as described above or with some slight variation, such as multiburst, window, and stair step. Some broadcasters use vertical interval reference signals. Figure 12.9 shows one in use currently in the United States.

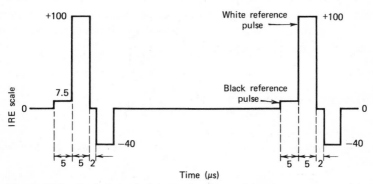

Figure 12.9 Typical vertical interval reference signal.

12.11.6   Test Patterns

Standard test patterns, especially those inserted at a point of program origination, provide a simple means of determining transmission quality. The distant viewer, knowing the exact characteristics of the transmitted image, can readily detect distortion(s). Standard test patterns such as the EIA test pattern used widely in the United States, with a properly adjusted picture monitor, can verify the following:

> Horizontal linearity
> Vertical linearity
> Contrast
> Aspect ratio
> Interlace
> Streaking
> Ringing
> Horizontal and vertical resolution

12.11.7   Color Bars

Color bar test signals are used by broadcasters for the adjustment of their equipment including color monitors. Color bars may also be sent over transmission facilities for test purposes. The color bar also may be used to test color transmission using a black and white monitor by examining gray densities of various bars depending on individual colors. A wide band A-scope horizontal presentation can show whether or not the white reference of the luminance signal and the color information have the proper amplitude relationships. The color bar signal may also be observed on a vector display oscilloscope (chromascope) which allows measurement of absolute amplitude and phase angle values. It also can be used to measure differential phase and gain.

REFERENCES AND BIBLIOGRAPHY

1. CCIR Xth Plenary Assembly, Volume V, Geneva 1963.
2. *Reference Data for Radio Engineers*, 5th ed., Howard W. Sams & Co., Indianapolis, 1968.
3. "Fundamentals of Television Transmission," *Bell System Practices*, Section AB 96.100, ATT, New York, March 1954.

4. "Television Systems Descriptive Information—General Television Signal Analysis," *Bell System Practices*, Section 318-015-100, Issue 3, ATT, New York, Jan. 1963.

5. "Engineering of Local Radio Television Links," *Bell System Practices*, Section R100.080, ATT, New York, July 1952.

6. "Television Systems—A2A Video Transmission System Description," *Bell System Practices*, Section 318-200-100, Issue 5, ATT, New York, Feb. 1962.

7. *Transmission Systems for Communications*, Bell Telephone Laboratories, 4th ed., ATT, New York, Chap. 29.

8. *Lenkurt Demodulator*, Lenkurt Electric Co., San Carlos, Calif.:Feb. 1962, Oct., Nov. 1963, Jan. 1965, March 1966, and Feb. 1971.

9. D. H. Hamsher, *Communication System Engineering Handbook*, McGraw-Hill, New York, 1967, Chapter 13.

10. J. Herbstreit and H. Pouliquen, "International Standards for Colour Television" (paper), International Telecommunications Union, Geneva, 1967.

11. K. Simons, *Technical Handbook for CATV Systems*, 2nd ed., Jerrold Electronics Corp., Philadelphia, 1966.

12. EIA Standard RS-250-A, Electronic Industries Association, Washington, D.C., Feb. 1967.

13. D. Kirk, Jr., *Video Microwave Specifications for System Design*, reprint of article appearing in *Broadcast Engineering*, Jerrold Electronics Corp., Philadelphia, 1966.

14. *Principles of Electricity Applied to Telephone and Telegraph Work*, American Telephone and Telegraph Co., New York, 1961.

15. John W. Rieke and R. S. Graham, "The L3 Coaxial System Television Terminals," *Bell Syst. Tech. J.*, July 1953.

16. W. von Guttenberg and E. Kugler, "Modulation of TV Signals for Combined Telephone and Television Transmission over Cables," *NTZ-CJ J.* No. 2, 1965.

GUIDE TO CCITT AND CCIR RECOMMENDATIONS*

CCITT and CCIR are two consultative committees operating under the auspices of the International Telecommunication Union (ITU) based in Geneva, Switzerland. More than 150 nations belong to the ITU and contribute to the consultative committees' work, as well as many of these countries telecommunication administrations and other common carriers. Likewise a large number of the major producers of communication equipment also contribute as nonvoting members.

The work of CCITT (Comite Consultif International Tèléphone Telegraph) and CCIR (Comite Consultif International Radio) deals with providing standardized practices and system specifications so that the world's nations can intercommunicate, whether by telephone, telegraph, facsimile, data, program channel, or television. CCITT is concerned mainly with wire communications and CCIR with radio, although there is considerable interplay and overlap between the two consultative committees' work, which is embodied in a series of recommendations.

CCITT

The CCITT Recommendations consist of nine volumes. The volumes are identified by color and the location in the world where the plenary assembly took place that authorized issuance of those recommendations, such as CCITT Green Books, Geneva, 1972.

CCITT uses the word "volume" with ambiguity. Some "volumes" contain several sections bound separately, identified by letters. Many volumes have separately bound parts identified by "stars."

* The basic tabular information presented in Appendix A is by courtesy of the International Telecommunication Union—CCITT, CCIR—Geneva.

Most recommendations are from 2 to 4 pages in length, although some are much longer. CCITT Recommendations (Rec.) and Reports (Rep.) are classified by letters A–V as follows:

Series Rec.	Contents*
A.	Organization of the work in CCITT, Vol. I
B.	Means of expression, Vol. I
C.	Blank
D.	Lease of circuits, Vol. IIA
E.	Telephone operation and tariffs, charging and accounting, statistics on international telephony and service quality, Vol. IIA
F.	Telegraph operation and tariffs, Vol. IIB
G.	Telephone transmission on metallic lines, radiolinks, satellite, and radiotelephone systems, Vol. III, 1 star
H.	Lines used for the transmission of signals other than telephone signals, such as telegraph, facsimile, data, etc., Vol. III, 2 star
I.	Blank
J.	Program and television transmission, Vol. III, 3 star
K.	Protection against interference, Vol. IX
L.	Protection of cable sheaths and poles, Vol. IX
M.	Maintenance: telephony, telegraphy, and data transmission, Vol. IV
N.	Maintenance: sound program and television transmission, Vol. IV
O.	Blank
P.	Quality of telephone transmission, Vol. V
Q.	Signaling and switching, including traffic, Vol. VI
R.	Telegraph (VFCT), Vol. VII
S.	Printing telegraph equipment, Vol. VII
T.	Facsimile, Vol. VII
U.	Telegraph switching, Vol. VII
V.	Data transmission, Vol. VIII

* Volume identification refers to "White Books," Mar del Plata, 1968.

**Contents of the CCITT Books Applicable after the
Fifth Plenary Assembly (1972)**

Volume	Green Book Contents
I	Minutes and reports of the Vth Plenary Assembly of the CCITT Resolutions and opinions issued by the CCITT General table of Study Groups and Working Parties for the period 1973–1976 Summary table of Questions under study in the period 1973–1976 Recommendations (Series A) on the organization of the work of the CCITT Recommendations (Series B) relating to means of expression. Recommendations (Series C) relating to general telecommunication statistics.
II-A	Recommendations (Series D) and Questions (Study Group III) relating to the lease of circuits. Recommendations (Series E) and Questions (Study Group II) relating to telephone operation and tariffs.
II-B	Recommendations (Series F) and Questions (Study Group I) relating to telegraph operation and tariffs.
III	Recommendations (Series G, H, J) and Questions (Study Groups XV, XVI, Special Study Groups C, D) relating to line transmission.
IV	Recommendations (Series M, N, O) and Questions (Study Group IV) relating to the maintenance of international lines, circuits and chains of circuits.
V	Recommendations (Series P) and Questions (Study Group XII) relating to telephone transmission quality, local networks, telephone sets equipment.
VI	Recommendations (Series Q) and Questions (Study Groups XI, XIII) relating to telephone signaling and switching.
VII	Recommendations (Series R, S, T and U) and Questions (Study Groups VIII, IX, X, and XIV) relating to telegraph technique.
VIII	Recommendations (Series V, X) and Questions (Study Group VII and Special Study Group A) relating to data transmission.
IX	Recommendations (Series K) and Questions (Study Group V) relating to protection against interference. Recommendations (Series L) and Questions (Study Group VI) relating to the protection of cable sheaths and poles.

Each volume also contains, where appropriate:
Definitions of specific terms used in the field of the volume.
Supplements for information and documentary purposes.

The G. Recommendations of CCITT have the most impact on transmission system design. Tables A.1–A.7 summarize those recommendations. The tables have been provided courtesy of the International Telecommunication Union. (Green Books, Geneva 1972.)

Table A.1a Summary of the Main Characteristics Specified by the CCITT for International Telephone Circuits[a] and International Connections[b]

Characteristic	For an International Circuit (1)	For a Complete Connection or for Its Parts (2)
Reference equivalent	G.111, B	For the connection and for the national systems G.111, G.121
Nominal four-wire equivalent (transmission plan, see G.101)	0.5 dB (G.141) Echo effects (G.131, B)	Four-wire chain national circuits G.101, B, b, G.121, G.122
Transmission stability	G.131, A	Balance return loss of national networks (G.122)
Band of frequencies effectively transmitted Limits (Hz)	at least 300–3400 (G.151, A)	From international centre to local exchange: 300–3400 (G.124) Four-wire chain of 6 circuits: 300–3400 (G.132)
Additional attenuation at limits of frequency	9 dB (G.151, A and G.132)	9 dB (G.151, A and G.132)
Attenuation distortion	G.151, A Fig. 1/G.151	Graph No. 1, desirable objective for 12 circuits (Fig. 1/G.132) For data see H.12
Group delay, t	G.114	For the connection (G.114) $t \leq 150$ ms, without reservation $t \leq 400$ ms, acceptable with conditions. For data see H.12

Phase distortion (from the group delay t) [c]	$t_m - t_{min} \leq 30$ ms[d] $t_M - t_{min} \leq 15$ ms[d] (G.133)	For the four-wire chain (G.133) $t_m - t_{min} \leq 60$ ms $t_M - t_{min} \leq 30$ ms For each national four-wire chain (G.133) $t_m - t_{min} \leq 15$ ms $t_M - t_{min} \leq 7.5$ ms
Variation of overall loss with time	Mean deviation from nominal $\leq \pm 0.5$ dB Std. dev. ≤ 1 dB or 1.5 dB (G.151, C)	Extension circuits: as (1) (G.151) For data see H.12
Linear crosstalk between different circuits (near- or far-end crosstalk ratio a)	$\Delta \geq 58$ dB (G.151, D, see Note 3)	Extension circuits: as (1) (G.151).
Near-end crosstalk ratio between the two directions of transmission	Ordinary circuits: ≥ 43 dB (G.151, D) With speech concentrator: ≥ 58 dB With echo suppressors: ≥ 55 dB (G.151, D — see Note 4)	Extension circuits: as (1) (G.151)
Circuit noise	See Table A.1b	
Impedance of the circuit		A single value for one trunk exchange (G.232, M)
Frequency difference at two ends of a carrier circuit	≤ 2 Hz (G.135, G.225)	G.135, G.225
Power at zero relative level point Telephony, mean power in busy hour	Speech currents, etc., 22 μW[e] (G.223) Electric signals + tones 10 μW[e] (G.223) (see G.224 for the power of signaling pulses)	

517

Table A.1a (*Continued*)

Characteristic	For an International Circuit (1)	For a Complete Connection or for Its Parts (2)
Voice frequency telegraphy. Maximum power per channel for vft systems having	Amplitude modulation. Power when sending continuous mark (H.23, A, a)	Frequency modulation mean power (H.23, A, b)
24 channels	9 μW	5.6 μW
18 channels	15 μW	7.5 μW
12 channels or less	35 μW	11.25 μW
Private wire telegraphy and telephony		
One or other	Sending continuous mark 0.3 mW maximum (H.31) [f]	
Both	Teleg. level $\leq$ −13 dBm0 (H.32) [f]	
Phototelegraphy	Amplitude modulation 1 mW, frequency modulation 0.1 mW (H.41)	
Maximum power for data transmission over leased circuits (H.51, A) [f]	1 mW on subscriber's line Frequencies $\geq$ 2400 Hz, see G.224 Frequency modulation: −10 dBm0 or −20 dBm0 Amplitude modulation: −6 dBm0 and 64 μW (mean for both directions in busy hour)	
Maximum power for data transmission over circuits in switched network (H.51, B) [f]	1 mW on subscriber's line Frequencies $\geq$ 2400 Hz, see G.224 Frequency or phase modulation: −10 dBm0 in simplex, −13 dBm0 in duplex Amplitude modulation: 64 μW (mean for both directions in busy hour)	

[a] Unless otherwise stated, circuits for voice-frequency telegraphy or phototelegraphy have the same characteristics.

[b] This very condensed table is not a recommendation, and reference should be made to the complete Recommendations.

^c m = nominal minimum frequency effectively transmitted.

M = nominal maximum frequency effectively transmitted.

min = frequency corresponding to minimum group delay time.

^d These values apply to the chain of international circuits.

^e Calculation target value or conventional value for a hypothetical reference circuit.

^f This recommendation contains restrictions of use. See also Recommendation H.34.

Table A.1b   Summary of Noise Objectives Specified

	General Objectives				
Types of Systems	Cable[b] or Radio Relay Link		Single Hop Satellite Link	Submarine Cable[b]	All Systems
Telephone circuits considered[a]	National four-wire extension circuits and international circuits from 250 to 2500 km	Circuits from 2500 to about 25,000 km	Circuits from 7500 to about 15,000 km	Circuits from 2500 to about 25,000 km	Chain of six international circuits
Recommendations Of the CCITT	G.152 G.212[c] G.222 G.226	G.153		G.153 G.143	G.143
Of the CCIR	391, 392, 393-1, 394, 395-1, 396-1, 397-2		352-1, 353-2		
Hypothetical reference circuit (HRC) or typical circuit considered	HRC of 2500 km[d] or similar real circuit	Circuit of 7500 km[d]	Basic HRC of at least 6500 km	Chain of about 25,000 km	Chain of more than 25,000 km
Recommended objectives Psophometric power Hourly mean Total power	10,000 pW				50,000 pW
Terminal equipment Line	2500 pW 7500 pW i.e., 3 pW/km	15,000 pw[i] 2 pW/km or better[e]	10,000 pW[i]	About 7000–9000 pW 1 pW/km[e] About 1.5 pW/km 1 pW/km for each section longer than 2500 km
For 1 minute exceeded during 20% of the month Line	7500 pW		10,000 pw[i]		
% of a month during which the psophometric power for one munite due to the line indicated can be exceeded 47,000 pW 50,000 pW 63,000 pW	0.1	0.3[i]	0.3[i]		
Unweighted power % of the month during which 10- pW (5 ms) can be exceeded	0.01	0.03[i]	0.03[i]		

[a] This very condensed table is not a recommendation and reference should be made to the complete Recommendation. Special objectives for telegraphy are indicated in Recs. G.143, G.153, G.222, and G.442. Objectives for data transmission are shown in Recs. G.143 and G.153.
[b] For these systems, it is sufficient to check that the objective for the hourly mean is attained.
[c] See, in this recommendation, the details of the hypothetical reference circuits to be considered.
[d] The objectives for the line noise, in the same column, are proportional to the length in the case of shorter lengths.
[e] Objective 3 pW/km for the worst circuits; if a real circuit has more than 40,000 pW, it should be equipped with a compandor.
[f] Except in extremely unfavorable climatic conditions.

by the CCITT and the CCIR for Telephone Circuits[a]

	In National Networks			Radio Relay Links				Tropospheric Radio/Relay Links in Special Conditions	Open-Wire Lines	
				Special Objectives						
				Composition of links very different from HRC						
Noise due to the national transmitter system	Circuits ≤250 km on FDM system	Circuits not very different from HRC $280 < L < 2500$ km	$50 \leq L < 280$ km	$280 < L \leq 840$ km	$840 < L \leq 1760$ km	$1670 < L \leq 2500$ km	One or two circuits at most in one world connection	Up to 2500 km	More than 2500 km
G.123	G.123							G.311	G.153
		395-1	395-1	395-1	395-1	395-1	396-1, 392-2		
Total length L in km of the long-line FDM carrier systems in the national chain							HRC of 2500 km[d]	HRC of 2500 km[d]	Circuit of 10,000 km
$(4000 + 4L)$ pW or $(7000 + 2L)$ pW[g]	1000 pW[h]							20,000 pW[i]	50,000 pW[i]
							2500 pW		
		$3 L$ pW	$(3 L + 200)$ pW		$(3 L + 400)$ pW	$(3 L + 600)$ pW		17,500 pW	
		$3 L$ pW	$(3 L + 200)$ pW		$(3 L + 400)$ pW	$(3 L + 600)$ pW	25,000 pW		
		$\dfrac{L}{2500} \times 0.1$	$\dfrac{280}{2500} \times 0.1$	$\dfrac{L}{2500} \times 0.1$	$\dfrac{L}{2500} \times 0.1$	$\dfrac{L}{2500} \times 0.1$	0.5		
							0.05		

[g] For planning purposes.
[h] 1000 pW is the average for all the channels of the system. Desirable value: 500 pW. Highest value for one circuit: 2000 pW.
[i] Provisionally.
Note: All the values mentioned in this table refer to a point of zero relative level of a telephone circuit set up on the system under consideration (of the first circuit, for the chain.) Furthermore (G.123), the psophometric e.m.f. of noise induced by power lines should not exceed 1 mV at the "line" terminals of the subscriber's station. The mean value of the busy-hour noise power through a four-wire national exchange: ≤ 200 pWp. Limits of unweighted noise through exchange: 100,000 pW.

Table A.2   Summary of the Main Characteristics Specified by the CCITT for Carrier Terminal Equipments[a]

	Systems Wholly in Cable (G.232)	Systems on Open-Wire Lines	
		3-Channel (G.361)	12-Channel (G.232)
Level of carrier leak on line			
Per Channel	−26 dBm0	−17 dBm0	−26 dBmp
Per group	−20 dBm0	−14.5 dBm0	−20 dBm0
Attenuation distortion	Fig. 1/G.232 and Fig. 2/G.232		
Group delay	Table 1 (G.232)		
Non-linear distortion	Fig. 3/G.232		
Amplitude limiting	Definition (G.232, H)		
Crosstalk ratio	≥65 dB for intelligible crosstalk (G.232, J) ≥60 dB for unintelligible crosstalk between adjacent channels (G.232, J)		
Near-end crosstalk ratio (A) between HF points	≥47 dB without echo suppressors (G.232, J) ≥62 dB with echo suppressors (G.232, J)		
Near-end crosstalk ratio (X) between audio points	≥53 dB without echo suppressors (G.232, J) ≥68 dB with echo suppressors (G.232, J)		
Relative levels	(G.232, L)		
Impedance	600 Ω (G.232, M)		
Protection and suppression of pilots	(G.232, N)		

[a] This very condensed table is not a recommendation, and reference should be made to the complete recommendations.

Note.—See Recs. G.234 and G.235 for 8-channel and 16-channel equipment, respectively.

Table A.3a Summary of the Main Characteristics Specified by the CCITT for Groups and Supergroups[a]

	Group	Supergroup
	At 84 kHz (G.242) (dB)	At 412 kHz (G.242) (dB)
Ratio between wanted component and the following components, defined on (G.242, p. 2), after through-connection		
Intelligible crosstalk [b]	70	70
Unintelligible crosstalk [b]	70	70
Possible crosstalk	35	35
Harmful out-of-band	40	40
Harmless out-of-band	17	17
Additional suppression to safeguard pilot frequencies (G.243)		at least 40 dB at 308 kHz ± 8 Hz at least 20 dB at 308 and 556 kHz ± 40 Hz relative to 412-kHz value)
Additional suppression to safeguard additional measuring frequencies (G.243)		at least 20 dB at 308 and 556 kHz ± 20 Hz at least 15 dB at 308 and 556 kHz ± 50 Hz (relative to 412 kHz) (see also Fig. 1/G.243)
Range of insertion loss over the passband for through-connection equipments	± 1 dB relative to 84 kHz (G.242)	± 1 dB relative to 412 kHz ≤ 3 dB for SG 1 and SG 3 (G.242)
Range of insertion loss over 10° and 40°C for through-connection equipments	± 1 dB at 84 kHz relative to the insertion loss at 25°C (G.242)	± 1 dB at 412 kHz relative to the insertion loss at 25°C (G.242)
		Absolute power level at zero relative level point (for tolerances, see G.241)

Table A.3a *Continued*

	Group		Supergroup
	Frequency (kHz) [c]	Accuracy (Hz)	(dBm0)
Pilot frequency for (G.241)			
Basic group B [d]	84.080	±1	−20
	84.140	±3	−25
	104.080	±1	−20
Basic supergroup	411.860	±3	−25
	411.920	±1	−20
	547.920	±1	−20

[a] This very condensed table is not a recommendation, and reference should be made to complete Recommendations.
[b] For telephony (G.242).
[c] See (G.241) for use of these frequencies.
[d] Also applies to 8-channel groups (G.234).

Table A.3b Summary of the Main Characteristics Specified by the CCITT for Mastergroups, Supermastergroups, and 15-Supergroup Assembly[a]

	Mastergroup	Supermastergroup	15-Supergroup Assembly
Ratio between wanted component and the following components, defined on (G.242) after through-connection:	At 1552 kHz (G.242)	At 11,906 kHz (G.242)	At 1552 kHz (G.242)
	(dB)	(dB)	(dB)
Intelligible crosstalk [b]	70	70	70
Unintelligible crosstalk [b]	70	70	70
Possible crosstalk	35	35	35
Harmful out-of-band	40	40	40
Harmless out-of-band	17	17	17
Variation of insertion loss in pass-band of through-connection equipment	± 1 db with respect to value at 1552 kHz (G.242)	± 1.5 dB with respect to value at 11,096 kHz ± 1 dB in each master-group (G.242)	± 1.5 dB with respect to value at 1552 kHz ± 1 dB in each super-group (G.242)
Variation of insertion loss between 10°C and 40°C of through-connection equipment	± 1 dB at 1552 kHz relative to insertion loss at 25°C (G.242)	± 1 dB at 11,096 kHz relative to insertion loss at 25°C (G.242)	± 1 dB at 1552 kHz relative to insertion loss at 25°C (G.242)
Relative levels at distribution frames (G.233)	(dBr)	(dBr)	(dBr)
Transmit	−36	−33	−33
Receive	−23	−25	−25
Return loss at modulator input (G.233)	(dB) ≥20	(dB) ≥20	(dB) ≥20

525

Table A.3b (*Continued*)

Master group, supermastergroup or 15-supergroup assembly pilots (G.241(in:	Mastergroup Frequency (kHz)	Supermastergroup Accuracy (Hz)	15-Supergroup Assembly Level (for tolerances, see (G.241) (dBm0)
Basic mastergroup	1,552	± 2	−20
Basic supermastergroup	11,096	±10	−20
Basic 15-supergroup assembly	1,552	± 2	−20

[a] This very condensed table is not a recommendation; reference should be made to the full recommendations on the pages shown in the table.

[b] For the telephony (G.242).

Table A.4 Summary of the Characteristics Specified by the CCITT for Carrier Systems on Open-Wire Lines[a]

	Systems Acting on Each Pair		
	3-Circuit Systems	8-Circuit Systems	12-Circuit Systems
Line frequencies For a single system	Fig. 1/G.361; (see also G.361, A, a; G.361, B. a; G.361, B. b; G.361, c)	Fig. 1/G.314	Fig. 1/G.311 or Fig. 2/G.311
For several systems on the same route	Fig. 1/G.361	(G.314, c)	See Fig. 3/G.311, and Fig. 4/G.311 for examples
Pilots Frequency	16.110 and 31.110 kHz or 17.800 kHz[b] (G.361, c) −15 dBm0	(G.314, d)	(G.311, e)
Level			−20 dBm0[c]
Terminal equipment and intermediate repeater output.	≤ 17 dBr (G.361, b)	≤ 17 dBr (G.314, b)	≤ 17 dBr ± 1 dBr (terminal equipment) ≤ 17 dBr ± 2 dBr (intermediate repeater equipment) (G.311, c)
Relative level per channel at 800-Hz equivalent frequency			
Frequency accuracy of pilot and carrier frequency generators	2.5×10^{-5} (G.361, c and h)	1×10^{-5} (G.314, d)	5×10^{-6} (G.311, f)

[a] This very condensed table is not a recommendation, and reference should be made to the complete Recommendations.
[b] Used only by agreement administrations.
[c] Provisional recommendation.

Table A.5 Summary of Characteristics Specified by the CCITT for Carrier Systems on Symmetric Pair Cables[a]

	System				
	1, 2 or 3 Groups	4 Groups	5 Groups	2 Supergroups
Line frequencies	Fig. 2a/G.322	Fig. 2b/G.322 Scheme 1 Scheme 1bis [b]	Fig. 2c/G.322 Scheme 2 Scheme 2bis [b]	Fig. 4/G.322 Schemes 3 and 4 Scheme 3bis [b]
Relative level at repeater output [c] (low-gain systems) (G.322, B.2, a)	−11 dBr	−14 dBr	−14 dBr	−14 dBr
Relative level at repeater output [c] (valve-type systems) (G.324, B, b)				
Nominal value	+4.5 dBr	+1.75 dBr	+1.75 dBr	+1.75 dBr
Tolerance	±2 dB	±2 dB	±2 dB	±2 dB

Return loss of repeater and line impedances (G.322, A, c)	$\leq 0.15 \sqrt{\dfrac{f_{max}}{f}}$ or	$\leq 0.08 \sqrt{\dfrac{f_{max}}{f}}$ or	$\leq 0.08 \sqrt{\dfrac{f_{max}}{f}}$ or ≤ 0.10 (paper-insulated cables)
	≤ 0.25	≤ 0.10	$\leq 0.10 \sqrt{\dfrac{f_{max}}{f}}$ or ≤ 0.17 (cable types IIbis and IIIbis [b] G.321)
Relative level at repeater input [c]	≥ -56.5 dBr (G.324, B, b)		
Pilots	For alternative methods see Fig. 5/322		60 kHz ± 1 Hz and 556 kHz ± 3 Hz (G.322, d, 2)
Monitoring frequencies (low-gain systems)	(G.322, B.2, b)		
Harmonic distortion (low-gain systems)	See Table (G.322, B.2, c)		
Harmonic distortion (valve-type systems)	See Table (G.325)		

[a] This very condensed table is not a recommendation, and reference should be made to the complete Recommendations. For 12 + 12 systems, see Recs. G.325 and G.327.

[b] Used only by agreement between administrations.

[c] Not applicable to power-fed repeaters.

Table A.6 Summary of Characteristics Specified by the CCITT for Carrier Systems on 2.6/9.5 mm Coaxial Cables[a]

	2.6 MHz Systems[b] (1)	4 MHz Systems (2)	12 MHz Systems (3)	40 and 60 MHz Systems (4)
Line frequencies	Fig. 1/G.337 and Fig. 1/G.338	Fig. 1/G.338 and Fig. 3/G.322	Figs. 1/G.332 to 4/G.332	Figs. 1/G.333 and 2/F.333
Pilot frequencies Line-regulating pilots	60 kHz ± 1 Hz or 308 kHz ± 3 Hz 2604 kHz ± 30 Hz (G.337, A, b)	60 kHz ± 1 Hz or 308 kHz ± 3 Hz 4092 kHz ± 40 Hz and see (G.338, b.1)	4287 kHz ± 42.9 Hz for valve-type systems (G.339, b.1) 12,435 kHz ± 124.3 Hz for transistorized systems (G.332, b.1)	4287 kHz ± 42.9 Hz 12,435 kHz ± 124.3 Hz 22,372 kHz ± 223.7 Hz 40,920 kHz ± 409.2 Hz 61,160 kHz ± 611.6 Hz (G.333, b.1)
Auxiliary line-regulating pilots	(G.337, A, b)	(G.338, b.1)	308 kHz ± 3 Hz and 12,435 kHz ± 124.3 Hz for valve-type systems (G.339, b.1) 308 kHz ± 3 Hz and 4287 kHz ± 42.9 Hz for transistorized systems (G.332, b.1)	
Frequency comparison pilots National	as (2)	60 or 308 kHz 1800 kHz[c] (G.338, b.2)	300 or 308 kHz (G.332, b.2)	

International	as (2)	1800 kHz (G.338, b.2)	308 and 1800 kHz 300 kHz[c], 808 kHz[c] and 1552 kHz[c] (G.332, b.2)	4200 or 8316 kHz (G.333, b.2)
Additional measuring frequencies	(G.337, A, c)	(G.338, b.4)	(G.332, b.3) and (G.339, b.3)	(G.333, b.3)
Level of line-regulating pilots and additional measuring frequencies Adjustment value	as (2)	−10 dBm0 ± 0.5 dB (G.338, b) −1.2 Nm0 for some systems (G.338, b)	−10 dBm0 ± 0.5 dB (G.332, b.1) −1.2 Nm0 for valve-type systems (G.339, b)	as (2)
Error in the level	as (3)	as (3)	± 0.1 dB (G.332, b.1)	as (3)
Variation with	as (3)	as (3)	± 0.3 dB (G.332, b.1)	as (3)
Impedance match between repeaters and line N (as defined on (G.332, c)	$N \geqq 40$ dB for $f < 300$ kHz (G.338, e) $N \geqq 45$ dB for $f > 300$ kHz G.338, e)		$N \geqq 48$ dB for $300 \leqq f \leqq 5564$ kHz (valve-type systems G.339, e) $N \geqq 48$ dB for $f = 300$ kHz and $N \geqq 55$ dB for $f \geqq 800$ kHz (transistorized systems G.332, e)	$N = 65$ dB[d] (G.333, e)
Relative level on line			(G.332, f) and (G.339, f)	(G.333, f)

531

Table A.7 Summary of Characteristics Specified by the CCITT for Carrier Systems on 1.2/4.4 mm Coaxial Cables[a]

	1.3 MHz Systems	4 MHz Systems	6 MHz Systems	12 MHz Systems
Line frequencies	Fig. 1/G.341	Schemes 1 and 2 of Fig. 1/G.343	Schemes 1, 2, and 3 of Fig. 1/G.344	(G.345)[b]
Pilot frequencies Line regulating pilots	1364 kHz ± 13.6 Hz (G.341, b.1)	See (G.343, b.1) and for Scheme 1 (G.338, b.1); for Scheme 2 (G.332, b.1)	308 kHz ± 3 Hz (G.344)	The provisions of this recommendation are provisionally those appearing in Recommendation G.322 (see Table 6), with the exception of the matching
Auxiliary line-regulating pilots	60 kHz ± 1 Hz or 308 kHz ± 3 Hz (G.341, b.1)	4287 kHz ± 42.8 Hz[c] (G.343, b.1)	4287 kHz ± 42.8 Hz[d] 6200 kHz ± 62 Hz (G.344, b.1)	
Frequency comparison pilots	60 kHz or 308 kHz (G.341, b.2)	Scheme 1 (G.338, b.2 and Scheme 2 (G.332, b.2)	Schemes 1 and 2 (G.338, b.2) Scheme 3 (G.332, b.2)	
Additional measuring frequencies	(G.341, b.3)	(G.343, b.3)	(G.344, b3.)	

Level of line-regulating pilots and additional measuring frequencies			
Adjustment value	−10 dBm0 or 1.2 Nm0 for some systems (G.341, b)	−10 dBm0 (G.343, b)	−10 dBm0 (G.344, b)	
Tolerances				
Impedance match between repeaters and line	$N \geqq 54$ dB for a 6-km repeater section $N \geqq 52$ dB for an 8-km repeater section (G.341, e)	(G.343, b) $N \geqq 50$ dB for $f = 60$ kHz $N \geqq 57$ dB for $f \geqq 300$ kHz (4-km repeater section G.343, e)	(G.344, b) $N \geqq 6.9$ Np or 60 dB for $f \geqq 300$ kHz $N = 50$ dB for $f = 60$ kHz (3-km repeater section G.344, e)	$N = 63$ dB for a 2-km repeater section (G.345)
Relative levels on line and interconnection	(G.341, f)	−9 dBr at 4028 kHz or −8.5 dBr at 4287 kHz (G.343, f)	−17 dBr[b] (G.344, f)	(G.332, f)

Table A.8 Principal Characteristics of Analogue Signals at Audio Frequencies, at Terminals of PCM Equipment[a]

Analogue Characteristics Measured at Input and Output Parts [b, c]	Test Signal			
	Signal	Frequency Range	Power Level dBm0	
Attenuation distortion			0	Figure 1/G.712
Envelope delay distortion			0	Figure 2/G.712
Idle channel noise:				
Weighted				−65 dBm0p
Single frequency				−50 dBm0
Due to receiving equipment				−75 dBm0p
Image frequency	sine wave	4.6–72 kHz	X	< X − 25 dBm0
Level of out-of-band image signals	sine wave	300–3400 Hz	0	< −25 dBm0
Intermodulation products:				
$2f_1 - f_2$ Any intermodulation product	two sine wave	f_1 and f_2 (Hz)	$-21 < X < -4$	< X − 35 dBm0
	sine wave	300–3400 Hz	−9	
	sine wave	50 Hz	−23	< −49 dBm0

Variation of gain:				
With input level (reference = gain at input level of −10 dBm0)	white noise		−60 < X < −10	Figure 6A/G.712
	sine wave	700–1100 Hz	−10 < X < 3	Figure 6B/G.712
	sine wave	700–1100 Hz	−55 < X < 3	Figure 6C/G.712
With time (stability)				± 0.2 dB in 10 minutes ± 0.5 dB in 30 days ± 1.0 dB in 1 year
Crosstalk:				
Interchannel	sine wave	700–1100 Hz	0	< −65 dBm0
	white noise		0	< −60 dBm0
Go and return	sine wave	300–3400 Hz		> 60 dB
Distortion	Gaussian noise		−55 < X < 3	Figure 4/G.712
	sine wave	700–1100 Hz	−45 < X < 0	Figure 5/G.712

[a] This very condensed table is not a recommendation, and reference should be made to the complete Recommendations.
[b] Parameters of input and output ports:
600 Ω balanced, 4 wire ports
return loss better than 20 dB over frequency range 300–3400 Hz (provisional recommendation).
[c] For correct application to the equipments, see p. 1 of Rec. G.712.

CCIR

The CCIR Recommendations (and reports, resolutions, and opinions) are numbered according to a system in force since the Xth Plenary Assembly. When a text is modified, it retains its original number, to which is added a dash and a figure indicating how many revisions have been made. For example, 396-1 inicates that the current recommendations now in force has been modified once from its original version. If Rec. 396-1 becomes 396-2, it is apparent that Rec. 396 has had two changes since the original. Table A.9 shows only original numbering; subsequent changes are not shown in the Table.

CCIR, New Delhi, 1970, has seven volumes as follows:

Volume	Contents
I	Spectrum utilization and monitoring (Study Group 1)
II (part 1)	Propagation in nonionized media (Study Group 5)
II (Part 2)	Ionospheric propagation (Study Group 6)
III	Fixed service at frequencies below about 30 MHz (Study Group 3). Standard frequencies and time signals (Study Group 7). Vocabulary (CIV)
IV (Part 1)	Fixed service using radio relay systems (Study Group 9). Coordination and frequency sharing between communication satellite systems and terrestrial radio relay systems (subjects common to Study Groups 4 and 9)
IV (Part 2)	Fixed service using communication satellites (Study Group 4). Space research and radioastronomy (Study Group 2)
V (Part 1)	Broadcasting service (sound) (Study Group 10). Problems common to sound broadcasting and television (subjects common to Study Groups 10 and 11)
V (Part 2)	Broadcasting service (television) (Study Group 11). Transmission of sound broadcasting and television signals over long distances (CMTT)
VI	Mobile services (Study Group 8)
VII	Information concerning the XIIth Plenary Assembly Structure of the CCIR Complete list of CCIR texts

Table A.9 shows the most commonly referenced CCIR Recommendations in the text, those referring to radiolink systems.

Table A.9 Index of CCIR Recommendations, Reports, Resolutions, and Opinions

Number	Volume	Number	Volume	Number	Volume
Recommendations					
45	VI	265, 266	V	374–376	III
48, 49	V	268	IV	377–379	I
77	VI	270	IV	380–393	IV
80	V	275, 276	IV	395–406	IV
100	I	279	IV	407–421	V
106	III	283	IV	422, 423	VI
139, 140	V	289, 290	IV	427–429	VI
162	III	302	IV	430, 431	III
166	I	304–306	IV	432, 433	I
182	I	310, 311	II	434, 435	II
205	V	313	II	436	III
214–216	V	314	IV	439–441	VI
218, 219	VI	325–334	I	442, 443	I
224	VI	335–340	III	444–446	IV
237	I	341	I	447–451	V
239	I	342–349	III	452, 453	II
240	III	352–359	IV	454–461	III
246	III	361	VI	462–466	IV
257, 258	VI	362–367	IV	467–474	V
262	V	368–373	II	475–478	VI
Reports					
19	III	226	IV	341–344	II
32	V	227–231	II	345–357	III
42	III	233–236	II	358, 359	VI
79	V	238, 239	II	362	VI
93	VI	241	II	362–364	III
106, 107	III	244–251	II	366	III
109	III	252	*a*	367–373	I
111	III	253–256	II	374–393	IV
112	I	258–266	II	394	VI
122	V	267	III	395–397	IV
130–137	IV	269–271	III	398–401	V
176–194	I	272, 273	I	403–407	V
195	III	275–282	I	409–412	V
196	I	283–290	IV	413–415	*a*
197, 198	III	292–294	V	416–423	I
200, 201	III	297–308	V	424–432	II
202	I	311–316	V	433–439	III

537

Table A.9 (*Continued*)

Number	Volume	Number	Volume	Number	Volume
203	III	318–320	VI	440	*a*
204–214	IV	321	III	441	III
215	V	322	*a*	442–456	IV
216	VI	324–334	I	457–498	V
218, 219	IV	335	III	499–515	VI
222–224	IV	336–339	II	516	V
		340	*a*		

Resolutions

Number	Volume	Number	Volume	Number	Volume
2–4	II	21–23	III	39, 40	VII
7, 8	II	24	VII	41–44	I
10	II	26, 27	VII	45–51	I
12, 13	II	30	II	52–54	III
14	III	33	VII	55, 56	IV
15, 16	I	36, 37	VII	57, 58	V
20	VI	38	{ V, VII }		

Opinions

Number	Volume	Number	Volume	Number	Volume
2	I	22, 23	II	32–35	I
11	I	24	VI	36, 37	III
13, 14	IV	26–28	III	38–41	V
15, 16	V	29, 30	I	42, 43	VI

a Published separately.

Table A.10 Characteristics of Radio Relay Systems Specified in CCIR Recommendations

		Maximum Number of Telephone Channels													
	Frequency Band	1	6	12	24	60	120	300	600ᵃ	900/960ᵃ	1200/1260ᵃ	1800	2700	Television	Trans-horizon
Occupied bandwidth	Band 8														} 388
Number of radio frequency channels	Bands 8 and 9														
Center frequencies and radio frequency channel arrangements	2 GHz						283-2			382-2					382-2	382-2
Polarization arrangements															
4 GHz								279-1		382-2					382-2	
6 GHz									373-1	384-1		383-1	384-1	383-1	
7 GHz							385								384-1	
8 GHz								386-1		386-1					386-1	
11 GHz										387-1					387-1	
Interconnection at:															
Audio frequencies								268-1								
Baseband frequencies								380-2								
Intermediate frequencies										403-2						
Video frequencies															270-1	
Hypothetical reference circuit					391					392						396-1
Allowable noise power in the hypothetical reference circuit									393-1						289-1 462	397-2
Noise in the radio portions of real circuits									395-1							
Frequency deviation								404-2							276-1	404-2
Preemphasis and deemphasis characteristics																
Line regulating and other pilots									275-2		401-2				405-1	275-2
Signaling and service channels											400-2					
Standby arrangements										305						
Auxiliary radio relay systems																
Residues of signals outside the baseband									389-1							
Maintenance measurements in actual traffic									381-2 398-2						463	381-2
Measurements of performance with the help of a signal consisting of a uniform spectrum									399-1							

ᵃ Or the equivalent.

B | STATIC NO-BREAK POWER SYSTEMS

In the modern communication plant, highly reliable power is assured by static no-break power systems. In some cases motor-generator sets are used as backup.

Static no-break power has come into its own, replacing flywheel dynamic systems, for a variety of reasons. One reason is the development of comparatively long lasting secondary cells, whether lead-acid or nickel-cadmium. A second reason, partially owing to the coming of age of the transistor, is the standardization of 48 VDC as prime power in both switching and transmission systems, rather than alternating current supplies.

In the common no-break arrangement the battery float is between charger and load as shown in Figure B.1. Today's chargers are almost exclusively silicon controlled rectifier (SCR) types. Given modern telecommunication plant life on the order of 15 years, lead-acid batteries remain very attractive. Older lead-antimony cells are still used. However, calcium-lead cells, costing somewhat near 10–15% more than the antimony cells, compete with Nicad (nickel-cadmium).

To assure long battery life, the system charger should provide a regulated output between 0.5 and 1%. Another important consideration in selecting a charger is output ripple. As can be seen in Figure B.1, the load is placed across the charger shorted by the battery. Indeed the battery provides good additional filtering, about 2000 μF/A-h. Nevertheless, ripple in any case should be below 0.50% of the dc output voltage to the load and preferably to 0.10% or below. To assure a low hum level, 50 mV is a good figure to specify for maximum ripple voltage.

Noise from the battery source should not exceed − 70 dBmp (100 pWp). Both ripple and noise measurements should be made during full-load conditions, ac mains connected.

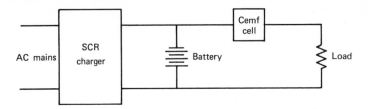

Figure B.1 Simplified diagram of static no-break power system.

The charging circuit should be able to withstand nominal variations of incoming ac line voltage of at least ± 10%. The charger should also provide the required dc output voltage under widely varying load conditions.

A telephone switch is a good example. During the busy hour a 10,000-line crossbar switch may draw 800 A at 48 V sometimes just before noon. At about 2 AM, it may draw only 20% of that amount, or about 160 A.

A radiolink, on the other hand, would show little variation in load and require much less capacity. Requirements on this sort of charger are much less.

The charger should use silicon controled rectifiers (SCR) because of their light weight, simplicity, and excellent operating characteristics. Load sharing should also be included in the charger specifications. Load sharing permits two or more chargers, not necessarily of equal capacity, to provide output to a common bus and supply load to that bus according to their rated capacity. This is done by the proper interconnection of the dc control signals and adjustment of the load-sharing potentiometers.

To properly size a battery and charger, the following expressions must be understood. Battery capacity in ampere-hours (A-h):

$$C_{\mathrm{ah}} = I_{\mathrm{L}} \times T_{\mathrm{R}}$$

where I_{L} = load current (A)

T_{R} = reserve time (h)

Reserve time is the time the system can operate off the battery supply exclusively, when the ac mains are shut down for whatever reason. Reserve time is usually given as 8, 12, 24, 48, or 72 h. The time selected is a design decision. It depends on the length of time an expected outage may last, or how long it would take to bring up a motor-generator set on-line. This time should include a technician's travel time to unmanned sites.

The relationship to determine the load on the charger is

$$I_{\mathrm{ch}} = I_{\mathrm{ba}} + I_{\mathrm{L}}$$

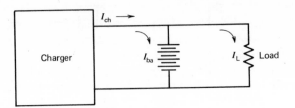

Figure B.2 Current, battery, load, and charger.

These values of current are shown in the simplified drawing, Figure B.2. Another parameter must be considered as well: recharge time, T_{ch}, expressed in hours. The average current required to recharge the battery can now be expressed:

$$I_{ba} = \frac{C_{ah}}{T_{ch}}$$

Combining these last two equations,

$$I_{ch} = \frac{C_{ah}}{T_{ch}} + I_L$$

Now combining this equation with the first equation,

$$I_{ch} = \frac{I_L \times T_r}{T_{ch}} + I_L$$

The energy required to recharge the battery is the term $I_L X T_r/T_{ch}$ and assumes a battery that is 100% efficient. However, some energy is required to compensate for chemical and heat losses; usually 10% is a good figure. Thus

$$I_{ch} = I_L + 1.10 \left(\frac{I_L \times T_r}{T_{ch}} \right)$$

Factored

$$I_{ch} = \left(1 + \frac{1.10\, T_r}{T_{ch}} \right)$$

With this equation, charger current, I_{ch}, can be determined. I_L will be given, sometimes called the house load or power budget. Reserve hours and recharge time are design parameters for us to determine.

The standard nominal 2-V lead-acid storage cells are considered fully discharged when cell voltage reaches 1.75 V at 77°F (23°C). This is called the final voltage and is used as a standard figure in the industry. However, some power system designers use a more conservative figure, 1.84 V. The final voltage is valid for an 8-h discharge rate (reserve hours = 8 h).

To adjust this 8-h reserve time capacity to other capacities, the following table is useful, and introduces a new aspect in battery sizing:

Rate (h)	Capacity per positive plate (A-h) final voltage 1.75	Capacity (%)
24	75	125
12	66	110
8	60	100
6	55	91
4	50	83
2	40	66

Batteries are sized (i.e., required ampere-hour capacity is established) in terms of single positive plates. Total cell rating is given by multiplying the single plate rating by the number of plates in a cell. It should be kept in mind that there is always one more negative plate than positive plates in a cell.

When specifying battery/charger installations, the following terms are used. The table below shows the terms with *example* values given for a 48-V dc installation. The terms are given with equivalent declining voltage order.

Term	Voltage per Cell	Voltage for 24 Cells
(3) Equalize	2.30	55.20
(2) Float	2.17	52.08
(1) Ac-off (load not connected)	2.05	49.20
(4) Ac-off, initial full load	1.97	47.28
(5) Ac-off, average full load	1.92	46.08
(6) Ac-off, final full load	1.75	42.00

Notes

1. A fully charged lead-acid battery under no load has 2.05-V potential difference between negative and positive plates of the cell. Its specific gravity will be 1.215 (i.e., specific gravity of the electrolyte). Some

automotive batteries using high density electrolytes have fully charged specific gravities up to 1.300.

2. The float condition permits a slow charge using a voltage just high enough to overcome internal resistance of the cell. Thus the charge voltage must be some tenths of a volt more than 2.05 V dc per cell.

3. The equalizing voltage is considerably higher than 2.05 V dc per cell and is applied for relatively short periods of time.

4. The initial full load voltage is 2.05 V dc $- I_{\text{L}} \times R_{\text{int}}$, where R_{int} is the internal resistance of the cell in ohms.

5. Average full-load voltage is the voltage that may be expected when the cell has reached a point in time halfway between full charge and the final voltage. A uniform discharge rate is assumed.

6. Final voltage is a standard parameter.

On sizing batteries and charger, the system designer must assure that the equipment which will present the load may withstand the usual ± 10% dc input voltage variation. This figure should be verified on the equipment specifications that consist the load.

Assume a 48-V dc supply. The high voltage would then be 52.8 V dc, and the low, 43.2 V dc. If the equalize charge were the 55.2 V from the table above, then on equalize the charger should also employ a regulating element, the cemf (counter emf) cell shown in Figure B.1. The voltage drop across that cell must be 55.2–52.8 V dc = 2.4 V dc. Provision is made in the cemf cell that it is dropped off-line when ac power fails so that the load will get the full battery voltage. The cemf cell must also be specified to withstand the full-load current.

The following example will illustrate these concepts.

Example.   Radiolink repeater, 48 V dc

8 transmitter-receiver groups	20 A
2 orderwire	1
1 fault alarm system	1
Tower lights, strobe/control	10
Interior lighting	3.5
Heat and ventilation	14.5
Total	50 A

The site is unmanned and comparatively isolated. A reserve time of 12 h is selected.

$$I_{\text{L}} = 50 \text{ A} \qquad T_{\text{R}} = 12 \text{ h} \qquad T_{\text{ch}} = 48 \text{ h}$$
$$I_{\text{C}} = 50 \,(1 + 1.10 \times 12/48)$$

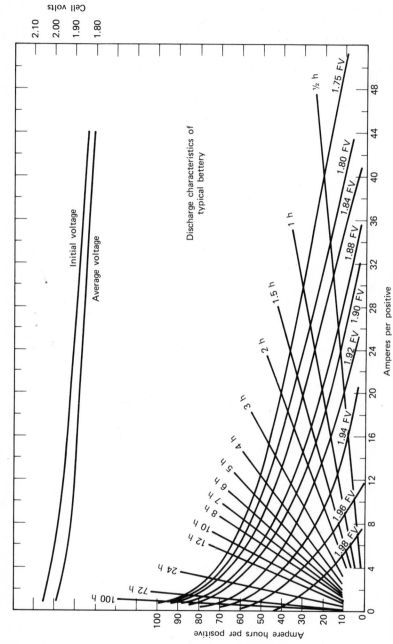

Figure B.3 Curves to determine amperes per positive plate. Courtesy Warren G-V Communications.

$$= 50\,(1\,+\,1.10\,\times\,0.25)$$
$$= 63.75\text{ A, the charger rating}$$

Battery, 48 V dc or 24 cells

Final cell voltage, 1.75 V dc

From Figure B.3, where 12 h crosses the 1.75-V curve, 5.5 A per positive plate.

Number of positive plates $50/5.5 = 9.9$ rounded off to *10*

Number of negative plates, $10 + 1 = 11$

Total plates per cell, 21

C SIGNALING ON TELEPHONE CIRCUITS

Signaling conveys the intelligence needed so that a switched telephone network can interconnect one subscriber with any other. Signaling tells the switch that a subscriber desires service, and gives the switch the data necessary to identify the distant subscriber desired and to properly route the call. It also provides certain status information (busy tone, dial tone, ringing tone, congestion, etc.). Metering pulses for call charging is also a form of signaling.

There are several methods of compartmentalizing signaling:

- Subscriber signaling.
- Interswitch signaling.

and/or

- Supervisory signaling.
- Address signaling.

It should be realized that on most telephone calls more than one switch is involved in the call routing. Therefore switches must interchange information between switches on a fully automatic dial-up call. Between modern switches address information is provided by interregister signaling and the supervisory function is carried out by line signaling.

Supervisory signaling provides information on line condition such as

- Calling party off-hook.
- Calling party on-hook.
- Called party off-hook.
- Called party on-hook.

Fundamentally, supervisory signaling is two-state signaling (e.g., the calling party may be off-hook or on-hook regarding the subscriber). Address signaling is more complex, consisting of sending a series of digits from the subscriber to the switch that serves him, similar information between switches, and finally to the destination switch, providing a connection to the called subscriber. To inform the called subscriber of the call, the destination switch sends ringing current down his line to ring his phone.

The aim of this appendix is to provide the reader with an appreciation of signaling and to underline its effect on transmission. To properly engineer a telephone transmission system, some knowledge of signaling is required. The transmission engineer must look at signaling from several viewpoints:

- Signaling over VF wire pairs.
- Signaling over carrier circuits.

The first item deals with signaling over subscriber pairs or over local interoffice trunks (junctions) where VF metallic pairs serve as the transmission medium.

Care must be taken in the design of these systems to assure that the proper signaling functions can be carried out over them, that signaling does not interfere with transmission, nor transmission with signaling.

One example of this "care" has been dealt with in Chapter 2. The subscriber's line condition is "on-hook" (the subscriber's telephone is in its cradle); no current flows in the loop connecting the telephone to the local serving exchange. When the telephone is lifted out of its cradle (e.g., goes "off-hook"), a switch is closed in the telephone set completing the loop to the local exchange. In this closed loop current flows, causing a relay to close in the local exchange, and a request for line seizure takes place. Upon seizure, a dial tone is returned to the subscriber. The dial tone is an audible "status" signal to indicate to the calling subscriber that the switch is ready to receive address information, his dialed digits.

The important item here for the transmission engineer concerned with the design of the subscriber loop is that the resistance of the loop will be low enough to permit sufficient current to flow to close the relay in the local serving exchange. The gauge of the VF pair comprising the loop, its length, and the resistance of other items in series in the loop (i.e., handset, load coils, relay windings, etc.) must be taken into consideration to assure that the minimum current will flow. Depending on switch design, certain maximum loop resistances will usually be encountered (Table C.1) and should not be exceeded in the design of the loop.

The concept of loop design is covered in Chapter 2. It should be kept in mind that the signaling limit on a subscriber loop may exceed the transmission attenuation limit, and conversely, the transmission attenuation limit may exceed the signaling limit.

Table C.1   Common Subscriber Loop Maximum Resistance

Switch Type	Usual Maximum Loop Resistance to Effect Line Seizure (Ω)
Step-by-step	800
Crossbar (standard)	1200–1300
Semielectronic	1900–2600

The second step in subscriber signaling after line seizure has been effected and dial tone received, is to convey the necessary address information to his local exchange. He does this with the telephone dial or with the key set on the subscriber telephone instrument. The dial simply makes and breaks the 48-V loop, forming a series of pulses corresponding to the digit dialed.

The number that the calling subscriber dials is a number that is unique to his called party, a series of digits that identifies the distant telephone that the calling subscriber wishes to reach. A telephone number also provides routing information for switches and may supply charging or metering information.

It is common practice today to use seven digits in a telephone number, although six, or even five digits are used in certain areas of Europe and other parts of the world. However, it is the seven-digit number we wish to discuss here. As an example, let us take the number 271-4568. The first three digits, 271, identify the exchange, and the last four digits identify the subscriber's telephone. The total numbering capacity of a four-digit block is 10,000. There are 10,000 different digit combinations between 0000 and 9999. Conveniently, 10,000 is the basic building block of most standard switches today.

The discussion above roughly outlines how supervisory and address information are handled on the subscriber loop. The same type of information must be sent from one exchange to another in a call setup, namely:

- Supervisory information.
- Address information, in this case, interregister signaling.

For interoffice trunks (junctions) in the local area, the transmission medium is VF cable. Supervisory signaling will be dc when distance (loop length) permits. Therefore, as in subscriber loop engineering, care must be taken to assure that maximum signaling distance is not exceeded for the wire gauge employed. The range of dc signaling can be increased by using a battery voltage over 48 V, by use of pulse link repeaters (a form of regenerator), by biasing line relays, or by increasing line relay sensitivity (see Table C.1).

When trunk lengths are in the intermediate range (e.g., 15–75 km), so-called low frequency signaling may be used with up to two repeaters in tandem. Signaling information is sent by pulses in this case. The length of the pulse distinguishes supervisory from address information. Common signaling frequencies for LF signaling are 50 Hz (Italy, France, Germany, Spain), 25 Hz (Germany), and 80 Hz (Spain).

On still longer trunks carrier systems are used, mostly FDM. Here supervisory signaling is either in-band or out-band (out-of-band). Out-band is favored in the long run, being simpler and more economical. It is also more reliable, as will be seen below.

The voice channel on an FDM carrier system occupies a nominal 4-kHz bandwidth (see Chapter 3). Speech is assigned the segment 300–3400 Hz. Signaling can take advantage of the narrow out-band slot provided above the speech segment and below 4000 Hz. The frequencies used in the slot are as follows: CCITT recommends 3825 Hz, ATT uses 3700 Hz, and Germany 3850 Hz. As we know, with supervisory signaling, two line conditions are possible, idle or busy. A 3825-Hz (or other out-band frequency) tone or lack of tone may indicate one condition or the other. There are advantages to each approach. Tone-on idle keeps a constant load on the channel when not in use. When the condition changes from tone-on to tone-off, a line seizure takes place. Thus any interruption such as a deep radio fade would cause a massive seizure, appearing to the switch as if all trunks on that particular circuit group required service at once. For tone-on idle proper precautions are taken to avoid this eventuality. One way of doing this is to sense group pilot tones from the FDM multiplex group equipment.

Tone-off idle (tone-on busy) provides a solid indication of a seized trunk and does not suffer the dropout problem that can cause massive seizure. The tone being on when busy does not affect the speech path whatsoever, the supervisory signaling channel being separate from the speech channel. Of course this cannot be the case for in-band signaling.

In-band signaling in this context refers to supervisory signaling carried on inside the speech band (e.g., between 300 and 3400 Hz). The big advantage of in-band signaling is that when patching of the channel is required, only the four-wire voice leads must be patched, whereas with out-band signaling an extra pair of leads must be patched as well. On the other side of the coin, in-band single frequency signaling requires special circuitry to help ensure that "talk-down" does not occur. "Talk-down" is the chance imitation of supervisory signaling tone(s) by speech tone or a combination of speech tones causing a freak disconnect. "Talk-down" was discussed in Chapter 3. Another disadvantage of in-band signaling is that active signaling cannot be carried out during the speech interval, but only during the time of setup and disconnect.

In-band signaling usually uses one or two tones above 2000 Hz. It is desirable to use frequencies above 2000 Hz because in normal conversation, less speech energy is concentrated in that portion of the voice channel spectrum (see Figure 1.2). This provides added assurance that talk-down will not occur. In-band signaling may be pulse or continuous (on-idle). Pulse is more desirable from a system loading viewpoint. Remember that out-band is always continuous.

In-band supervisory signaling frequencies may be of the 1VF or 2VF types. Simply speaking, this means that it could be one-tone, called SF (single frequency) signaling, or two-tone, used either simultaneously or singly. For instance, CCITT No. 4 signaling uses 2040 and 2400 Hz in both directions, and it is a pulse type. The North American R-1 system uses 2600 Hz in both directions SF, continuous. The British SS-AC11 uses 2280 Hz. The Germans with their IKZ50 uses in-band signaling as an option, 3000 Hz, occasionally 2280 Hz, in both directions.

Before we get deeper into the transmission aspects of supervisory signaling, a short description of E and M signaling is in order. All out-band signaling is E and M. It is a two-state signaling system operating in both directions of a call setup (and disconnect). Elongating the signaling system into carrier equipment, on-hook is equivalent to idle, off-hook to busy. Consider now Figure C.1. Only the originating and terminating exchanges (offices) are considered in the figure.

In signaling terminology, the devices that generate and put tones on the line are called senders; those that receive tones and actuate relays as a consequence are called receivers. These devices may be incorporated in the carrier equipment, particularly in the case of out-band signaling, or in separate signaling converters, or right in the switch registers. From the transmission aspect, we want to speak of sending levels, receiver sensitivity, and receiver dynamic range.

For the moment, let us consider only the supervisory function in this regard. Table C.2 gives values of some typical operational systems. Another important consideration is that of noise bursts or impulse noise. If the signaling receiver is very sensitive, an impulse noise hit may cause actuation or deactuation. Still, enough sensitivity must be provided for actuation when a real signal is impressed on the receiver. For instance,

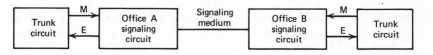

Figure C.1 E and M signaling conditions.

Signal A to B	Signal B to A	Condition at A		Condition at B	
		M Lead	E Lead	M Lead	E Lead
On-hook	On-hook	Ground	Open	Ground	Open
Off-hook	On-hook	Battery	Open	Ground	Ground
On-hook	Off-hook	Ground	Ground	Battery	Open
Off-hook	Off-hook	Battery	Ground	Battery	Ground

Table C.2   Certain Transmission Characteristics of Signaling Systems

System Type	In-Band/ Out-Band	Sand Level (dBm0)	Receive Level (dBm0)	Dynamic Range (dB)
CCITT No. 4	In-band 2VF	−9	0 to −18	18
CCITT No. 5	In-band 2VF	−9	−2 to −16	14
R-1	In-band 1VF	−8 for 300–550 ms, then drops to −20	−1 to −27	27
R-2	Out-band 3825 Hz	−20	Recognition tone off −33 Recognition tone on −27	

[a] CCITT Rec. Q.144.

if the send level is − 9 dBm0 and the receive level is set at − 18 dBm0, only 9 dB of loss is permitted from sender to receiver. This is an important item that the transmission engineer must take into consideration. Suppose that the signaling receiver is adjusted for a threshold of − 18 dBm; a 9-dB loss exists between sender and receiver; and stability of the transmission link connecting them is such that it drifts negative (e.g., more loss is added due to deterioration in amplifiers). It is obvious that the signaling receiver will not actuate properly with only a fraction of a decibel of increased loss. In this hypothetical case, there is not sufficient level to actuate the signaling receiver. Such an eventuality probably would not take place for the case of supervisory signaling, but could well be the case for end-to-end interregister signaling. These points will be covered in greater detail during the ensuing discussion of interregister signaling.

By definition, interregister signaling is "machine-to-machine." There is no intervention of man. Hence we work exclusively with machine intelligence. There are a number of ways of exchanging this intelligence, which, as mentioned previously, consists of digits identifying the· distant subscriber's telephone, and at the same time, provides routing information and often charging information.

There are several ways of sending this information:

- By dc pulses on systems connected by metallic pairs.
- By ac pulses of the 1VF or 2VF types.
- By multifrequency (MF) pulses.
- By multifrequency (MF) compelled systems.

Pulse sending may comprise simple pulse trains, binary codes (often four-element) or there is significance in the frequency makeup of the pulse. There also may be meaning in the pulse length and where it fits in the sequence of sending. In this discussion we concentrate more on MF signaling methods.

Before proceeding further in the discussion of signaling the terms "compelled" and "noncompelled" signaling should be distinguished. In most cases interregister signaling is bidirectional. Information is sent from the initiating exchange to a receiving exchange, and information is also sent from that receiving exchange back to the initiating exchange.

Noncompelled signaling is when the signals in one direction are essentially independent of signals in the other direction. Of course there is some ambiguity here in that when a line seizure is sent in the forward direction and a "proceed to send" is not forthcoming in the backward direction, no forward address information will be sent. An example of noncompelled interregister signaling is CCITT No. 5 (Table C.4). It should be noted that pulse length is important for noncompelled signaling systems.

Compelled signaling requires acknowledgement signals from the receiving exchange as the signaling sequence proceeds. With compelled signaling each forward signal continues to be sent until acknowledgment is received. Likewise, the acknowledgment signal continues to be sent until the initial (forward) signal ceases.

Noncompelled signaling is faster than compelled signaling. By faster we mean that it takes less time to complete the signaling process. However, noncompelled systems may require timing adjustments on long circuits owing to delay caused by the fact that it takes time for a signal to traverse a specific link (or system) and this time, of course, is a function of the velocity of propagation. Compelled signaling systems adapt automatically to long links (or systems) because pulse length is not a key feature as in noncompelled systems. The call setup is simply delayed a little longer.

In MF systems, used widely today for address signaling on automatic networks, the signaling information is contained in the *pairs* of frequencies sent. As a consequence, two signal receivers are required to operate a corresponding register relay. One receiver is provided for each frequency, consisting of a bandpass filter, detector, and some form of signal amplifier or relay driver.

Some MF systems use five frequencies and are referred to as 2-out-of-5 systems (e.g., SOCOTEL used in France and Spain). Others use six frequencies (i.e., R-2) in both directions. Still others use 2-out-of-6 in the forward direction and 2-out-of-4 in the reverse (backward) direction (i.e., Italian OOb-MFC).

Besides the 10 digits used for "selection" (switch acutation), at least two special signals with their own two-frequency groupings are used: KP, sent in the forward direction "start of pulsing," and ST, sent in the forward direction "end of pulsing." There are numerous other special signals that may be used. A listing of those accepted for international use may be found in CCITT Rec. Q.140.

Several of the more common MF codes are described below. Table C.3 shows the R-1 MF code used widely in North America; Table C.4 shows the CCITT No. 5 MF code.

Table C.3   MF Code   North American R-1

Digit	Frequencies (Hz)	Digit	Frequencies (Hz)
1	700 + 900	7	700 + 1500
2	700 + 1100	8	900 + 1500
3	900 + 1100	9	1100 + 1500
4	700 + 1300	0	1300 + 1500
5	900 + 1300	KP	1100 + 1700
6	1100 + 1300	ST	1500 + 1700

Source: CCITT Rec. Q.317.

Table C.4   MF Code CCITT No. 5

Signal	Frequencies (Hz)	Remarks
KP1	1100 + 1700	Terminal traffic
KP2	1300 + 1700	Transit traffic
1	700 + 900	
2	700 + 1100	
3-0	Same as table C.3	
ST	1500 + 1700	
Code 11	700 + 1700	Code 11 operator
Code 12	900 + 1700	Code 12 operator

Source: CCITT Rec. Q.151.

The R-2 system is rich in coding possibilities, taking full advantage of 2-out-of-6. 15 possibilities are available, but this number is doubled in each direction by having meanings Groups I and II in the forward direction and Groups A and B in the backward direction. Groups I and A are said to be of primary meaning; Groups II and B, secondary. The change from primary to secondary meaning is commanded by a backward signal A-3 or A-5. Secondary meanings can be changed back to primary only when the original change from primary to secondary was made by use of the A-5 signal. Table C.5 shows the frequency makeup of the R-2 MF code.

The 10 digits to be sent in the forward direction in the R-2 system are in Group I and are index numbers 1 through 10 in Table C.5. The *ST* signal is Group I, index 15.

Primary meaning backward signal 1140 + 1020 Hz means "send next digit." The index 15 signal, Group A, indicates "congestion in an international exchange or at its output." This is a typical "backward information" signal giving circuit status information. Group B consists of nearly all "backward information," and in particular, deals with subscriber status.

A complete description of the R-2 system may be found in CCITT Rec. Q.361. R-2 is a regional code for Europe.

SOCOTEL, as used in Spain and France, is a 2-out-of-5 frequency system in both directions with 700-, 900-, 1100-, 1300-, and 1500-Hz tones. 1900 Hz serves as the acknowledgment signal.

For toll (long-distance) systems one or many tandem (transit) exchanges may be required in the call path to effect a connection. How is the call setup carried out at these tandem exchanges? It may be on a *link-by-*

Table C.5   MF Code, European R-2 System

	Frequencies (Hz)						
Index No.	1380	1500	1620	1740	1860	1980	Forward Direction I/II
for Groups							
I/II and A/B	1140	1020	900	780	660	540	Backward Direction A/B
1	x	x					
2	x		x				
3		x	x				
4	x			x			
5		x		x			
6			x	x			
7	x				x		
8		x			x		
9			x		x		
10				x	x		
11	x					x	
12		x				x	
13			x			x	
14				x		x	
15					x	x	

link basis or on an *end-to-end basis.* For instance R-1 is link-by-link, and R-2 can be end-to-end (or link-by-link). All line signaling is carried out on a link-by-link basis. Interregister signaling may be link-by-link or end-to-end.

Link-by-link signaling may be defined as a signaling system where *all* interregister address information must be transferred to the subsequent exchange in the call setup routing. Once this information is received at this exchange, the preceding exchange control unit releases. This same operation is carried on from the originating exchange through each tandem (transit) exchange on a link-by-link basis to the terminating exchange of the call.

End-to-end signaling abbreviates this process such that tandem (transit) exchanges receive only the minimum information necessary to route the call. For instance, the last four digits of a seven-digit telephone number need only be exchanged between the originating exchange (e.g., the calling subscriber's local exchange or the first toll exchange in the call setup) and the terminating exchange in the call setup. With this type of signaling fewer digits are required to be sent for the overall call setup. The advantage is that the signaling process may be carried out much more rapidly, decreasing the postdialing delay. Intervening exchanges on the route need work much less, handling only two or three digits of call routing information. The key to end-to-end operation is the originating register which stores the entire call signaling information in memory. As the call progresses on the call setup, it is called upon to repeat various digits upon request from down-route tandem exchanges. As a consequence, to effect end-to-end signaling a considerable backward information capability is required to provide signaling digit requests to the originating register (sometimes called leading register). The R-2 signaling system is particularly fitted for this function.

For the transmission engineer end-to-end signaling may present some special problems. With link-by-link signaling we had to be concerned only with signaling levels and losses on a link basis. On end-to-end signaling we are concerned with levels and losses across the whole system from originating register to the terminating exchange. Stability of the system becomes a very important consideration.

Previously, an example of a 9-dB difference was cited between send level and signaling receiver threshold. On most link-by-link systems the 9 dB or even less, to cover loss and loss variation would be sufficient (see Chapter 2). On end-to-end systems the signaling receiver dynamic range may well have to be considerably increased. This will depend much on the transmission plan and on how well it is adhered to.

Consider the R-2 system, which is typical end-to-end:

Transmit level: $- 11.5 \pm 1$ dBm0 per frequency (CCITT Rec. Q.364)
Receive sensitivity range: $- 5$ to $- 35$ dBm (CCITT Rec. Q.365)

Without allowing any tolerance for circuit level stability, assuming a transmit level of $- 12.5$ dBm per frequency, the maximum end-to-end loss permitted would be 22.5 dB. Now allow four links in tandem in the call setup with 1-dB stability per link; considering the worst case, the signaling level could then drop 4 dB. Thus the maximum loss on the connection could be no more than 18.5 dB. From the strictly transmission viewpoint such a loss is very high but may well be possible when reference equivalents are above 30 dB.

Keep in mind that improving sensitivity by lowering threshold of signaling receivers makes them more vulnerable to noise "hits" and improper operation. Special care must be taken in this regard in environments where there is a high level of impulse noise, such as with step-by-step switches.

A major new development in signaling is CCS (common channel signaling). Up to this point in our discussion the transmission of signaling information has been associated physically with the telephone channel it controls. With CCS the signaling information, both supervisory and address, is separated (as it is with the 30 + 2 PCM system mentioned in Chapter 9) and sent as a serial data bit stream on a specially dedicated channel for signaling information only. Two systems are now proposed and, to some extent, operative. These are the CCITT No. 6 and the North American CCIS (common channel interoffice signaling). With the CCITT No. 6 system one CCS channel can handle the signaling information for up to 2048 telephone speech channels, whereas CCIS can handle 8192 telephone channels. CCS systems require teleprocessors at each switch for implementation.

The extensive use of such systems may become attractive once large, stored program controled (SPC) switching networks come into existence. With SPC each switch will have its own processor (computer), and such a processor can easily accommodate CCS as well. However, with the loss of the CCS channel, all signaling capability would be lost for the telephone trunks it serves. Thus highly reliable channels will be required to carry the CCS information with efficient alternative routing.

GLOSSARY

AC	Alternating current
AF	Audio frequency (as distinguished from RF, radio frequency)
AFC	Automatic frequency control
AGC	Automatic gain control
ALC	Automatic load control
AM	Amplitude modulation
AMI	Alternate mark inversion
Ant	Abbreviation for antenna
APL	Average picture level (TV terminology)
ARQ	Automatic repeat request, a method of correcting errors on data/telegraph systems by means of a back channel
ASCII	American Standard Code for Information Interchange
ATT (AT & T)	American Telephone and Telegraph (Co.)
Autodin	Automatic digital network. A digital data and telegraph network with message and line switching
Autovon	Automatic voice network. A telephone network operated by the U.S. Department of Defense
AWG	American Wire Gauge
AZ	Azimuth (refers to orienting earth station antenna systems, or any antenna system)
B factor	The penalty factor used in the VNL concept
BB, bb	Baseband
BCD	Binary coded decimal. A code where the 10 digits, 0–9, are conventionally binary coded from 0001 through 1010 and all other numbers are built up from these basic digits
B_{if}	IF bandwidth
bit	Binary digit

BNZS   Bipolar but with N zeros substitution (PCM terminology)

BPO   British Post Office (the telephone administration of the UK)

bps   Bits per second

BTM   ITT's large Belgian affiliate Bell Telephone Manufacturing Company

busy hour   A traffic engineering term. CCITT definition: "The uninterrupted period of 60 minutes during which the average traffic flow is maximum"

BWR   Bandwidth ratio. Used in determining S/N from NPR

$$\text{and} = 10 \log \frac{\text{(occupied baseband bandwidth)}}{\text{(voice channel bandwidth)}}$$

CARS   Community antenna radio service

CATV   Community antenna television, sometimes called cable television. It is a private line method of bringing broadcast (and other TV/communication) services into the home. Often called "pay TV"

CCIR   International Consultive Committee for Radio. It is organized under the auspices of the ITU (International Telecommunications Union), Geneva

CCITT   International Consultive Committee for Telephone and Telegraph. It is also organized under the auspices of the ITU

CEPT   Conference European des Postes et Telecommunication. The regional European telecommunication conference committee

cm   Centimeter

cm region   As designated by the ITU, that band of frequencies from 3 to 30 GHz

C/N Carrier-to-noise ratio

CNT   Canadian National Telephone Co.

codec   Coder-decoder. This is an acronym used in PCM terminology

Compandor   Acronym for compressor-expandor. See Section 3.13

CRC   Communication relay center (HF terminology)

CRPL   Central Radio Propagation Laboratory operated by the US Department of Commerce in Boulder, Co.

C/T The ratio of carrier to thermal noise power, usually expressed in dBW per degree Kelvin

CT   International routing term. Centre de Transit, usually followed by a number, such as CT1, CT2, or CT3. The

	number indicates its hierarchical rank, 1 being the highest rank. Also called transit exchange.
CU	Crosstalk unit (refer to equation 1.14)
CW	Continuous wave. Sometimes is used to mean "Morse operation"
D1, D2, etc.	ATT-Bell System series of PCM systems.
dB	Decibel
dBa	Decibels adjusted, a noise measurement unit using F1A weighting (now obsolete)
dBm	Decibels referred to 1 mW
dBm0	An absolute power unit in dBm referred to the 0 TLP (zero test level point)
dBmp	A noise measurement unit based on the dBm using psophometric weighting
dBmV	An absolute voltage level measurement unit based on the dB referred to 1 mV across 75 Ω
dBr	Decibels referenced. The number of decibels of level above or below a specified reference point in a system. A minus sign indicates the level to be below or less than than at the reference, the plus, above or more
dBrnC	A noise measurement unit using C-message weighting. Usage limited essentially to North America
dBW	Decibels related to 1 W. 0 dBW = $+30$ dBm
dBx	The crosstalk coupling in dB above reference coupling, which is a crosstalk coupling loss of -90 dB
DC	Direct current
DCA	Defense Communication Agency (an agency under the U.S. Department of Defense)
DCS	Defense Communication System (US)
di-bit coding	Coding for transmission at 2 bits at a time such that each transition on transmission represents 2 bits
dN	Decineper or 1/10 of a neper (see neper)
DSB	Double sideband. A form of AM (modulation) where the same information is sent on each sideband. Conventional AM broadcast is a form of DSB. If the term DSB is used, it should be stated whether the carrier is suppressed or not. On conventional AM it is not
DSBEC	Double sideband, emitted carrier
EBCDIC (code)	Extended binary coded decimal interchange code
EDP	Electronic data processing
entropy	A term used in thermodynamics. In data transmission it defines the measurement of uncertainty

EIRP	Effective isotropically radiated power
EL, el	Elevation (refers to the orientation in a vertical plane of an earth station antenna)
ERL	Echo return loss, term used in VNL concept. A single, weighted figure for return losses in the band of 500–2500 Hz
Erlang	The international dimensionless unit of traffic intensity. One erlang (E) is the intensity in a traffic path continuously occupied, or in one or more paths, carrying an aggregate traffic of 1 call-hour per hour, 1 call-minute per minute, etc.
exhaust, exhaustion	When all wire pairs in a cable are considered, for planning purposes, to be used up (assigned), on any installed system is considered to have reached its full capacity. Systems, particularly wire pair, never really are filled (used, assigned) to 100%. Many are considered exhausted for planning purposes at 80%, 70%, etc. The remaining unassigned pairs are bad pairs, emergency spares, etc.
exhaust date	The date for planning purposes that the exhaustion is supposed to take place
F1A (weighting)	An obsolete weighting used in the United States. Noise units used with this weighting network are dBa
FCC	Federal Communications Commission, the U.S. federal telecommunications (and radio) regulatory authority
FDM	Frequency division multiplex
FDMA	Frequency division multiple access
FM	Frequency modulation
FOT	An HF propagation term from the French, "fréquence optimum de travail." We more often call it OWF or optimum working frequency
frame	In PCM the assembly of digits which together constitute the repetitive cycle of the multiplex. Also used in TV
FSK	Frequency shift keying
ft	Feet
GBLC	Gaussian band-limited channel
GHz	Gigahertz $\text{Hz} \times 10^9$
G/T	The "figure of merit" of an earth station. Refers to the earth station receiving system. It is usually expressed in dB and the expression may then be stated $$G/T = G_{\text{dB}} - 10 \log T$$ where G is the gain of the antenna at the receiving fre-

	quency and T is the effective noise temperature in degrees Kelvin of the receiving system
HDB3	High density binary 3 (a PCM term)
HF	High frequency. The radio frequency band from 3 to 30 MHz
highway	A common path over which signals from a plurality of channels pass with separation achieved by time division
Holbrook-Dixon	The two Bell Telephone Laboratory scientists who developed an empirical formula for the loading of multichannel amplifiers. When referring to "Holbrook-Dixon," reference is actually being made to their formulas or curves
HPA	High power amplifier
Hz	Hertz, the basic measurement unit of frequency. We used to call it cycles per second
IBM	International Business Machine (Co.)
ICSC	International Communications Satellite Consortium
IEEE	Institute of Electrical and Electronic Engineers (US)
IF	Intermediate frequency
IM, IM distortion, IM products	Intermodulation, intermodulation distortion, intermodulation products
inside plant	See outside plant
Intelsat	International Telecommunication Satellite (under the control of ICSC)
I/O	Input-output (device). A generic group of equipments that serve as input and output for a telecommunication system. Teleprinters, and telephone handsets are input-output devices
IRE	Institute of Radio Engineers (U.S.), which, now joined by the AIEE, has become the IEEE (Institute of Electrical and Electronic Engineers)
ISB	Independent sideband. A form of AM modulation where the information sent on the upper sideband is different from that sent on the lower sideband
ITA	International Telegraph Alphabet
ITT	International Telephone and Telegraph (Co.)
ITU	International Telecommunications Union based in Geneva, Switzerland
junction	See trunk

K factor	Used in troposcatter and radiolink path profiling to determine the amount and type of bending a radio beam may undergo in a particular set of circumstances
	If *K* is greater than 1, the ray beam is bent toward the earth; if less than 1, it is bent away from the earth
kHz	Kilohertz; Hz $\times$ 10^3
km	Kilometer
kW	Kilowatt
λ	Lambda (the Greek letter) referring to wavelength
L-carrier	The SSBSC FDM series developed by the Bell System for long-haul transmission over broadband media. We may expect to find the L1, L2, L3, L4, or L5 systems
lay-up	Assembly of tubes inside a single sheath (coaxial cable systems)
LBO	Line build-out, a method of extending the length of a line electrically, usually by means of capacitors
LNA	Low noise amplifier
LOS	Line-of-sight, usually referring to radio systems where relay stations operate within line-of-sight of each other. See Chapter 5
LP	Log periodic (antenna)
LPA	Linear power amplifier
LSA mode	Limited space-charge accumulation mode
LSB	Lower sideband
LUF	An HF propagation term meaning lowest usable frequency
M out of N code	An error detection code
Mb/s	Megabits per second
mesh connection	A telecommunications routing term describing the situation where telephone exchanges are interconnected directly by trunks. *Full mesh* is a term which is more explicit, meaning that all exchanges in an area are each and every one connected to all other exchanges in an area. It implies that tandem operation is not used at all. See star connection.
MHz	Megahertz Hz $\times$ 10^6
mi	Mile; in this work, unless otherwise specified, we refer to the statute mile (i.e., 5280 ft)
mm	Millimeter(s)
mm region	As designated by the ITU, the frequency band from 30 to 300 GHz. In this text, from 13 to over 100 GHz

modem	acronym for modulator-demodulator
ms	milliseconds
MUF	An HF propagation term meaning maximum usable frequency
NARS	North Atlantic Radio System. A radio system, principally troposcatter, connecting the UK via the Faeroes, Iceland, Greenland, Baffin Island, and Canada, for U.S. and North Atlantic defense
NBFM	Narrow band frequency modulation. Maximum deviation is 15 kHz for commercial operation, 3 kHz for amateur operation
NBS	National Bureau of Standards (U.S.)
neper	A logarithmic measurement unit expressing a ratio similar to the dB but the logarithm is based to the natural log base e. 1 db = 0.1151 nepers; 1 neper = 8.686 dB
NL	Nonloaded (in reference to VF telephone cables)
NLR	Noise load ratio (or noise load factor)
NPR	Noise power ratio
ns	Nanoseconds
NTSC	National Television Systems Committee (U.S.)
OCL	Overall connection loss, term used in the VNL concept
off-hook	A telephone handset that has been taken off-hook, closes the loop, and "busys" the line
on-hook	A telephone handset that is on-hook is not in use. Placing a handset on-hook opens the loop, making the line "idle"
ORE	Overall reference equivalent. ORE is the sum of the TRE (transmit reference equivalent) plus the RRE (receive reference equivalent) of a telephone connection
OSC, osc	Abbreviation for oscillator
outside plant	A telephone operating company term, or derived from telephone operating companies. It has a varied usage. Of course, it is the opposite of "inside plant." We may well say that all major operating equipment located indoors may be considered inside plant, but what about the subscriber's subset and protection equipment? We may equally well say that all equipment located out-of-doors is outside plant. For the discussion in this text, outside plant is that part of the telephone plant that takes the signal from the local switch to the subscriber

	as well as the local trunk (junction) plant. Many telephone engineers lump all civil engineering activities as outside plant
OW	Orderwire, service channel
PAL	Phase alternation line. One of the two European color television systems. The other system is SECAM
PCM	Pulse code modulation
PEP	Peak envelope power
Periscopic antenna system	Instead of using a long waveguide run up a tower, some radiolinks resort to an "optical mirror," similar to that used with a submarine periscope. In this case a passive reflector is installed at the top of the site antenna tower (or whatever operating height) and the parabolic reflector antenna is installed* oriented upward with its ray beam in near field at 45° angle with the reflector in such a way that the reflected ray is directed at the distant antenna. Depending on near-field location and passive reflector size, a periscopic system may display up to a 3-dB gain or loss over simple parabolic antenna systems with their waveguide run losses not included.
P-P, p-p	Peak-to-peak. A voltage measurement from positive voltage peak to negative voltage peak
pps	Pulses per second
PRF	Pulse repetition frequency
PSK	Phase shift keying. A form of phase modulation in which the modulating function shifts the instantaneous phase of the modulated wave between predetermined discrete values
psophometric	A noise weighting used in Europe with a noise meter called a psophometer. See Section 1.9.6 and CCITT Rec. G.223.
PSV	Pair shield video
pW	Picowatt(s), 10^{-12} W or 1 pW = -90 dBm
pWp	Picowatts psophometrically weighted
QPSK	Quaternary phase shift keying, four-level phase modulation. See PSK
QSY	International "Q" signal indicating that a change in frequency is required or is being carried out
RC	Resistance-capacitance product, determines the time constant of a circuit

* Usually at the tower base.

RCVR	Abbreviation for receiver
reference equivalent	A measurement in dB of telephone subscriber satisfaction. In broad terms reference equivalent considers only the level of a telephone connection
Rep.	Report, and in particular, a CCIR (or CCITT) report
RF, rf	Radio frequency (as distinguished from AF or audio frequency)
RFI	Radio frequency interference
rms	Root mean square (math)
RRE	Receive reference equivalent. The reference equivalent of the receiving portion of a telephone connection
scintillation	A random fluctuation of a received signal about its mean value, the deviations being relatively small. It should be noted that the word is an extension of the astronomical term for the twinkling of the stars, and the underlying explanation may be similar (taken from the IEEE Standard Dictionary). It should also be noted that we are often redundant when we say scintillation fading. Another fact that separates scintillation from other types of fading·is that it is usually rapid, with many fades occurring inside of a minute of time
S/D	Signal-to-distortion ratio. Term used in PCM
SECAM	Sequential with memory. One of the two Europran color television systems. The other is PAL (phase alternation line)
SHF	Super high frequency. That band of frequencies encompassing 3000–30,000 MHz
SID	Sudden ionospheric disturbance. An HF propagation term
S/N	Signal-to-noise ratio
SPADE	Single channel per carrier multiple access demand assignment equipment
SSB	Single sideband. A form of amplitude modulation where information is transmitted on one sideband only. Suppressed carrier operation is usually implied
SSBSC	Single sideband suppressed carrier
star connection	A telephone/telecommunications routing term implying tandem operation to interconnect exchanges which have a comparatively low level of traffic flow giving a savings of the additional trunks that would be required for full mesh connection

STC	Standard Telephone and Cable, ITT's large British affiliate
STL	(1) Studio-transmitter link. A radiolink connecting a broadcast studio (usually TV) to its associated radiating transmitter site. (2) Standard Telephone Laboratories (ITT-UK)
SWL	Shortwave listener, one who listens to HF signals as a hobby
sync	Synchronization, synchronizing
T1, T2, etc.	ATT-Bell System series of PCM line arrangements
TASO	Television Allocation Study Organization (U.S.)
TDA	Tunnel diode amplifier
TDM	Time division multiplex
TDMA	Time division multiple access
TE_{11} mode	TE for transverse electric wave. In circular waveguide the TE_{11} mode is the dominant wave
throughput	The amount of useful information that a data transmission system can deliver end-to-end. Parity bits, stop and start elements, etc., reduce throughput
TRE	Transmit reference equivalent. The reference equivalent of the transmit portion of a telephone connection
tropo	Tropospheric scatter (transmission)
trunk	In the local area those circuits connecting one exchange to another. This is called a junction in the UK. In the long-distance (toll) area, any circuit connecting one long-distance exchange to another. Toll connecting trunks connect the local area to long-distance (toll) service
TWT	Traveling wave tube, a broadband amplifier
TWTA	Traveling wave tube amplifier
UHF	Ultra-high frequency, 300–3000 MHz
UK	United Kingdom (Great Britain)
USB	Upper sideband
USITA	U.S. Independent Telephone Association
variolosser	A balanced attenuator whose attenuation is controlled by a signal obtained by rectification of the signal which passes through it. It can be connected so as to increase or decrease the attenuation with an increase in signal strength
VF	Voice frequency, encompassing the band of frequencies from nearly dc to about 30,000 Hz. In this text we nominally consider VF to be the voice channel or 0–4000 Hz; for CCITT carrier application, from 300 to 3400 Hz

VFCT, VFTG	Voice frequency carrier telegraph
VHF	Very high frequency. The band of frequencies encompassing 30–300 MHz
VNL	Via net loss. The method used in North America of assigning minimum loss in a telephone network to control echo and singing
VNLF	Via net loss factor. Term used in computing VNL, and the unit is dB/mi
VOGAD	Voice operated gain adjust device
VSB	Vestigial sideband
VSWR	Voltage standing wave ratio
VU	Volume unit. A measure of level, usually used for complex audio signals such as voice or program traffic. The unit is logarithmic. For a continuous sine wave across 600 Ω, 0 dBm = 0 VU. For complex signals in the VF range we say that

$$\text{Power}_{dBm} = VU - 1.4 \text{ dB}$$

w/g, W/G	Abbreviation for waveguide
XMTR	Abbreviation for transmitter
Z_0	Characteristic impedance
0 TLP	Zero test level point. A single reference point in a circuit or system at which we can expect to find the level to be 0 dBm (test level). From the 0 TLP other points in the circuit may be referenced using the unit dBr or dBm0 such that

$$dBm = dBm0 + dBr$$

INDEX

Absorption, oxygen and water
vapor, 270, 298
Absorption attenuation, 238-
239, 270, 432-434, 444
Activity factor, 92-95, 98
Address signaling, 547, 549,
554
Advantage, FM over AM, 204-205
Alarms, 101, 102, 199
radiolink, 248-250
see also Supervisory systems
A-law companding, 456, 465
Alternate mark inversion (AMI),
471
Aluminum as a wire conductor,
72
AM, conventional compared to
SSB, 147-148
conventional compared to FM,
204
AM-DSB, 371, 385
American Wire Gauge (AWG), 64,
67, 68
Amplitude modulation, see AM
Amplitude response, see Atten-
uation distortion
AM-VSB, 371-372, 385
Analog transmission versus
digital, 339
Angle of incidence (HF),137-138

Antenna cost trade-off, 301
Antenna gain, mm region, 439,
442
Antenna gain degradation, see
Aperture-to-medium coupling
loss
Antenna gain nomogram, 278
Antenna height, 273
Antenna noise, 303-305
Antennas, HF, 155-156
radiolink, 207-209
tropo, 278, 281, 283
Antenna towers and masts, 250-
253
Antenna tracking subsystem,
319-321
Aperture-to-medium coupling
loss, 270-271
ARQ, 355-356
Artificial line, 44
ASCII, 342, 344, 346-348
Aspect ratio, 483
Asynchronous (start-stop)
transmission, 360-361
Atmospheric absorption, 296-297
Atmospheric noise, 296-297,
303
Attack and recovery times, 128,
152
Attenuation, atmospheric, due

to oxygen and water vapor, 270, 298
Attenuation distortion (frequency response), 13, 14-16, 38, 76, 79-80, 108, 375, 376, 393, 402, 414, 418, 423, 426, 467, 492, 493, 503, 506
Attenuation per unit length (coax), 402
Aural channel, *see* Program channel
Auroral zones (HF transmission), 136
Automatic equalization, 389
Automatic frequency control (AFC), 149-150
Automatic gain control (AGC), 128, 170, 419
Automatic level control, 405
Automatic load control (ALC), 152
Automatic tracking, 320-321
AUTOVON, 43, 52
Average talker level, 99

B6ZS code, 471
Babble, 22
Back-off power, 308, 327
Back porch, 487, 497
Backward channel for error control, 391, 392
Balance, 45
Balance return loss, 47, 52, 53
Balancing network, 44-46, 47, 49
Balun, 159
Band limited, 21
Bandwidth, 13, 143, 476
  IF (B_{if}), 287; *see also* Carson's rule
  PCM, 448, 455, 476
Bandwidth assigned, 140, 145
Baseband, 201, 227, 313, 403, 485, 503
Baseband repeater, 241, 243

Battery, telegraph, 358
Battery sizing, 541-545
Baud, 366-367, 385
Baudot teleprinter code, 342
BCD code, 353
BCD interchange code (with Hollerith), 352
Beacon, satellite, 327
Beamwidth, antenna, 208-209, 282
Bending of radio waves, 183-186
B factor, 59
Bias distortion, 365-366
Binary, 339
Binary coding techniques, *see* Coding
Binary convention, 340-341
Binary transmission and concept of time, 359-367
Binomial distribution, 10
Bipolar transmission, 449, 471
Bit (binary digit), 339, 340, 342, 366
Biternary, *see* Bipolar transmission
Bits, bauds and words per minute, 366-367
Blanking pulse, 486-487
Block, (data transmission), 354-355, 357
BNZS code(s), 471
Boltzman's constant, 20, 33
Break point (overload), 224-225
Bridged tap, 71, 124
BWR (bandwidth ratio), 230

Cable characteristics (coaxial cable), 401-403
Calcium-lead cells, 540
Capture effect, 194-195
Card punch, card reader, 367, 368, 369, 370
Carrier (equipment), 13, 15, 17, 18, 23, 42, 43, 46, 49, 57-58, 82-130, 374-378, 388-389, 550-551

Carrier leak, 86
Carrier suppression, 306
Carrier-to-noise ratio (C/N),
  194-197, 275
Carson's rule, 195, 244, 290
Cassegrain feed, 303
CATV, 399
CCIS, 558
CCITT, CCIR guide to, Appn. A,
  512-539
CCITT hierarchy, 59-62
CCITT loading, 98-99
CCITT No. 5 code, 348-350
CCITT pilots, 102
CCS signaling, 558
Central battery, 34
Central tendency, 11
CEPT 30+2 PCM, 459, 460, 462,
  464, 469
Channel bank (FDM), 92, 125
Channel capacity, 382-384
Channel-end, 124-126
Channel noise (in radiolink
  systems), 230-231
Character, 339
Characteristic impedance, 155,
  165
Charger, battery, 540-543
Chrominance, 497
Circuit conditioning, see Con-
  ditioning
Circulator, 210, 222
Class 1, 2, 3, 4 and 5 offices
  (North American switching
  hierarchy), 54, 55, 58, 61,
  62
Clock, 362-364
C-message line weighting, 25,
  28
Coaxial cable, 60-MHz systems,
  420-424
 transmission lines, 219-221
Coaxial cable carrier systems,
  115-121
Coaxial cable communication
  systems, 399-429

Coaxial cable-radiolink, the
  decision, 424-428
Code, 339
Codec, 463, 465
Coding, 330, 341-353
 for load coil spacing, 66
 levels, (see 342), 452
 PCM, 448, 457-461
Codes, specific data/telegraph,
  343-353
Color bars, 510
Color television, 482, 497
Combiner, see Diversity com-
  biners
Common equipment, 125
Common volume, see Scatter
  volume
Communication subsystem, earth
  stations, 316-319
Companding, PCM, 455, 456
Companding improvement, 457
Compandor, 13, 126-128
Comparison, coaxial cable-
  radiolink, 425
Compelled signaling, 553
Composite noise baseband power
  (for NPR measurements), 227
Composite telegraph, 100
Composite television signal,
  485-488
Compression, 194
Compression ratio, 127
Compromise balance (network),
  47, 49
Compromise rhombic antenna,
  156-157
Concentration, telephone, 61,
  69, 72
Conditioned pairs, video trans-
  mission, 505-506
Conditioning, 73, 375, 387-389
Construction design, coaxial
  cable systems, 400
Contents, CCITT Recommendations,
  514-515
Conversion factors, 32-33

Coordination contour, 246,
  336-337
Coordination distance, 336-337
Counter emf cell, 541, 544
Critical frequency (HF propa-
  gation), 135
Critical video parameters, 491-
  493
Critical video transmission
  impairments, 488-490
Crosstalk, PCM, 449, 466, 467
Crosstalk, 20, 22-23, 76, 79,
  92, 376
Crosstalk coupling loss, 23
Crosstalk index, 22-23
Crosstalk unit (CU), 23-24
CRPL (Central Radio Propaga-
  tion Laboratory), 134, 136
Cryogenic cooling, 302, 306,
  318
C/T (ratio of carrier to ther-
  mal noise power), 306-308
CU (crosstalk unit), 23
Current ratio, 5
Customer satisfaction, see
  Subscriber satisfaction
CW (continuous wave), 140
Cyclic distortion, 366

D1 PCM, 462, 464, 473
D2 PCM, 460-461, 462, 463, 464,
  473
Data interface, 367-369
Data loading, 100
Data modems versus critical design parameters, 384-387
Data on PCM, 475
Data rate, 383, 385, 389; see
  also throughput
dBa, 25, 28, 29
dBm, 5, 25, 92, 93, 95
dBmO, 19
dBmp, 26, 28
dBrn, 24
dBrnC, 25-26, 28
dBmV, 7-8

dBr, 19
dB sum/difference, 9-10
dBV, 506
dBW, 6-7, 31
dBx, 23
DCA pilots, 102
DC nature of data transmission,
  357-359
Decibel, 2-10, 38
Dedicated plant, 71
De-emphasis, 199-201; see also
  Pre-emphasis
Delay, 17-18
Delay distortion, see Envelope
  delay distortion
Demand assignment, 315
Deviation, 195, 205-206, 329
Deviation sensitivity, 205
Di-bit coding, 385, 393
Differential phase and gain,
  492, 493, 504, 508
Digital data transmission (over
  telephone networks), 339-398
Digital transmission (PCM),
  448-481
Direct trunk group, 61
Dispersion (the degree of vari-
  ation), 11
Dispersion loss, 76
Dissipation line, 156
Distance to radio horizon, 273-
  274
Distortion in digital data
  transmission, 364-366
Diversity, path, 437
Diversity (diversity reception),
  210-216, 279-280, 281, 286-
  288
Diversity combiners, 172, 211,
  212, 214-218, 287-289
dN (decineper), 8
Domestic satellite systems, see
  Regional communication satel-
  lite systems
Double sideband, emitted car-
  rier (DSBEC), 113, 126, 143

Down link, *see* Satellite-earth
  link
D region, 132
Drop and insert, 109
Dynamic range of talker levels,
  126, 452-453

E and M signaling, 551
Earpiece (telephone), 1, 34-35
Earth bulge, 183-185
Earth-space communication,
  millimetric, 445-446
Earth-space window, 295-298
Earth station, definition of,
  295
Earth station technology, 295-
  337
EBCDIC, 350-351
Echo, 13, 45-49, 50-53, 59
Echo delay, 48
Echo return loss (ERL), 47, 54
Echo suppressor, 48, 51, 55,
  58, 80, 391
Echo tolerance, 54-56
Economics of carrier trans-
  mission, 124-126
Effective noise temperature,
  30-31
EIRP (effective isotropically
  radiated power), 31-32, 299,
  307, 308, 310-311, 329, 330,
  335
Elastic store, 476, 477
E layer, 132
Elements, tv, 483-485
Emission bandwidths, 145
Emission types, 140-143
End office, *see* Local office
End-to-end signaling, 556-557
Entropy, 341
Envelope delay distortion
  (EDD), 17-18, 108-109, 374,
  377, 384, 386, 387-389, 393,
  426, 492, 493
Equalization, *see* Conditioning
  of coaxial cable systems,

414-418
of pair shield video, 506
Equalize (battery), 543, 544
Equalizers, fixed (coaxial
  cable systems), 415-416
  variable (coaxial cable sys-
  tems), 416-418
  TV coaxial cable, 501
Equiprobability (coding), 456
Erect sideband, 84, 146
ERP, 32; *see also* EIRP
Error detection and error cor-
  rection, 353-357
Error rate, 25, 100, 353, 377-
  380, 386, 449, 467, 471-472
Errors, digital, 359
ESRO, 335
European 12 MHz coaxial cable
  tv system, 501-503
Excess attenuation, *see* Attenu-
  ation
Expansion ratio, 127
Extended unigauge, 70-71

FlA (weighting), 25-26
Facility layout, HF, 167
Fade margin, determination of,
  mm links, 440, 441
  determination of, radiolinks,
  232-239
  determination of, on tropo
  links, 276-279
Fading, on millimetric links,
  430
  on radiolinks, 197-198, 232-
  239, 426
  on tropo links, 266-267, 276-
  277
Far field, 257-259
Fault alarms, radiolinks, 248-
  250
  PCM, 469, 470
FDM, *see* Frequency division
  multiplex
FDMA (frequency division multiple access), 300, 311-313,

314, 315, 332
Feed, antenna, 282
Feed bridge, 62
Field period, 485
Figure of merit, earth stations, see G/T
Final route, 61
Final voltage, 543, 544, 545
First Fresnel zone radius, 185, 189-191
Flat noise, 258
F layer, 132-135
Flicker, 484
Floating, 541, 543, 544
Flux density, see Illumination level
Flyback, 485, 486
FM improvement threshold, 193, 194-195, 205, 275-276, 287
FM receiver, 210, 285-287; see also Receiver group
FM transmitter, 201-206
Forbidden regions, 434
Format, PCM, 450, 463
Fortuitous distortion (digital), 365-366
FOT (frequence optimum de travail), 137
Four-phase shift keying, see Quadrature phase shift keying
Four-wire transmission, 43-45, 48, 51, 57, 73, 75, 76, 77, 80, 82-84
Frame, 452, 461-463
Frame rate (tv), 483-484
Free space path loss, 181-182, 268, 433, 439, 442
Frequency assignment, basis of, 247-248
tropo, 281, 292-294
Frequency bands, satellite communication, 297
Frequency deviation, see Deviation
Frequency diversity, 211-213

in-band, 172-173
Frequency division multiplex (FDM), 13, 43, 82-130, 321-325, 380, 381, 399, 403, 449, 476
Frequency frogging, 113
at radiolink repeaters, 247
Frequency generation (FDM), 103-104
Frequency inversion, 146
Frequency planning, radiolinks, 244-248
tropo, 281
Frequency reinsertion, 149-150
Frequency reuse, 330
Frequency shift (keying) modulation (FSK), 140, 372, 385, 389-391, 478
Frequency synchronization, 102-103, 149-150
Frequency synthesizer, 149-152
Frequency tolerance, end-to-end, 102, 150
Frequency translation errors, 381
Fresnel clearance, 185-191
Fresnel zones, 185-191
Front porch, 487
FSK, see Frequency shift modulation
Full duplex, 43, 91, 95
Functional block diagram, data transmission system, 396-397
Functional operation of a "standard earth station," 316-325

Gain, antenna, 208-209
Gain-frequency characteristic, see Attenuation distortion
Gain stability, see Stability
Galactic noise, 296-297, 305-307
Gating, 452
Gaussian band limited channel (GBLC), 383-384

Gaussian distribution, 10, 20, 412

Gaussian noise, see Thermal noise

Geostationary satellite, see Synchronous satellite

Glossary Appn D, 559-569

Granularity, PCM, 456, 465

Great circle bearing and distance, 167-168

G. Recommendations CCITT (index), 516-535

Group, FDM, 100; see also Standard CCITT group

Group delay, 108-109

Group regulation, 102

Ground wave, 131

G/T (figure of merit of an earth station), 302-306, 335

Gunn oscillator, 439

Harmonic distortion, 76

Harmonics, 21, 412-413

HDB3 code, 471

Heterodyne, see Mixing

Hierarchy, PCM, 473-475
  telephone network, 59-62

Higher order PCM, 473-476

High frequency (HF) radio, 131-176

Hits, noise, 24, 376, 378-379

Holbrook and Dixon, 19, 229, 230, see also 91-101

Hollerith code, 350-352

Homodyne detection, 501

Horizontal and vertical polarization (HF), 166

Horizontal synchronizing interval, 487

Hot-standby operation, 240-242

Hourly median values (signal level), 267

House load, see power budget

Human interface 368; see also I/O devices

Hum level, 540

Hybrid, 44-45, 51, 74; see also Term set

IBM data transceiver code, 344-346

Idle channel noise, PCM, 465

Idling signal, 364

IF, carrier inserted, 147

IF bandwidth, 287; see also Carson's rule

IF frequencies, 210

IF interference, 247

IF repeater, 241-244

Illumination level, 308-309

IMPATT, 439

Impedance, coaxial cable, 402, 409, 427

Impedance mismatch, 46-48

Impulse noise, 24, 76, 376-380, 393, 469, 552, 557

Inband signaling, 129, 550-551

Independent sideband (ISB), 143, 168-171

Information baseband, see Baseband

Input-output (I/O) devices, 34
  data, 368-370

Intelligible crosstalk, 22-23, 79

Intelligibility, 14

Intelsat II, 300, 312

Intelsat III, 300

Intelsat IV, 325-330

Intelsat IVA, 330

Intensity range, voice signals, 126-128, 452-453

Interaction crosstalk, 113

Interface, CCITT, 75-80
  with end instrument, 367-369

Intermodulation distortion, 153

Intermodulation noise, 20, 21-22, 23, 92, 225, 227, 231, 290, 376-378, 407, 408, 412-414

Intermodulation products (IM products), 20, 21, 98, 313,

412-414
Interleaving, 484
Interregister signaling, 547, 549, 553-558
Interstices, 400
Interswitch signaling, see Interregister signaling
Intersymbol interference, 383-384
Intertoll trunks, 54, 57-58
Inverted sideband, 84, 146
Ionogram, 174-175
Ionosphere, 132
Ionospheric refraction, 131
Ionospheric sounder, oblique, 131, 136, 173-175
ISB, see Independent sideband
Isolation, polarization, 209, 246, 289
Isotropic, 32, 208
Isotropic antenna, 181
ITA No. 2 code (CCITT), 343
ITU (International Telecommunication Union), 38-39

J-carrier, 112
Jitter, see Phase jitter; timing jitter
Johnson noise (thermal noise), 20-21
Jumbo group, 423-424
Junction (trunk), 12, 39, 40, 72-74

K-carrier, 112-113
Keyboard sending units, 369
Keying, neutral and polar, 358-359
K factor, 184-191
Klystron, power amplifier (high power), 284

Land area, minimum required for towers, 252-253
Lay-ups, basic coaxial cable, 400-401

L-carrier, 121, 500, 501
L-3 carrier, 502
L-coaxial cable systems, 406, 415, 416, 418, 419
L-5 coaxial cable system, 422-424, 476
Lead-acid cells, 540
Level, 13, 18-19, 38, 52, 93, 101, 102
 absolute, 6
 coaxial cable systems, 418-419
Level regulating pilots, 101-102, 418, 419
Level stability, 419, 424
Levels and level variations (data transmission), 380-381
"Lincompex," 171
Linearity, 153
Linear polarization, 209
Linear quantization, 455
Line buildout (LBO), 74-75
Line buildout networks, 415, 416
Line code, PCM, 469
Line frequency (FDM), 90-91, 96-97, 110-121, 403, 421
Line-of-sight, 177
Line parameters, ITT/BTM PCM, 469
Line signaling, 547
Line engineering (radiolinks), 177-180
Link-by-link signaling, 555-557
Loading, 91-101
 HF transmitters, 171
 inductive, 15, 46, 48, 57, 65-68, 69, 124
 trunks, 73-74, 467
 radiolink systems, 225-231
 single channel, 99
 FDM systems, 91-101
 with constant amplitude signals, 99-101
Load isolator, ferrite, 222, 282-283
Load on charger, 541-542

Local area trunks (junctions), 72-74, 466, 479

Local office (switch), 54, 59-60

Logatom, 39

Log normal fading (tropo), 267

Log periodic (LP) antenna, 154, 162-166

Long distance transmission via PCM, 476-477

Loop, DC (data transmission), 357-359

telephone, 35-37, 39, 46, 51, 61-72, 123-124, 548-549; see also subscriber loop

Loop extender, 65, 69-71

Low noise receiver (low noise amplifier), 302, 305, 306, 316, 317, 318

LPA, see Power amplifier, linear

LPSS, 424

LSA mode, 439

LUF (lowest usable frequency), 137

Luminance, 482, 483, 484, 493, 497

Magnetic activity, 133-134, 138

Magnetic storms, 138

Manual pointing (earth station antennas), 319-320

Mark (and space), 340, 359, 360, 361, 362, 364, 365, 389-390

Mark-space convention, see Binary convention

Mark-to-space transition, 361, 363, 364, 366

M-ary techniques, 372

Master frequency generator, 103, 104

Master group, 121-122; see also CCITT mastergroup

Maximum feasible median path

loss (tropo), 290-291

Mean, arithmetic, 10-11

Median signal level, 438

Median path loss, 267-270, 292, 293

Median received signal level, 266; see also Signal level, median received

Mesh connection, 60-61, 72

Message service, 319

Message traffic, 91

MF codes, 554-557

Microwave (radiolink) frequency bands, 178-179

Microwave (LOS) line-of-sight, see Radiolink systems

Millimeter wave transmission, propagated, 430-447

Minigauge, 71-72

Mismatch, 46; see also Impedance mismatch

Mixing, 21, 82-86

Modem, 368, 371

Modems, for bandwidths in excess of 4 kHz, 395

high data rate, 392-393

medium data rate, 391-392

practical application, 389-393

Modulation-demodulation schemes (data transmission), 371-373

Modulation index, 203-204

Modulation plan, CCITT, 86-91, 389-391

Modulator-exciter (tropo), 284-285

Molecular resonance, 432

Monitored functions, radiolink alarm and supervisory systems, 248-249

Monkey talk, 149

Monochrome television, 482

Monopulse, 320-321

Moore ARQ code, 355-356

Morse code, 140, 144

Mouthpiece (telephone), 1, 34

m out of n code, 354

MUF (Maximum usable frequency),
  137, 174
Multipath, 174, 191, 198, 211,
  427
Multipath distortion, 380
mu law, 456, 460, 465
Multiburst, 508
Multuple access, 311-315
Multiples (dB), 4
Multiplex, see Frequency divi-
  sion multiplex; Time divi-
  sion multiplex
Multiplexer, HF, 169
mVp, 28

NBS method (tropo), 268-270
N-carrier, 113
Near field, 256-259
Near field boundaries, 224
Near sing, 53
Negative impedance repeater,
  75
Neper, 8
Network, 2, 29, 30, 52, 59-61
Neutral transmission (keying),
  358-359
Nickel cadmium (NiCad) cells,
  540
NLR (noise load ratio), 230
No-break power, 444, Appn B
  540-546
Noise, 13, 31-33, 76, 77, 115,
  375-380, 449
Noise, atmospheric (sky), 296-
  298
  PCM, 449, 465
Noise bandwidth (B_{if}), 20, 285,
  439; see also Carson's rule
Noise budget, earth stations,
  329
Noise calculations, 105-108,
  126, 426
  CCITT approach, 258-262
  US Mil approach, 106-108
Noise contributors, earth sta-
  tions, 304

Noise figure, 27, 30-31, 285-
  286, 301
Noise measurement units, 24
Noise power ratio (NPR), 225-
  232, 258, 290
Noise temperature, 20-21, 30-
  31, 285, 302-306
Noise threshold, 275, 285, 439;
  see also Threshold, thermal
  noise
Noncompelled signaling, 553,
  554
Nonuniform (non-linear) quanti-
  zing, 455, 456
NOSFER, 39
NPR, see Noise power ratio
NPR measurement frequencies,
  228
NTSC, 483, 494, 497, 508
Numbering, telephone, 549
Nyquist, 383-384, 392

Off-hook, 12, 35, 37, 59, 69,
  129, 547, 551
Off-line devices, 370
Offset, frequency (synthesizer),
  150-151
On-hook, 12, 71, 129, 548, 551
On-off telegraphy, 371
Open wire carrier, 112
Operator sequence(s), 342
Optical horizon, 180
Orderwire (also called service
  channel), 199, 285, 324
Outage time, 193
Out(-of-)band signaling, 106,
  129, 550-551
Out-of-band testing, 231-232
Outside plant, 68-69
Overall connection loss (OCL),
  54-57
Overall reference equivalent
  (ORE), 40, 73, 80
Overload, 98
  and margin, coaxial cable sys-
  tems, 409-412, 414

Overload point, 410, 418
Over-the-horizon(OTH), 264
Overshoot, radiolinks, 246
OWF (optimum working frequency,
    137; see also FOT

Pair shield video (PSV), 505
PAL, 497
PAM (pulse amplitude modula-
    tion) wave, 450, 451, 452
Parallel-to-serial conversion
    for improved economy of cir-
    cuit usage, 394-395
Parallel transmission (at in-
    terface), 367
Parameters, critical for data
    transmission, 373-382
Parametric amplifier, 286, 302,
    305, 306, 439
Parity, 344, 348, 349, 354,
    357
Path calculations, 180, 193-
    198, 232-239, 274-279, 301
Path diversity, 437, 446
Path loss (free space), 181-
    182, 298-299, 336
Path profiling, 180, 187-191
Path reliability, see Propaga-
    tion reliability
HF, 131
PCM, see Pulse code modulation
PCM higher order multiplexers,
    474-476
PCM, practical application,
    466-473
    to interoffice trunks (junc-
    tions), 466-467
PCM switching, 479-480
PCM transmission via radiolink,
    477-479
Peak deviation, 205-206
Peak distortion, 366
Peak envelope power (PEP), 152
Peak factor, 98
Percentage of modulation, 203
Perforator, 370

Periscopic antenna systems,
    223-226
Phase jitter, 382
Phase shift, 12, 17-18
Phase shift (keying) modulation
    (PSK), 372-373, 385
Pilot carrier, 146, 149-150
Pilot tones, 101-103, 216-218,
    285, 418-419, 422, 505
Plane reflectors, 253; see also
    Periscopic antenna systems
    as passive repeaters, 253,
    256-258
Polarization, earth stations,
    327
HF, 166
radiolink, 209, 439
tropo, 280, 283
Polar transmission (keying),
    358-359, 367
Post equalization, 388
Power amplifier, linear (LPA),
    152
tropo, 284-285
Power budget, 542
Power fading, 198
Powering, coaxial cable sys-
    tems, 420, 422, 424, 428
Powering PCM systems, 470
Power ratio, 4
Power series, 412
Power split(ter), 223, 301,
    318-319
Practical transmission of
    speech, 12-13
Precipitation attenuation, 238-
    239, 297, 335, 433, 434-437,
    440, 441, 442
Precision balancing network,
    74-75
Preemphasis (de-emphasis), 98,
    199-201, 203, 232-235
Preemphasis, coaxial cable sys-
    tems, 403
tv transmission systems, 504
Preselector, 210

Pressurization, waveguide, 220
Printer, 370
Program channel, 331
Program channel transmission
  (tv), 499, 503
Programmed pointing, tracking,
  319-320
Propagation, 181
Propagation, basic HF, 131
  millimetric, 430, 431-434
Propagation reliability, 193,
  211, 238
Propagation time, 80, 362, 363,
  475
Prove-in-distance, 125-126,
  450, 466, 480
Psophometric weighting, 26, 28
PSV, see Pair shield video
Pulse code modulation (PCM),
  443, 448-481
Pulse link repeater, 549
Pulse stuffing, 475
pW (picowatt), 28, 76
pWp (picowatt, psophometrically
  weighted), 26, 28, 76, 77

QPSK, see Quadrature (quaternary) phase shift keying
  (modulation)
QSY, 136
Quadrature phase shift keying
  (modulation), quaternary
  phase shift keying, 373, 439,
  443, 478
Quantization, 448, 452-455,
  460
Quantizing distortion or quantizing noise, 449, 453, 463-
  465

Radiation efficiency, 207-208
Radiation resistance, 207-208
Radio continuity pilots, 216,
  218
Radio frequency interference,
  see RFI

Radiolink-coaxial cable, the
  decision regarding, 424-428
Radiolink frequency bands, 178-
  179
Radiolink systems, 177-263
Rainbarrel effect, 53
Rainfall absorption, see Pre-
  cipitation attenuation
Rainfall, millimetric link,
  430, 434-437, 440-441
  radiolink, 238-239
Random noise, see Thermal noise
Range extender, see Loop ex-
  tender
Ratio, current, voltage, 5
  power, 2-4
Rayleigh distribution, 172, 198,
  213, 216, 237, 266
Receiver, HF, 147-148, 151
Receive reference equivalent
  (RRE), 39, 40
Receiver group, FM 210, 285-287
  telephone, 35
Receivers, signaling, 551, 554,
  557
Recharge time, 542
Redundancy, digital, 353
Reference coupling, 23
Reference equivalent, 13, 38,
  72-74, 80, 480
Reference frequency (US vs
  Europe), 13, 16, 37, 38-41
Reference level point, 18-19
Reflection, 198
Reflection point, 191-192
Reflectors, passive, 223-226,
  253, 256-258; see also Peri-
  scopic antenna systems
Refractive index, 183-184, 187-
  191
Refractivity, 183-191
Regeneration, digital, 448, 449
Regenerative repeaters, PCM,
  467, 468, 469, 477
Regional satellite communica-
  tion systems, 334-337

Reliability, equipment, 240–241, 444
  propagation, see Propagation reliability
Remote control, radiolinks, 250
Repeater, terminal, 403
Repeater spacing, 428
Repeater spacing, PCM, 468–469
Repeaters, coaxial cable, main (power feeding), 400, 403
  secondary (dependent), 400, 403
  passive, 256–257
  radiolink, 241–244
Reperforator, 370
Reserve time (batteries), 541, 542, 543
Residual noise, 225, 231
Resistance design, 37, 62–63
Retrace time, see Flyback
Return loss, 13, 46–48, 51, 52, 54, 71, 73, 76
RFI (radio frequency interference), 167, 245–246, 427, 438
RF repeater, 241–244
Rhombic antenna, 154–161
Ringing, impulse noise, 379
Ringing current, 548
rms deviation, 205
Round trip delay, 57
RS-232(EIA), 367–369

Sampling, 448, 451
Sampling sequence, 461
Sampling theorem, 448
Satellite, 295
Satellite-earth link, 229–310
Saturation, 194–195
Scanning sequence, PCM, 461
  video, 483–486
Scattering attenuation, 238–239
Scatter angle, 265–266, 268
Scatter loss, 268–270; see

also Median path loss
Scatter volume, 265–266, 271
Scintillation fading, 437–438, 440, 441
SCR charger, 541
SECAM, 497
Segmented curve, 457–461
Seizure, massive, 550
  switching, 12, 36, 37, 548, 550
Self-adaptive equalizer, 389
Sender, 551
Sense of transmission (data, telegraph), 340–341
Serial-to-parallel conversion (at data interface), 367
Serial-to-parallel conversion for transmission on impaired media, 393
Service channel, see Order wire
SF signaling, 551
Shannon, 341, 382–384
Short hop concept, 443–444
Sidelobes, antenna, 282, 336
Sidetone, 35
Signaling, 12, 36–37, 62, 68, 69, 73, 75, 129–130, Appn C 547–558
  on carrier systems, 129–130
  on PCM systems, 462, 463, 464, 479
Signal level (HF), median receiver, 136–138
Signal level (tropo), median receiver, 266
Signal-to-distortion ratio, 456
Signal-to-noise ratio, 13, 27, 28, 92, 100, 128, 193–197, 204, 230–231, 275, 377–378, 384–386, 449, 471–472, 473, 491, 493, 499, 504
Silicon controlled rectifier (SCR), see SCR
Simple telephone connection, 12
Sine-squared test signal, 507–508

Singing, 13, 45-49, 52, 53, 55, 59, 75
Single frequency (SF) signaling, 129
Single sideband (SSB), 82, 143, 147-153
Single sideband suppressed carrier (SSBSC), 82, 86, 126, 143-147
Sink, 1-2, 12, 369
Site shielding factor, 337
Skew, 11
Skip zone, 135
Sky noise, 305; see also Atmospheric and galactic noise
Skywave, 131-139
Slope, attenuation distortion, 16, 80
Slope pilot, 112
Smooth earth, 185-186
Snell's law, 183
SOCOTEL, 554, 555
SOJ carrier, 112
Source, 1-2, 12, 368
Space diversity, HF, 171-173 microwave (including radio-link), 191, 211-213
Space (and mark), 340, 359, 360, 361, 362, 364, 365
SPADE, 315, 327, 331-334
Span, span line, 467
Specific gravity, 543, 544
Speech energy, 14
Speech measurement, 92-96
Speech plus derived, see Voice plus
Speech transmission, human factor, 13-14
Spillover, antenna, 304
Splatter, 293-294
Sporadic E, 138-139
Spreading waveform, 318, 319
Spurious emission, 245
Squarewave tilt, 504-505
SSB, see Single sideband
Stability, 13, 47, 52, 53, 54,

69, 77-78, 419, 480, 557
frequency, 150, 206
Stair steps, 508-509
Standard CCITT basic master-group, 89-90
Standard CCITT basic super-mastergroup, 90
Standard CCITT (FDM) group, 86-88
Standard CCITT supergroup, 89-90
Standard deviation, 10-11, 54, 55, 99
Standard distribution, 10-11, 52, 54
Standard refraction, 185
Standards, video transmission (criteria for broadcasters), 494-499
Star cable carrier, 113-115
Star connection, 60-61
Start space, 360-362
Start-stop transmission, see Asynchronous transmission
Static no-break power, Appn B, 540-546
Station carrier, 121, 123-124
Station margin (earth stations), 309-310, 319
Stop mark (stop element), 360-362
Subscriber carrier, 121, 123-124
Subscriber loop, see Telephone loop
Subscriber satisfaction, 38-39
Subscriber signal level, 12, 38
Subscriber signaling, 547
Subset, telephone, 11-13, 19, 24, 34-35, 39, 40, 43, 45, 68, 71
Sudden ionospheric disturbance (SID), 138
Sum and difference, decibels, 9-10
Sunspot activity, 138

Sunspot cycle, 136
Sunspot number, 131
Sunspots, 132, 138
Supervision (signaling), 12, 37, 69, 129
Supervisory, 419-420, 422, 469, 470; see also Fault alarms
Supervisory signaling, 547, 549, 550, 552
Surface efficiency, antenna, 208
Switch, 12, 35, 37, 38, 42, 46, 52, 59-62, 69, 73-74, 75-76, 77, 92, 123, 129, 479
Switching, PCM, 479-480 transmission considerations, 75-76
Syllabic, 92, 126
Synchronization, 359-364 frequency, 149-150 tv, 485-488
Synchronous (binary) transmission, 362, 392, 393
Synchronous earth satellite, 295, 312-313, 319
Synthesizer, see frequency synthesizer
System design, coaxial cable, 403-407

T1, T2, T3 PCM, 473-474
Takeoff angle, HF, 138
Takeoff angle, tropo, 271-273
Talk battery, 34-36
Talk-down, 129, 550-551
Talk-listen effect, 95
Tandem, maximum number of circuits in, 77
Tandem switching center, 61
Tangential noise threshold, see Noise threshold,
Tape recorder (mag tape), 370
Tape punch, 370
Tape reader, 369
TDMA (Time Division Multiple

Access), 311, 313-315
Telegraph loading, 100, 101
Telegraph transmission, see Data transmission
Telephone connection, 11-13
Telephone networks, 42-59
Telephone sets (Bell System types), 14-15
Telephone transmission, 34-81
Television transmission, via satellite, 311, 327, 330-332, 335
Temperature variation loss, coaxial cable, 416-418
Termination (terminating resistance - HF antennas), 155, 159-161; see also Dissipation line
Term set, 44, 48, 51-53; see also hybrid
Terrestrial link subsystem (earth stations), 323-324
Terrestrial station, 295, 336
Test level point, 18-19
Test patterns (tv), 510
Tests for video quality, 507-510
Test tone, 52
Thermal noise, 20-21, 225, 375-377, 408, 409, 471-472; see also residual noise
Threshold, FM, 193-194, 205, 287, 301
thermal noise, 20-21, 193-194, 275, 409
Threshold extension, 279, 287, 309, 319
Through-group, through-supergroup techniques, 109-110
Throughput, 353, 354
Time block assignments, 267
Time division multiplex (TDM), 82; see also PCM
Time slots, 463
Timing, binary transmission and the concept of time, 359-367

Timing error, 360, 362
Timing jitter, 473, 476-477
Toll-connecting trunk, 40, 56, 58, 73
Tone-on idle, tone-off idle, 550
Topo maps (for path profiling), 187
Total noise (coaxial cable systems) and its allocation, 414, 415
Towers and masts, 250-253
Tracking, *see* Antenna tracking subsystem
Transit time (tv), 485
Transitions, binary data, 363
Transition timing, 363
Translation, frequency, 84-91, 103
Transmission design, 37, 48-49, 63-65
Transmission limit, 63
Transmission line devices, 222-223, 283, 284
Transmission lines, 161-162, 219-222, 285
Transmission loss planning, 49-59
Transmission of fault and supervisory information, 249-250
Transmission plans, 40, 62, 68, 557; *see also* VNL
Transmit-receive separation, radiolinks, 246
Transmit reference equivalent (TRE), 39, 40
Transmitter (telephone), 34
Transmitter-distributor (T/D), 370
Transponder, 325-329
Transversal filter, 387
Tropo scatter parameters, typical, 291-292
Tropospheric scatter (tropo), 32, 264-294

Trunk (junction), 12, 46, 55-57, 72-74, 466, 467, 549-552
Trunk group, 61
Twist and sway, radiolink towers, 251-253, 254-255
Two-cable working, 467
Two-tone test, 153
Two-wire transmission, 43-45, 48, 51, 62, 73
TWT (travelling wave tube), 203, 305, 306, 315, 318, 325-327

Unbalance (against ground), 76
Unigauge design, 66, 68-71
Unintelligible crosstalk, 22, 23
Up-link considerations, 310-311

Variation of transmission loss with time, 77-78; *see also* Stability
Variolosser, 127
Velocity of propagation, 17, 48, 50, 51, 57, 65
Vertical interval, 487, 488, 509
Vertical interval test signals (VITS), 509
Vestigial sideband (VSB) modulation, 371-372, 385, 500-501; *see also* AM-VSB
VFTG (VFT, VFCT), 149, 389-391, 394
Via net loss factor (VNLF), 57-58
Video standard, transmission, interconnection at IF and video frequencies, 499
Video transmission, 400, 482-511
  over coaxial cable, 500-503
  over conditioned pairs, 505-506
  over radiolinks, 503-505
Video transmission standard-

level, 491
Visual display, 370
VNL (via net loss), 47, 49, 50, 53–59
VNL penalty factor, 59
VOGAD, 170
Voice channel, 13, 83–84, 86–88
Voice plus, 324, 390
Voice frequency carrier telegraph, *see* VFTG
Voice terminal, 13
HF, 170
Voltage ratio, 4
Volume unit (VU), 19, 92–95
VU, *see* Volume unit

Waveguide, 219–223
Weighting, 25
Weighting factor, 26
White noise testing, 153
Windows, 434
Window signal (tv), 507
Wire systems, 34
Wolf sunspot number, 138

Yeh method (tropo), 268–269

Zero relative level point (O TLP), 77, 86, 98
15-supergroup assembly, CCITT, 91